AILING STEEL

Ailing Steel

The Transoceanic Quarrel

WALTER H. GOLDBERG

St. Martin's Press   New York

© Walter H. Goldberg, 1986
All rights reserved. For information, write:
St. Martin's Press, Inc., 175 Fifth Avenue, New York, NY 10010
Printed in Great Britain
Published in the United Kingdom by
Gower Publishing Company Limited.
First published in the United States of America in 1986

ISBN 0-312-01502-X

Library of Congress Cataloging in Publication Data

Goldberg, Walter, 1924–
  Ailing steel.
  Bibliography: p.
  Includes indexes
  1. Steel industry and trade — Europe — Addresses, essays, lectures. 2. Steel industry and trade — United States — Addresses, essays, lectures. 3. Steel industry and trade — Japan — Addresses, essays, lectures. 4. Competition, International. I. Title
  HD9525.E752G64   1984   338.4'7669142   83-40527

ISBN 0-312-01502-X

Contents

Figures

Tables

List of contributors

János Ács

János Ács is a lecturer at the University of Technology in Vienna. He has also taught at the University of Duisburg (West Germany), Linz (Austria) and at the University of Technology in Budapest. He has practical experience in international marketing of complete equipments, especially of steel plants, in various countries. His present research interests focus on decision analysis issues in marketing and innovation management. He is author of numerous articles in these fields and has been consulting with private corporations and public agencies.

Anthony Cockerill

Anthony Cockerill teaches Economics at the University of Manchester Institute of Science and Technology, having previously held posts in the Universities of Leeds, Salford and Cambridge. His research interests are in industrial organisation analysis and policy and he is the author of a number of publications on the steel industry. He is economic adviser to the Industry and Trade Committee of the House of Commons.

Sara Cole

Miss Sara Cole B.A., M.Sc., read Economic and Social History at the University of Nottingham and gained a Masters degree in Management Sciences at the University of Manchester Institute of Science and Technology in July 1982.

The British Steel Corporation's Corporate Plan was the subject of her M.Sc. research. She is currently registered for a part-time Ph.D. and is again studying the UK steel industry.

Miss Cole is now a director of Primco Ltd., a small group of companies which produces high-technology plastics.

Robert W. Crandall

Robert W. Crandall is a Senior Fellow at The Brooking Institution where he specialises in the economics of regulation, industrial organisation, and antitrust policy. His publications include *The Steel Industry in Recurrent Crisis, Controlling Industrial Pollution*, and *The Scientific Basis of Health and Safety Regulation* (with L. Lave). Prior to joining Brookings in 1978, he was Deputy Director of the US Council on Wage and Price Stability. From 1966 through 1974 he taught economics at the Massachusetts Institute of Technology. He has also taught at Northwestern University, George Washington University, and the University of Maryland.

Bela Gold

Bela Gold has just been named to the Fletcher Jones Chair in Technology and Management at the Claremont Graduate School in Claremont, California, after having served for many years as the Timken Professor of Industrial Economics and Director of the Research Program in Industrial Economics at Case Western Reserve University in Cleveland. He has had extensive experience as an economic adviser and as a consultant to managements and to government agencies on the productivity, technology, growth and international competitiveness of a wide array of industries, including steeel. His ten books and more than one hundred papers have been published in fifteen countries and have been translated into seven languages. He is also president of Industrial Economics and Management Associates, Inc. Further details are available in *Who's Who in America* and *Who's Who in the World*.

Walter H. Goldberg

Walter H. Goldberg, Professor of Management, Graduate School of Business Administration, University of Gothenburg (since 1963); Director, International Institute of Management, Science Center Berlin, 1973-1979, has published over forty articles in international journals about issues of industrial stagnation/crisis and strategies of coping with related problems. Recent books by Walter Goldberg are *The Multi-national Firm and National Policies* (with M. Z. Brooke), (1975); *Governments and Multinationals: The Policy of Control versus*

Autonomy (1983); *Mergers: Motives, Modes, Methods* (1983); *The Instrumentality of Innovation Policy* (forthcoming).

Carlo Maria Guerci

Carlo Maria Guerci is Professor of Industry Economics of the University of Genova (Italy). He was Economic Adviser of the Ceos of Nuova Italsider, Montedison and at present of the ENI Group. He has written important books and articles regarding the economics of industry.

Joel S. Hirschhorn

Joel S. Hirschhorn is a Senior Associate at the Office of Technology Assessment, the analytical arm of the US Congress whose basic function is to help legislators anticipate and plan for the positive and negative impacts of technological changes. During the five years he has been at OTA he has directed several major assessments. His work on the international competitiveness of the US steel industry is generally regarded as the premier government analysis of the problems facing this industry. It was this study that brought national attention to the restructuring of the domestic steel industry and the rise of the scrap-based 'mini-mills'.

Shigeyoshi Horie

Shigeyoshi Horie was born in Japan in 1926 and graduated from the First Faculty of Engineering, Tokyo University, in 1948. In the same year he commenced his career with Nippon Kokan K.K. His career progressed over the years and he became Director of the Steel Division in 1981, Director of Technical Research in 1982 and achieved the post of Managing Director in 1983.

Klaus R. Kunzmann

Klaus R. Kunzmann has, since 1974, been Director, Institute of Spatial Planning, University of Dortmund. He studied architecture and urban planning at the Technical University of Munich and got his Ph.D. from the Technical University of Vienna. He had done practical work as an urban planner in Sweden, Austria, Germany and Thailand. Since 1974, he has carried out extensive research and numerous publications on development in urban agglomerations, both in industrialised and in developing countries. He is an occasional consultant to the German Ministry of Economic Cooperation, the Council of Europe and the European Commission.

Stanley Y. Margolin

Stanley Y. Margolin is a senior staff member who has been associated with Arthur D. Little, Inc. for the past 30 years. He is a graduate of MIT with a Bachelor of Science and Master of Science in chemical engineering. His activities cover a wide range of chemical engineering, process metallurgy, technical, economic, and environmental studies dealing with basic industries. He has been responsible for a number of studies on various aspects of the steel industry in both North America and internationally. A registered professional engineer, Mr Margolin is a member of a number of professional societies.

Manfred Neumann

Manfred Neumann Dr. rer.pol. 1963 and Habilitation 1967 from the University of Cologne. He has been Professor of Economics at the University of Erlangen-Nürnberg since 1969 and a member of the Economic Advisory Committee to the Federal Ministry of Economics of West Germany since 1977.

Gerhard Rosegger

Gerhard Rosegger is the Frank Tracy Carlton Professor of Economics at Case Western Reserve University, Cleveland, Ohio. His main field of interest is the economics of technological change. He is the author of *The Economics of Production and Innovations: An Industrial Perspective* (Oxford: Pergamon, 1980), a co-author of *Evaluating Technological Innovations: Methods, Expectations and Findings* (Lexington, MA: Lexington Press, 1980) and of *Technological Progress and Industrial Leadership: The Growth of the American Iron and Steel Industry, 1900–1965* (Lexington, MA: Lexington Press, 1983), as well as of numerous chapters in symposia and papers in professional journals. He has served as consultant to producers and customers in the iron and steel industry. During 1983–84 he was Visiting Professor at the Institute for Innovations Research, University of Innsbruck, Austria.

Lutz Schröter

Dr. Lutz Schröter was until recently active at the Institut für Raumplanung, Universität Dortmund. He has a long standing interest in regional problems connected with faltering industries. To this extent he has not only delved into the crisis of the German Ruhr District (steel and coal) but also participated in Anglo-German comparative research, comprising the Liverpool — Merseyside region of England.

Hiroshi Takano

Hiroshi Takano was born in 1914 and graduated from the Faculty of Engineering of Tohoku University in 1940. His long career at Nippon Kokan K.K. commenced the following month. In January 1967, he was promoted to General Manager of Tsurumi Works and was appointed a member of the Board of Directors in November of the same year. In April 1968, he became Assistant General Manager of Fukuyama Works and General Manager of Keihin Works in 1972. He became a Managing Director in 1971, a Senior Managing Director in May 1974 and, two months later, he was appointed Executive Director of the Steel Division. Mr. Takano was appointed Executive Vice President in June 1968, a position he held until June of this year, when he became Executive Counsellor.

Sergio Treichler

Sergio Treichler is the Corporate Planner at MaxMeyer Duco S.p.A. (Italy). When he wrote his contribution to this book, he was the Chief Economist at Nuova Italsider, the biggest steel company in Italy. He represented Italy during 1980 and 1981 on the IISI's Economic Studies Committee, participating in the working group on long-term steel demand forecasting, of which the results have now been published. He was joint-relator to the National Commission for the Steel Industry for the Ministry of Industry in 1975. He is, at present, member of the Eurofabe Council, the European Federation of the Business Economists.

Helmut Wienert

Helmut Wienert born in 1948, is a Research Economist at the Rheinisch-Westfälisches Institut für Wirtschaftsforschung (Institute of Rhine-Westphalia for Economic Research).

Foreword

None of the crises in large-scale industries — shipbuilding, construction, automobile and truck production, etc. — which are shaking the highly industrialised countries of the world since the mid-1970s, has had as deep and serious consequences, at regional, national and international levels, as the steel industry crisis. It is not only the question of collapsing big workplaces which cause severe long-term unemployment problems hitting large, monostructured regions. It is also the question of the repercussions of various measures of support — protection, subsidisation etc. — on regional and world trade, which are most cumbersome, as they apparently contribute to postponing the desperately needed recovery, not only of the western economies, but of the world economy at large. This postponement however accentuates the steel crisis, as steel is one of the commodities gaining in demand in the early stages of a recovery, as more than two-thirds of the demand for steel is for investment goods — the traditional forerunners of business cycle improvement.

There is urgent need for a thorough analysis, not only of the roots and causes of the crisis, but, above all, for stocktaking and assessment of means and instruments to improve the situation of the ailing steel industry. As the crisis is complex and by no means uniform, the investigation must cover a wide range of instruments and measures, as to their applicability and potential efficiency at various decision levels: the firm, the region, the national and supranational levels.

In order to provide a differentiated view, a panel of the world's most reputed experts on steel market and steel industry problems has been asked to contribute to the analyses and recommendations.

As a most timely point of departure, the acute conflict over world steel market problems has been chosen, mainly involving the United States of America on the one hand, the European Community, and

Japan, each acting on its own, on the other.

The analyses and conclusions are of relevance to crisis management in other large-scale industries as well. This volume is one in a stream of publications by the author on issues of industrial stagnation in highly industrialised countries.

The editors of the Wissenschaftszentrum — Science Center — Berlin extend their most sincere thanks to Professor Walter H. Goldberg for his contributions as author and editor.

Meinolf Dierkes
Professor, President
Wissenschaftszentrum Berlin

Acknowledgements

At the International Institute of Management, Science Center Berlin, Dipl.-Kfm. Manfred Fleischer has provided mental and financial support to the project. Without the patience, perseverance and cleverness of Brigitte Schmidt, in taking the manuscript through the different stages of production, the project would hardly have succeeded. I owe most sincere thanks to both.

Copyright permits and permissions to quote have been obtained from the International Iron and Steel Institute and the Japan Iron and Steel Federation. The contributors to the volume (see List of contributors) to whom I owe the most sincere thanks for their preparedness to cooperate and also to discuss and penetrate the issues, have also made it possible not only to use materials from their organisations, but have also been kind enough to obtain copyright permits for material quoted from different sources:

Arbeitsamt Dortmund
The Bell Journal of Economics
Cambridge University Press
Prof. Dipl.-Ing. Dr Mont. Herbert Hiebler, Institut für Eisenhüttenkunde, Montan-Universität Leoben
Industrie- und Handelskammer zu Dortmund and the other Chambers of Commerce in the Ruhr District
Krupp Stahl Aktiengesellschaft
Lexington Books
Fred Margulies, IFAC, International Federation of Automatic Control, Laxenburg
Pahl-Rugenstein Verlag
The Quarterly Journal of Economics, John Wiley & Sons
Springer-Verlag,
Dr-Ing. Michael Wegener, Universität Dortmund

In preparing the manuscript I have enjoyed the most competent support from Professor Hans Mueller of Middle Tennessee State University, Murfreesboro, Tennessee, who has generously not only put his printed and mimeographed material at my disposal, but also given very valuable advice.

Without the support of the Wissenschaftszentrum — Science Center — Berlin this volume could not have been written. I owe the sincerest thanks to its President, Professor Dr Meinolf Dierkes.

PART I
INTRODUCTION

1 The problems

The transpacific/transatlantic quarrel

The steel industries of the western industrialised world outside of the USA have become used to assembling all their top brains at their headquarters in early June of each year. Likewise governments keep trade secretaries and steel experts alerted at the same time. It seems to have become a rule that the government of the United States of America proclaims wars against its allies' steel industries only a few weeks after the annual summits of the political leaders of the major western nations. At each of those meetings, pathetic speeches in favour of free world trade and against the evils of protectionism are held.

As a rule, the President of the United States of America acts as the strongest protagonist of free world trade. But only a few days after the summits are closed, the US Government announces severe cuts in steel imports and/or punishments to be applied, sometimes retroactively, against the western world's steel industries.

This has been the case now for three consecutive years:

1   The 1981 Ottowa summit was followed by bans against the producers of steel tubes and pipe.

2   The 1982 summit meeting at Versailles was to be accompanied by the announcement of excises of up to 40 per cent upon imports of bulk steel to the United States of America.

3   A fortnight after the Williamsburgh summit of 1983, the foreign speciality and alloy steel suppliers of the US market received their share of the regular(?) annual steel industry punishment'.[1] Speciality steel imports to the USA will hereafter be regulated by quota, implying quite drastic cuts compared to regular deliveries.

At least in the cases of 1982 and 1983, the US steel industry had urged its government to protect the industry against foreign imports. It is not the first time protectionist measures have been required by the industry. Earlier protective measures, aimed at restructuring and strengthening the industry to better withstand foreign competition, had failed to give the expected results (as demonstrated by this book). However, this time all the steel industries of the industrialised world were more or less in deep crisis, and this time the President of the USA did act as the foremost engineer of free world trade, whereas his administration, a few days later, demonstrated quite the opposite policies, with great emphasis and harshness.

It is not the aim of this book to one-sidedly criticise the US government or steel industry. The aims are rather to analyse the causes of the present situation of the steel industries of the mature industrialised countries, to assess the measures taken in the past in order to restructure the industry, and to demonstrate at which levels decisions need to be taken in order to overcome one of the most severe industrial crises of our time.

The transatlantic and transpacific quarrels about steel are indicators not only of how severe the crisis is, but also of the fact that solutions can hardly be found at the expense of others.

Definitions

The steel industry consists of corporations producing steel products from ferrous materials like ore, sinter, pellets, sponge or other direct reduced iron; further from pig iron, recycled iron, steel scrap and comparable wastes. The iron and steel industry does not comprise the iron and steel foundry industry, although there are considerable overlaps.

The industry is, in terms of structure, the operations and products of the firms, as well as the marketing strategies made up of three rather different segments:

a) Integrated steelmaking companies;
b) Non-integrated steelmaking companies; and
c) alloy or special steel producers.

Integrated plants comprise processes from extracting of iron from ore (by means of blast furnaces or other processes for gaining iron from various types of raw materials – see above), steelmaking to finishing. They serve larger markets, within their ranges of specialisation.
Non-integrated mills do not extract iron, but buy iron in more or less refined form (scrap, waste and so on). They operate melting, casting

4

and usually a limited range of finishing equipment. They usually serve regional markets. They tend to be located in markets with a high demand for certain products (for example construction material) or where scrap is available (often both demand and supply are good in metropolitan areas, where the non-integrated mills thus tend to be located).

Non-integrated mills are frequently called mini-mills because of their, originally, rather limited sizes, that is under 500 tonnes capacity per year. As will be shown in this book, there is a tendency towards larger, 'midi-mill', plant sizes, even beyond the million tonnes per year, which approaches the average size of older integrated mills.

Special or alloy steel producers make special products of different types. They may or – more often – may not engage themselves in reduction of iron from raw materials. This segment of the industry is not dealt with in this volume but will, however, be mentioned repeatedly.

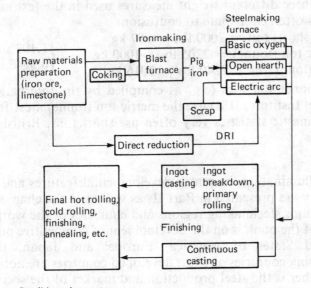

Possible major routes:
Integrated: coking-blast furnace-basic oxygen-ingot casting-finishing
Non-integrated: scrap-electric furnace-continuous casting-finishing
Semi-integrated: direct reduction + scrap-electric furnace-continuous casting-finishing

Source: Office of Technology Assessment, 1980

Figure 1.1 Schematic flow chart for integrated and non-integrated steelmaking

5

To provide a brief schematic overview over the production flows in integrated and non-integrated steelmaking, a flow chart representing the major processes is reproduced in Figure 1.1.

Frequently used abbreviations are:

DR: direct reduction (i.e. not using blast furnaces for ironmaking)

DRI: iron produced by direct reduction

BOP: the basic oxygen process will often be mentioned as the major and widely used post-World War II steelmaking process, in which oxygen-injected steelmaking furnaces (BOF) are used.

EAF: electric furnaces, which are often used in mini-mills.

OHF: Open hearth furnaces, which are an older technology for the reduction of steel from iron. It has largely been replaced by BOF or EAF in most modern steelworks.

CC: continuous casting process, which is a major innovation in finishing operations of steel.

TON: Three different weight measures used in the text and in steel statistics often contribute to confusion:

U.S. (short) ton = 2000 lbs = 907 kg
Metric ton (tonne) = 2205 lbs = 1000 kg $\begin{cases} 1,103 \text{ short ton} \\ 0,9843 \text{ long ton} \end{cases}$
G.B. (long) ton = 2240 lbs = 1016 kg

International statistics (e.g. as compiled by the International Iron and Steel Institute, IISI) use the metric ton (tonne) most frequently. U.S. domestic statistics very often use short tons, British statistics long tons.

Statistics illuminating the major developmental features and trends of the industry are presented in Part II, as well as in the chapters dealing with the major steelmaking regions and countries of the world. As the emphasis of the book is on the development of competitive problems in the United States of America, Europe[2] and Japan, the newly industrialising countries and less developed countries are not treated in depth, neither is the steel production and market of the socialist bloc, since it is nearly self-supplying and since, compared to its size, only marginal quantities cross its borderlines to and from the rest of the world.

Change in the steel industry

The steel industry is, like any other industry, undergoing change. The situation of today has its historical roots, and is caused by a number of factors, which will be briefly exemplified. As one may distinguish between several levels at which the various factors influencing the iron and steel industry development are working, there will be reason to return to those in more detail when discussing the steel industries of the major regions or countries.

The levels referred to above may be shortly defined here.

The *world steel market* is a statistical construct used for the measurement and recording, for statistical purposes, of the trade of steel between different regions and countries. It is often also specified by quality and, furthermore, can embrace flows of raw material and fuels used in steelmaking.

There is no decision making body at 'the top of the world' which, as such, can influence and control the market and the trade flows.[3] Certainly there are, however, phenomena active or discernible at this level, for example crisis or boom periods.

The *regional level* is of greater interest, as there are in some cases bodies active in decision making, influencing the market and/or the production system. Regions, however, are not necessarily clearly defined entities. When, for example, talking about the European market, it is *not* the region geographically specified as Europe that one refers to. First, the socialist countries are usually excluded (although, when they are included, in most cases the USSR is treated as a European country, which it is only, however, to a minor degree). Europe is occasionally interpreted as being identical with the European Communities; sometimes as the EEC of the 'Nine', or the original six members of the European Coal and Steel Treaty (see the section below on the European Community), sometimes as the EC plus affiliated countries.

For our purposes – in particular when referring to, or discussing, the Davignon or EC plan for restructuring the European steel industry – interest will most frequently focus on the European Community.

The socialist bloc will refer to the COMECON countries. Mainland China (PR of China) is not a COMECON member. (China ranks number five in the world in steel production capacity, close to 40 million tonnes. Together with North Korea, it is a rather closed market, however, importing approximately 5 to 6 million tonnes a year, which gives the two countries the rank number 6 amongst the importers of steel.)

The term *region* will be used in another meaning as well, viz. to define a geographical area, in which steelmaking is a, or *the*, dominant industry and which, because of this monostructural dominance, is heavily affected by the steel crisis, as few other job openings are available.

The level of the *country* is a very relevant focus of interest for the following reasons:

1   The country makes up a political, social, cultural and historical

constituency, an environment of substantive importance to industry and to economic activities as a whole. There are country-specific factors influencing the status of an industry. Even if those factors are largely intangible, they influence individual companies' or firms' behaviour.

There are differences – visible as characteristics in competition, for example between the steel industries of Great Britain, the United States, France, Germany, Belgium, The Netherlands – which have little to do in the first place with differences in political systems and governmental policies for the steel industry. (It should not be forgotten that there have been long periods of nil or minor governmental direct impact on, or monitoring of, the steel industry.)

2   The country is a decision level, potential or actual, on industrial policy matters. The government can influence the conditions for an industry to a greater or lesser extent. And as steel is a so-called basic industry, and as it operates many large units, employment-wise, the likelihood is greater than for other industries that steel will be monitored in various ways by political bodies. The cases of Japan and the socialist countries are most obvious examples of this.

Even in countries in which government's basic policy is not to interfere with the industry directly – other than either to create a business climate favourable for growth (industry being regarded as one of the major motors of growth) or to constrain potential or actual negative side effects of industrial activity – the industry is requested to become active in crisis situations (see as a specific example of this the section on the Federal Republic of Germany).

3   In some cases national governments are the sole or principal owners of the steel industry, for various political or historical reasons.[4]

The country also regularly represents the *level of the 'industry'*, but the industry does not represent a decision making body – except in countries where the industry is state-owned. The industry level of countries is one of the main focuses of this study.

The *firm* is the actual level at which responsible decisions are made, and is not the central focus of the study. The actions or reactions of firms or cases of firms will, however, be used quite frequently to illuminate interactions between the state and the industry, for example when contemplating or implementing industrial or crisis policies. Examples of firms' strategic moves to cope with crisis without public intervention or support are also given.

It should be pointed out that the *plant* in many cases is a more relevant unit of analysis than the firm, as individual plant characteristics ultimately make up the characteristics of the firm. However, plant statistics are only rarely available. Thus the *plant* can only exceptionally be the level of analysis.

Finally, a point of caution is raised: in illustrating the text with statistical figures, tables or statements, often industry or country averages are used. Statistical averages, however, are fictions, as they are extracts from mixtures of high and low performance etc. Averages represent 'mediocre' behaviour or performance, hiding both the best and the worst levels.

Change in capacity and cost

Devastation during World War II – rebuilding

World War II left behind ruins and run-down factories around the world, particularly in the iron and steel industry of the old world and in Japan. The steel industries of the axis powers were put under allied control, which dissolved the Japan Steel Co. merger and privatised the Company. In Europe, the allied powers kept control over the steel industries of Germany and Austria until almost the middle of the 1950s.

The only steel industry which was fully intact and running at high capacity utilisation was that of the *USA*, which was enjoying the big rush for steel to build up devastated European industries – including the iron and steel industries. Strangely enough, the only additions made to US steelmaking capacity during the post-war period used pre-war technology, and rather little new capacity was added.

In hindsight it may not have been a poor strategy, if one puts oneself in the role of steel industry management acting in the post-World War II context, when the experience from the post-World War I period was still in the minds of many – politicians and economists as well as management.

The devastation of war had then left behind spoilt economies, which were not able to enter the growth path. Now there were expectations of a similar depression to freeze down economic and industrial activities after World War II. The Marshall Plan was designed to prevent such a new disaster – and it succeeded. There were also other reasons for a stronger growth to follow the much more severe devastations during and after World War II, but, given the situation immediately after the war had ended, and also given the fact that the throughput time for investment in new integrated steelmaking capacity is six to eight years,

9

a conservative, but not unmotivated, 'stay low' sequence of decisions seems to have been taken, which from today's point of view may be harder to understand as a rational strategy.

Against this conservative US steel industry management strategy, the situation in Europe was, at least in some respects, quite different.

The *British* steel industry was not in as bad a shape as one would have expected to be the case after German air raiding over Britain in the beginning of the 1940s. The industry was operating modestly modern plants, some of them with quite good efficiency. Another feature was a rather split-up industry, which had proven to be an advantage during the war in that the Luftwaffe had never been able to knock out the British iron and steel industry to an extent which would have mattered.

Of course, British mills were using pre-war technology and some of them were quite worn out. What perhaps mattered more was the uncertainty during the 1950s about the fate of the steel industry. Would it be nationalised or not? This uncertainty undoubtedly stopped many plans for restructuring and modernisation as well as investment in new up-and-coming technology. In the mid-1950s the industry was almost nationalised, then a political shift turned the fate of the industry again for some time. When nationalisation finally came it was high time for reconstruction and modernisation. Unfortunately, the programme, however, came too late and was never implemented to the planned extent.

The *French* iron and steel industry was mainly located in the border regions towards Germany and in northern France towards Belgium, both traditional industrial regions in coal districts or with access to medium and low Fe-content ore[5], and suffered substantially during the occupation. Subsequently it was modernised with participation from the state, and, essentially, the traditional locations were chosen for post-war reconstruction. During the 1960s, the move to coastal – deep harbour locations – was picked up (see the section on France below), and the old locations were severely hit by the crisis of the recent years with drastic reduction in employment and consequently booming regional unemployment.

The war and post-war development of the *Belgian* iron and steel industry is similar to the French one – except that the state did not enter into the industry, until during the recent crisis. The reconstruction efforts of the 1970s seem to have been in vain, and the Belgian iron and steel industry is amongst the biggest losers in Europe.

Italy, which had a tiny steel industry at the end of World War II, only decided to build up a highly modern steel industry with strong government participation after the war, as a measure in the course of industrialising Italy. The locations were chosen mainly in the under-industrialised south. The highly capital-intense new industry has suffered heavily from high cost of capital and a slackening market. At the same time, the mini-mill sector grew perhaps faster in Italy than elsewhere, not least because of its access to both reasonably priced energy and nearby local growth markets. Because of its low average age, the Italian iron and steel industry is less prepared to take part in the European capacity reduction attempts.

Germany's steel industry suffered most from war damage. The allied powers' measures against the steel industry to begin with were rather strict. The industry practically had to start from scratch. It was quite successful in building up a modern industry, although some traditional steel locations later became burdensome. It is off-Rhine steel regions which were hit by the crisis and there the recent restructuring efforts are most painful but also difficult to monitor. The share of government involvement in the industry is marginal.

When the political insight matured that Europe could not afford the suffering and pain of another war and that thus one should cooperate rather than fight, the first major material opportunity to develop and pursue a joint policy programme was in the field of coal and steel. A treaty – the first cornerstone of the forthcoming European Communities – was signed in early 1951. This meant elimination of customs barriers and quotas on coal and steel, promotion of competition in order to improve the structure of the industries, anti-trust measures etc. in the long term. Further subsidies were to be eliminated. The agreement also included freight rates, indirect taxes and excises. Thus, the long-term objectives were structural improvement to further competitiveness and, consequently, growth of sales and production.

When the treaty was concluded, nobody could then imagine that the administrative structure created would soon have to be employed with the task of finding and applying solutions to the emerging crisis in coal-mining, but also, some twenty years later, in the iron and steel industry.

The development of the *Japanese* steel industry deserves to be illustrated in its historical perspective at somewhat greater length, because of the rather massive government–industry interaction. The Ministry of Agriculture and Commerce, MITI's predecessor, took the initiative by founding the first large government-owned steel industry, Yawata Steel Works, in 1896. During the crisis of the 1920s, the Ministry of Commerce and Industry took active steps to support both the public and private steel industries in order to protect them from collapse. In 1934, the (old) Japan Steel Corporation was formed by the merger of Yawata and Fuji, and the 1930s and 1940s saw a strong state leadership in industry. A number of industrial laws, amongst them the Steel Industry Law, were promulgated, implying governmental support in financing, through tax reductions, protective measures and a mixture of state and private enterprises, in order to build up a strong industrial backbone for imperial Japan.

The Supreme Command of Allied Powers (SCAP) dissolved the Japan Steel Co. in 1945 and reprivatised it. The immediate post-war activities of MITI, during the allied occupation, were aimed at minimising its effects on Japan's industry and economy as a whole. One of the first steps taken was to build up the steel industry, together with coal and electrical power. MITI used its influence and cleverness to provide the industry with the necessary financial support (Reconstruction Finance Bank, Japan Development Bank and World Bank loans – 1953). In the years 1956–58 came the end of the US occupation and also of the SCAP-induced anti-monopoly laws. In 1958, MITI caused the Federal Trade Commission to accept the so-called public sales system for the steel industry, an ingenious system of price rigging. Also in 1958, MITI arranged, during a slack period in demand, a price maintenance cartel for steel which was also accepted by the Federal Trade Commission (FTC). Between 1948 and 1950, and in 1960, MITI exercised detailed control over investment in the steel industry through the plans of the Industrial Rationalisation Council and the Foreign Capital Law. After 1960, it exercised 'self-coordination'. Steel has been called the 'honour student' of Japanese government–business relationships. However, in 1965, during a slack period in demand, problems occurred both in the coordination of new blast furnaces and converters and in the coordination of production in order to maintain prices at reasonable levels, and the self-regulation broke down because of a combination of over-capacity and recession. MITI again stepped in and recommended a reduction in production to prevent a collapse of steel prices. In July to September it ordered a 10 per cent cut of output, which was accepted by the industry. Its prolongation for another three months in 1965 resulted in the Sumitomo incident, in which Sumitomo refused to obey, on account of its superior export performance. The Sumitomo incident laid open MITI practices and had some long-lasting repercussions. Indirectly, the incident triggered the second Yawata–Fuji merger into (new) Nippon Steel Corporation, which became operative in 1970 with the objective to create a genuine industrial hierarchy in steel, resulting in a stable oligopoly.

The Capital Liberalisation Act, which was introduced at the request of the United States in 1966, was made invalid for the steel industry (as for a range of other industries) by acts of MITI.

During the late 1970s, MITI again created cartels for the 'structurally depressed industries' (including steel) in order to allocate market shares to be abandoned, to indicate the number of employees to be retained or to be put on early pension schemes and the scrapping of excess facilities, and finally to create investment limiting cartels. This whole plan was opposed by the Federal Trade Commission.

The average capital outlays in the Japanese steel industry at 1980 prices were per one five-year period approximately:

in 1954 to 1959 200 billion Yen
in 1960 to 1964 400 billion Yen
in 1965 to 1969 600 billion Yen
in 1970 to 1974 1,120 billion Yen
in 1975 to 1979 950 billion Yen

Japan is the clearest case of a most efficient cooperation – coordination between state and industry resulting in an incomparable growth and efficiency. Essentially two models had been applied, namely state planning for large industry, until 1945. In the post-war period the state–industry interaction was rather of indicative type or administrative guidance type, although of rather different degrees at different points or periods in time. Certainly the influence of MITI has been stronger than admitted in public by Japanese authorities. In the case of the steel industry it should be remembered that many rather strong personal ties have existed between MITI and the industry ever since the beginning of large steel in Japan at the turn of the century. Of the large steel corporations only Sumitomo had not been participating in the fraternity. Otherwise, the Sumitomo incident would never have occurred. Sumitomo, however, learnt from this lesson and joined the fraternity shortly after the quarrel had been settled.

The steel industry in newly industrialised and less developed countries; the four 'blocs'

The newly industrialised countries (NICs) had, during the immediate post-World War II period, hardly played any role as steel producers, but some of them were importers of steel. From the mid-1960s onwards, they have increasingly been appearing as sellers on the world market. First, they took their home markets, then they penetrated nearby markets and those in less developed countries. Since the mid-1970s they have appeared to an increasing degree on the world market, even if, to begin with, only in the low grade quality range, sometimes explicitly so, in order to creep under import obstacles in highly developed countries, which often leave out the inferior quality ranges. To be mentioned are the steel industries of Brazil, Mexico, South Korea, Taiwan, India, Australia, South Africa and Argentina.

The post-war development as far as production capacity and production is concerned is gradually showing a pattern of blocs with intra-bloc trade and a gradual decline of the interregional trade between the blocs. The discernible 'blocs' or areas are the highly industrialised countries, the USA, Japan, and Europe. The above statement is valid also for this grouping: the inter-trade is gradually declining, more and more being confined to special grades.

The 'bloc' of the newly industrialised countries is self-supporting but also exporting, however with limited inter-regional trade. The most aggressive countries in Asia are South Korea, Taiwan and to some extent also India. In Latin America both Brazil and Mexico are the dominating steel suppliers. South Africa is essentially attempting to become independent but is also serving some neighbouring markets.

It is obvious that some newly industrialised countries like Taiwan and South Korea are trying to gather larger shares of foreign markets. They had even to some extent been entering the Japanese home market (approximately 6 per cent in 1982). Also Brazil has been rather active in exporting, for example to European countries.

Despite the fact that the newly industrialised countries can produce steel at lower cost than US or European mills, and in some cases also cheaper than Japanese mills, they are quality-wise competitive only in rather narrow segments. Still, they are much better equipped and run modern equipment, compared to many less developed countries (LDCs).

On the whole, there is little evidence that NICs and LDCs will be able to acquire larger shares of the steel market in the highly industrialised countries, but certainly more nations are participating in steel production. This tendency has been going on since the 1930s, with an accelerating speed during the last decades. This also means that the exchange of steel is increasing, as most countries are not in a position to cover the entire product range.

The COMECON-countries constitute a rather closed bloc, with marginal quantities only traded externally.

One central conclusion on market development for steel is that steel is essentially being produced in the same region where it is used. If one examines the statistics of production of raw steel (in million tonnes) and the development of shares of world raw steel production, one can easily see that world production figures have constantly been on the decline (almost linearly) since the 1960s. The decline for the United States and for Western Europe is practically parallel.

The development for Japan is, first, rapid growth from the mid-1950s up to 1974. Thereafter (Japan had by 1974 a share of only 3 per cent less than the United States) the trend for Japan is almost parallel to that of Europe and the United States.

The COMECON share and the share of the developing and newly industrialised countries were almost constant between 1960 and 1974. Thereafter they increased: the NIC–LDC share goes up from approximately 11 to 20 per cent, that of COMECON from about 25 to nearly 30 per cent.

If one looks at the absolute figures, one sees that the growth in the NIC–LDC is accelerating, and that within COMECON is growing at a slower pace, after a rapid growth period until 1974, whereas for the western world the growth ends after 1974.

There are a number of reasons behind this development.

1   If one relates steel consumption to the gross national product per inhabitant, in constant dollars or value, one finds that steel consumption per inhabitant is growing rapidly as development takes off, but the growth soon culminates and turns into a decline, which, to start with, is rather fast but then levels out slowly. Steel intensity is highest in early phases of development, which is to be explained by the fact that approximately two-thirds of steel consumption is investment-related. In later stages of development and in the maturing stages the steel intensity of investment declines. More productivity gives less steel-intense investment.

2   World trade in steel in relation to total world production is stagnant or slightly decreasing as NICs and LDCs produce more of their own needs – and they are in steel-intense phases of growth.

3   Over time the ratio of sophisticated steel products (the so-called value-steel) is increasing because the processes are difficult to learn and because rather limited quantities are bought, which leads to a specialisation in few products, which then are sold 'worldwide'.

The most drastic change in post-World War II markets came at the end of 1974, when great shifts occurred in steel-intense industries. Shipbuilding faltered, and construction activities in industrial buildings, office buildings, housing, but also in road construction began to level off as early as the beginning of the 1970s and then quickly dipped. Other steel-intense production also started to decline, such as machinery construction and automobile production. The only steel-intense growth in this sector was in the offshore and oil pipeline sector (but also in energy production, particularly the nuclear power sector, which, however, is special steel).

The development of the apparent steel consumption over gross private investment in the highly developed countries has been on the decline during the entire post-World War II period, and will continue to decline.

Another secular trend is that production follows consumption (see the initial general statement).

The less developed countries can, and do, produce their steel for domestic consumption. In fact, it has become easier for them to do so with the advent of the mini-mill technology, which requires only approximately one-tenth of the investment capital needed for integrated capacity. The LDCs will remain marginal markets for special steels and for special carbon steel makes.

In summary, the steel market in the 1980s will be marked by a continuously slow growth in consumption and production and, simultaneously, a continued shift of the growth in demand and supply to the newly industrialised and less developed countries. The highly industrialised countries will have to struggle with their steel industry's adaptation to lower demand, whereas newly industrialised and less developed countries probably still want to increase their capacity further.

Some forecasters predict a steel shortage for the second half of the 1980s and the beginning of the 1990s. Most of those forecasts, however, date back to the late 1970s and the beginning of the 1980s. The first three years of the 1980s have, however, brought a further slump in demand for steel, now also including the recent growth sector in the field of construction for energy, including oil.

The COMECON countries and the newly industrialised countries and LDCs have also been affected by the long economic crisis in the highly industrialised countries. Thus, even their demand of steel is today growing at lower rates than during previous periods.

The forecasts for crude steel capacity made by the International Iron and Steel Institute and other agencies, like the OECD, have repeatedly been adjusted downwards. The principal causes for this are, on the one hand, the drastically reduced investment activities in the world and in the highly industrialised countries in particular, and, on the other hand, technological improvements, particularly the increased utilisation of continuous casting, which is gradually reducing the amount of raw steel needed per unit of finished steel output. Both developments make the possibility of a steel shortage, as mentioned above, most unlikely. There may, however, be occasional bottlenecks appearing in narrow ranges of quality, but this is a quite natural phenomenon. The history of steel shows that there are sometimes sudden drastic changes in demand, which cannot possibly be met by adaptation of production equipment in the short term. Nor would it be economically justifiable and possible to keep reserve capacity available to meet the temporary bottlenecks.

Trade and location

Steel has traditionally been an internationally traded commodity for a range of reasons, for example, industrial tradition, access to energy deposits (charcoal, coal, electricity, natural gas). During the 1960s and the beginning of the 1970s the availability of cheap oil caused the energy factor of location to become unimportant. Further factors of location are access to ore deposits, often in the neighbourhood of coal deposits, access to means of bulk transportation (during the 1960s and onwards this became the major factor of change, however, due to the appearance on the market of large bulk tonnage, the so-called OBO – ore-bulk-oil carriers). This, in combination with cheap oil, gave great cost advantages to steel plants with immediate access to deep sea terminals. At the same time, large deposits of high quality ore and coal and oil became available at many new sites.

The examples given of the supply side factors are matched by few demand side factors: steel is a most important material in high demand during the build-up phase of a national industrial production capacity. As more and more countries enter into the phase of modernisation, industrialisation grows and thus the demand for steel. During modernisation, the investments are steel-intense. Over time, the steel intensity of investments is gradually declining.

Steel is not a homogeneous product but it is used in a wide and steadily growing range of 'qualities'. As producers in a process industry like steel by necessity have to concentrate on a limited range of grades there is, and will remain, a market demand for variety. This demand will be the backbone of world trade, that is trade between the large steel-producing regions.

Present-day problems: the birth of a crisis

1974 – the best steel year ever...?

The peak demand of 1974 was, to some extent, artificial. A major steel buyer of that time was the world shipbuilding industry, and there was an extremely intense but rather short-lived peak demand for ships. In the shipyards, order books overflowed with orders mainly for large-scale tankers – very large crude carriers (VLCC), ultra-large crude carriers (ULCC), parcel tankers, ore – bulk – oil carriers (OBO), roll-on-roll-off carriers (RoRo ships etc.). A number of factors contributed to the boom in ship demand. The Suez Canal had been closed because of an act of war, forcing oil to be carried around the Cape of Good Hope. As the oil carriers in principle had been built to use the Suez

Canal and thus were of quite limited size, there suddenly emerged a shortage of tanker tonnage. Japanese shipyards had been successful in substantially increasing the size of oil carriers. Thus as freight rates for oil boomed, a great number of orders for carriers in the larger ranges were placed with Japanese and, to some extent, even European shipyards.

As shipyard technology had to be changed – conventional berths and docks had become far too small for the new giant-size carriers – a number of new shipyards were established, but they often were placed in countries with low labour costs, as the Japanese competition, to begin with, was mainly felt to be a labour cost competition. New shipyards emerged in, for example, Mediterranean countries and other new industrialised countries. Thus, the shipbuilding boom had also triggered a shipyard building boom.

The good years of the 1960s had been pushing not only consumption but also, consequently, investment in production capacity, and in office building and housing. Never before had so many tall, steel and glass-intense bank, insurance, industrial headquarters and office buildings been erected at so many places all over the world as during the mid-1960s to the mid-1970s. The order books were full in many steel-intense industries, and it became a custom to sign long-term contracts for steel plant capacity in order to ensure supplies of the steel needed according to the company forecast, and to specify the qualitative details when the true demand was known. Thus, the business cycle for the steel industry was actually overheated and even speculative.

About six to ten months after the oil price crisis, its repercussions became visible in the order books. Many forward buying contracts were cancelled because the drastically increased oil prices had suddenly reduced the amounts of money available for consumption and, soon, for investment as well.

The industry which was hit most severely in the short term was probably the shipbuilding industry. Many 'speculation orders' were withdrawn, inducing the shipbuilding industry to cancel forward orders with the steelworks. Soon a number of shipyards collapsed and, by 1976, even the Japanese shipyards were reducing their workforce quite drastically.

The crisis then hit the automobile industry and the construction industry. By the end of the 1970s the seemingly chronic shortage of housing had suddenly disappeared in many industrialised countries. Vacant housing and office space suddenly became available on the market. Again, orders and long-term contracts with the steel industry were cancelled, and the industrialised world experienced its first really deep business cycle crisis since the war.

As demand for steel slumped, new steel capacity became available, according to decisions taken under conditions of ever-increasing demand for steel at the beginning of the 1970s. This new capacity grew, particularly in Japan, but also in other parts of the world. In the steel industry as well as in manufacturing industry, the slackening in demand triggered dismissals at an accelerating pace, the purchasing power available in society slumped, which triggered contraction of the sums of money available for spending and thus also business activities. Gradually it became obvious that the economies of different countries of the world were facing not a business cycle problem but rather a range of structural crises.

As the crisis progressed, the steel corporations gradually consumed their reserves. As practically the entire western world was affected by the rather new phenomenon of stagnation, older and modern mills were affected almost equally by the crisis, in so far as firms operating older, more labour-intense equipment had to carry through their labour force for a considerable time. The much less labour-intense highly-automatic modern factories, on the other hand, were hit by unusually high levels of interest on debt. US mills may have a greater degree of freedom when it comes to laying off labour during times of slackening demand for steel; however, the bonus for steelworkers over the average income for workers in manufacturing industries in the United States is unusually high, growing to approximately 75 per cent by the beginning of the 1980s.

The Japanese industry (where the steelworker bonus is approximately 50 per cent over the industrial manufacturing average) was in a slightly better situation than its European or American competitors, for two reasons: a) it is much less labour-intense and b) as big Japanese steel usually belongs to Zaibatsus, that is conglomerate trade, industrial, and financial groups, it could in many cases shift its superfluous labour to other industrial firms belonging to the same group. One should remember, however, that even shipbuilding and somewhat later also the Japanese automobile industry have been hit by employment problems. Groups with strong involvement in those three industries were thus rather constrained in their flexibility. The burden therefore to a considerable extent had, and still has, to be shared by employed persons sent into early retirement and by sub-suppliers (of which steel has rather few, but shipbuilding and carmaking rather many). Interest rates in Japan are still much lower than elsewhere in the industrialised world. Against this must be held that any of the Japanese integrated steelworks are of large size and rather recent vintage. The capital cost in real terms is thus considerable indeed.

Once it started to become clear that the crisis was not a business cycle problem, restructuring efforts were made. From 1979 onwards the steel industries of many highly industrialised countries were faced with an awkward situation in that forward strategies and retrenchment strategies collided, often within the same firm, which sometimes led to absurd consequences.

The forward strategies had essentially three roots:

1   Because of the boom years 1973–74 and as a consequence of generally optimistic forecasts, many capacity increase and modernisation decisions taken during the boom time now matured in the shape of new highly efficient production capacity. The industry had been programmed for growth.

2   Many firms had come to the conclusion that the best strategy to overcome the crisis would be the establishment of superior, highly modern, high quality and labour-, as well as energy cost-saving, equipment.

3   At the same time, and independent of the decisions taken in big steel, the advantages of little steel had become obvious, while resulting in a quick growth of this segment of iron and steel, which was capable of taking market shares from big steel and at good profitability. In the United States of America the entire growth in demand during the 1970s was taken care of by the mini-mills. The same has certainly been the case in several European countries.

Strategic responses to structural crisis

The restructuring efforts made in different parts of the world were of quite different nature.

The US steel industry had, like the steel industries of other countries, experienced slackening demand during the second half of the 1960s. It had then managed to get governmental support for protective measures – which certainly must have been a great achievement after the steel industry had been subjected to severe public criticism for its inflation-pushing behaviour ever since the beginning of the 1950s. The protective measures in the form of the so-called Voluntary Restraint Agreement (VRA) were intended to restrict imports from abroad in order to give the US steel industry an opportunity to restructure and modernise, whilst being protected against excessive imports.

When the order books started swelling, the VRA was abandoned at the initiative of the steel industry, not least because some of the large producers were also big buyers of foreign steel.

When it had become obvious that the crisis of the mid-1970s would become deep and be of longer duration, big steel, supported by the steelworkers' union, succeeded in convincing the government of the necessity of new protective measures. The basic idea of the so-called trigger price mechanism (TPM) was an inducement to modernise and rationalise by setting artificial prices for various steel qualities which had some relation to Japanese prices at US destinations.

The American Iron and Steel Institute undertook to demonstrate how desperate, in the meantime, the situation for the steel industry had become and would remain, unless the government took steps in order to support the steel industry in different ways and by different means. The US Congress asked its Office of Technology Assessment to undertake a study of the steel industry (OTA: *Technology and Steel Industry Competitiveness*, 1980). The investigation proposed a programme called 'An Industrial Policy for the US Steel Industry' with the aim of restructuring and modernising the industry for future growth. The government had also taken a number of steps in immediate response to the obviously deepening crisis, such as a change in the depreciation rules for the steel industry.

As the crisis, against expectations, continued to deepen and a number of companies, even some of the big eight, were driven to the brink of bankruptcy, new and stronger protective measures were requested. The TPM (which in the meantime had been revised) was terminated at the same time as anti-dumping suits were filed against many European steel producers.

A few days after a summit meeting of the political leaders of the free world at Versailles in early June of 1982, at which unanimous strong pleas for the maintenance of a free world trade were raised, not least by President Reagan, the Commission of the European Communities was confronted with US claims for steel quota and subsidy-neutralising US import excises of up to 40 per cent on the value of certain steel imports. A hot summer weather-wise had also become a hot summer in transatlantic political relations: the transatlantic steel quarrel was starting to glow, rapidly escalating to white heat, as will be shown later.

In spite of the existence of the European Community, the nation states of Europe initially reacted individually to the steel crisis of the 1970s.

An extensive reconstruction programme was under-way in Great Britain, although it had been delayed for political reasons. From this point of view, the British example demonstrates that public ownership in a crisis-stricken industry is not necessarily an ideal prerequisite for thorough and also badly needed restructuring, providing an ideal decision context because of its unitary ownership. When the restructuring efforts finally got under-way with a programme implying both thorough rationalisation and growth, the industry, at that time heavily concerned with internal problem solving, was hit by the drastic decline in steel demand. The rationalisation and modernisation programme had to be adjusted repeatedly, and its capacity targets were adjusted downwards over and over again.

In France planification of almost Japanese type was in full bloom during the mid-1960s and for a few years to come. The centralised structure of France gave the government considerable means to influence industry toward structural reorganisation, concentration and rationalisation. And, as the state had large holdings in steel, the government had great freedom to get its plans implemented.

The French steel industry had been laggard in several measures of productivity. It was now to be brought into line with the best steel industry of the world. But steel industry restructuring does not render big results in the short or medium term, not even in France. The industry was in the midst of its restructuring efforts when hit by the boom in demand, the potential yield of which could not be harvested. Because of the heavy monostructural geographical concentration of the steel industry in a few regions, the adaptation had to be revised to take longer than originally envisaged as desirable. Traditional steel regions are very difficult to change to other economic activities. (With hindsight the restructuring of the textile industries in many countries, however painful the experience when it took place, was much easier to monitor than that of the steel industry – and this observation is not constrained to France.) Also, the French plan for restructuring had the combined characteristics of forward growth and thorough rationalisation, as well as weeding out of obsolete capacities.

In Italy, as already mentioned, new capacity had been established at a rapid pace, becoming operational towards the end of the 1960s and the beginning of the 1970s. At the same time, without state participation, mini-mills popped up in Northern Italy, almost like mushrooms. Italy had brand new capacity – considerable overcapacity – when the crisis came.

According to its post-World War II non-state interventionist strategy, the Federal Republic of Germany did leave its industry on its own to cope with growth or retrenchment. The case of the newly integrated coal and steel region of the Saar is an exception, described in some detail in Chapter 7.

The German steel industry adapted in many ways. The traditional steel giants, Thyssen, Krupp, Mannesmann, differentiated, quite successfully on the whole, away from steel, still maintaining considerable steelmaking capacity. Some of the new lines of industry entered into a type of downstream integration: (steel) construction industry, shipbuilding, production of turnkey factories, in all gradually reducing the importance of steel proper.

Other firms – less affluent than the big ones – chose innovative steel strategy; for example Hoesch merged with Hoogovens, attempting to establish a most modern huge sized plant on the coast near Rotterdam. A few other examples will be discussed in Chapter 7.

Although independent and unguided by state authorities, the German steel industry pursued similar strategies to the other European ones: growth, modernisation and abandonment of plants which soon would become obsolete. Some differences should, however, be underscored: the uncoordinated German steel industry was largely running at higher productivity and efficiency compared to the US industry or to other European industries, its plants being highly, or fairly, modern. Some of the major firms, although investing heavily in steel, gradually reduced their dependence on the steel sector and thus were able to build up some useful'shock absorbers', as would become obvious a few years later. The better profitability, but also the differentiation, gave the German steel industry considerably greater freedom of action, but also reserves when it came to meet the long slack in demand. The industry not only knew that it had to expect little help from government; it did not even want to be subjected to political influence.

A few companies met the deep crisis at the beginning of the 1980s in a rather bloodless shape.

Hoesch, after carrying the burden of reorganisation as a consequence of the German–Dutch merger, decided, after the Rotterdam project had failed to obtain the necessary permissions and thus could not be realised, that it should go back home whilst it was still in reasonable shape. As a matter of fact, it had been able to attract a new bridegroom who, however, deserted the bride at the church gate. (See Chapter 7.)

Kloeckner, which had chosen a different strategy similar to Krupp and Thyssen but which had organised it in a different way, had also started to invest in brand new super-modern finishing capacity plant by the seaside at Bremen. The large new plant was just coming on stream when the steel market collapsed. It then not only had to carry the heavy burden of high capital service costs, but, as it had no market share at the reference point for the steel quota later to be established by the European Commission, it had essentially no permission to sell. In the meantime, it had accumulated tickets from the European Community for violating the production quota regulation to amounts which it probably never will be able to meet.

Not to be forgotten in the German context is the innovator–entrepreneur, Dr Willy Korf, pursuing an aggressive growth strategy in the mini or perhaps rather midi-mill sector. Korf's mistake in the long run may have been that he took a rather long, strategic view on his new road into steel: in order to protect his mills against the consequences of scrap shortage, he engaged himself in equally innovative direct reduction plants, essentially based on then available, rather cheap, natural gas. When the prices for natural gas skyrocketed much faster than oil prices, Dr Korf could only consider his investment in direct reduction as sunk cost. But the blows he had to take forced him ultimately to take the full count.

The European Community for Coal and Steel was founded in April 1951 on the initiative of the French Minister of Foreign Affairs, Robert Schuman. Its aims, in brief, were:

1. A foreign policy goal: the control power of the allied forces over the German industrial concentration at the Ruhr was taken over, within a long-range objective implying a peaceful uniting of the West European states, ultimately creating a political union.

2. The economic political target envisaged the rational supply of coal, iron and steel products to the consumers and users by a coordination of national markets to a common market for coal and steel.

3. From the social political point of view the working and living conditions for all persons employed in those industries should be harmonised as well as improved.

The major and difficult tasks of the union were in the field of crisis management, during the late 1950s and early 1960s, in coalmining, and, towards the end of the 1970s and the beginning of the 1980s, in the restructuring of the West European steel industry. The activities of the communities in coping with the steel crisis are based on a legal obligation, as laid down in the original contract, as well as on the political concept underlying the formation of the union and the social aspects which comprised the original treaty. When the parliaments of the member nations signed the treaty, they agreed not only to the fact that a super-national agency could become active with regard to the concerned national industries, but also that it had an obligation to do so under certain circumstances. It is thus inherent in the concept of the Community to become active in crisis situations and to see that its pertinent programmes will be emulated in meaningful ways, from the economic medium and long range point of view, that they be politically acceptable and that they pay due attention to the social conditions for the employed, but also for the inhabitants of crisis-stricken regions.

It is also inherent in the basic agreement that the necessary change has to be supported not only by the individual states but by the Community as a whole. The programmes have to be coordinated and harmonised between the member nations. The burdens incurred through their implementation have to be shared by the members.

The general opinion in the steel industries of the world, not only within the European Community, was that the slump of 1975 was principally caused by two factors: insufficient production capacity to meet the market demand, as well as temporary shortage in the supply of certain raw materials. Nobody could foresee then that the growth of the last 25 years was coming to an end, to be replaced by stagnation or even decline of demand. The general expectation was that 1978 would be another boom year for steel. It became slowly clear that the industry was actually facing a structural crisis. A number of factors had contributed to it, leading to combined effects, which struck the industry almost simultaneously: the energy crisis, inflation, new environmental legislation, the rather sweeping changes in international currency markets (at least compared to previous periods), certain lack of capital, at least for major projects, gradual decline in productivity, and tougher competition, not only from Japan but now also from new producers in newly industrialising countries.

The highly industrialised countries (the USA, the European Community and Japan) saw their share of world production reduced by 8 per cent, or 55 million tons, between 1976 and 1979. It was then the Communiy began to become active in view of a number of factors, like for example:

a) growth of overcapacity in combination with stagnant demand;
b) the emergence of new capacity in Japan as well as in newly industrialised and in less developed countries, all of them operating at lower costs than Europe; and
c) significant changes in the structure and location of the consumer markets.

The objectives of the Community's restructuring programme were:

a) an adaptation (read: reduction) of productive capacities;
b) a programme aimed at restoring the competitive capacity of the industry;
c) financial restructuring of the ailing integrated steel producers; and
d) a concentration (but not restriction) of marketing objectives to the Common Market, which in the longer term would be the only profitable market to work at.

The Community chose to concentrate its activities on certain objectives:

a) breaking down trade barriers and protectionism wherever met;
b) counteracting and abandoning subventions;

c)   initiating agreements on voluntary restrictions in the trade with steel products;
d)   organisation of a price and quantity cartel within the Community; and
e)   temporary neutralisation of competitive advantages enjoyed by certain steel producers within the boundaries of the Community in favour of those who because of the crisis had lost their competitive capacity.

The European Community's programme is treated in more detail below.

The EC's activities concern big steel. The growing mini-mills sector feels at disadvantage, particularly as it has no voice so far in the EC's political bodies. The large European steel producers are often heard individually, but are always represented by their organisation, EUROFER. This led, in 1981, to the formation of the European Independent Steel Producers Association (EISPA), representing 68 member firms, under the Presidency of Dr Willy Korf.

Regional and labour market problems

Integrated steelworks are large employers. They also put their hallmarks on the physical environment because of their size and their very typical requirements for transportation facilities (access to waterways, harbours, railway yards or turnpikes); their ore yards and coke yards; their huge production facilities, cooling towers, coke ovens, blast furnaces; their exhaust products like dust and phosphorus fumes, sulphur gases, steam and other pollutants.

The influence of big steelworks is not confined to the physical and natural environment. It also concerns the region's economic structure and the infrastructure. Steelworks, like other large employers, tend to put their imprint, willingly or unwillingly, on the economic structure in so far as other industry and, in particular, medium-and small-size enterprises are repelled from the region.

There are several reasons why the neighbourhood of dominant enterprises should become unattractive to other industry with the exception of directly steel-related or steel-dependent enterprises.

One such reason can be found in the relatively higher cost of labour that large and, in particular, heavy industries carry or can afford to carry, either because the working tasks are heavy, dirty, noisy, unpleasant, or because the industry is profitable above the average, or because the industry is relatively easily blackmailed by unions, on account of the high or even prohibitive cost of having it idle on the occasion of a strike. Consequently other industries, unable to compete with the steel industry for high wages or high costs of employment, will not find the necessary labour and thus not be able to operate within a reasonable distance of steel or other heavy industry.

In addition, steelworks and comparable industries usually provide a great deal of their own infrastructure within the ranges of their own organisation, which consequently can lead to lack of attractiveness to certain trades or service enterprises.

Coal and steel industry enterprises usually are also big landowners and thus have a large influence on the utilisation of land. There are several reasons why they should become big landowners:

1   Because they need space for their activities, they may plan for expansion and thus keep land in reserve for this purpose.
2   They may have to acquire land because the neighbourhood of heavy industry is unattractive, sometimes to the degree that it could be an impediment.
3   As steel-producing firms are subject to cyclical variation in demand and thus also in profitability, holding of land also may be used as a means of monitoring variations in profitability over time.

Often heavy industry provides housing for its employees, so it may be unattractive for private house owners to provide rented living space in a region dominated by one, or a few, big industries.

If the industry is old, it may have developed a patriarchal tradition for better or worse. Big steel has often been successful over time in creating a kind of family atmosphere and has even been anxious to do so, for a variety of reasons.

Big firms may want to keep competitors out of reach, for example, competitors for labour and other resources which are available on a regional basis.

As a consequence of several of the above trends or characteristics, big steel, like other heavy industry, gradually tends to develop a dominance in the local and regional administration, which, if the industry is large enough, sometimes 'degenerates' to a type of 'appendix' organisation to the dominant firm. Thus regional and labour market issues become intertwined, not only when it comes to big enterprises, but they take on a particular character when bigger enterprises dominate the region. If big steel is growing, it will have to recruit and import labour into the region (and also to provide the necessary housing) with the result that immigration to the region is very often comparable to the immigration of foreign labour.

In times of recession, the steel industry firstly tries to keep its labour force busy with maintenance. If the recession continues and particularly if it turns to decline, labour will be dismissed. Lay-offs are then a burden to the region, as little or no alternative employment is available, because of the specific characteristics of the region, as indicated above. There is no 'safety net' of medium and small-scale business available, which temporarily, or rather in the long run, could be developed to provide alternative employment. Experience demonstrates that it is most difficult to attract other business to regions dominated by heavy industry, mainly because of the high labour cost and the unattractive physical environment.

The role of the politicians and of the state

The changes in the steel industry during the last six to eight years have provided ample examples of regional and labour market problems. They have also been one of the major reasons for political and/or state intervention. Politicians, irrespective of being elected for the regional or state level, cannot neglect large pockets of unemployment in monostructured regions. It is not only that the history of industrialisation has provided many examples of unrest emerging from such regions. It is that nowadays voters look upon the politicians as being the guarantors of employment, security, welfare growth and protection against economic hazards. This has come about because the politicians have been labelling their political programmes in similar terms and have very often taken credit for having induced growth and corresponding improvement in the standard of living of the broad masses. They have also proclaimed that they possess the political and administrative instruments to prevent similar catastrophes to those which occurred during the 1920s and 1930s. Their credibility is now at stake. If they fail, the voters will leave them and desert them, most likely for more militant programmes. No wonder then that structural change in heavy industry where big workplaces and dominant regions are concerned will trigger political activities, often implying government participation, involvement or even take-over of industry in a certain region. The examples of France, Great Britain, Japan and some others are indicators of early state involvement of different types and degree. Other political systems avoided becoming committed in responsible terms for a considerable time. Obviously, however, the long and deep crisis has ultimately also brought traditional non-interventionists like the United States and the Federal Republic of Germany into a situation where the State or the Federal Government must sit down together with industry, and together with the unions, to negotiate ways and means to reduce capacity, to restructure the industry and to provide secure jobs in, as well as outside the steel industry.

The mere fact that the state enters into the steel industry, as a shareholder or through nationalisation, has so far never been a solution to the problem. The clearest example of this statement is the British case, where the degree of political intervention in restructuring and rationalisation plans has been very high. At the other extreme, the case of Austria may be cited, where the state-owned steel industry (transferred by the allied powers into public ownership by the Federal Republic of Austria) has been run almost as if it were a privately owned corporation. The Austrian Government-owned steel industry has a remarkable record of innovativeness – the basic oxygen process was developed there to industrial applicability.

The case of Belgium deserves to be recorded since it demonstrates that, once the state has entered into the industry, it may be committed to further participation until complete or almost complete takeover without essentially solving the crisis.

Belgium increased its steel production during the 1960s by more than 80 per cent (compared to 36 per cent for the entire European Community). In 1972 Belgium accounted for 10 per cent of the Community's production, or approximately 16 million tonnes. Belgium's export dependence at that time was 80 per cent. In terms of the per capita steel production quota, Belgium took the second place in the world to Luxembourg, with 1,200 kilogrammes per inhabitant per year.

An essential feature in the Belgium case is that the steel industry is located in the French-speaking part of Belgium. In 1975 every third job was in, or dependent on, the steel industry. The export value of iron and steel, as well as iron and steel goods and machinery, accounted at that time for about one-third of Belgium's total export value.

When the crisis came in 1975, the Belgian Government had to help the steel industry with state loans. A treatise, presented in 1978, pinpointed the major problems of the Belgium steel industry: weak market position in exports and on the domestic market, caused by low productivity compared to other European countries, very high wage levels and overaged plants. On the basis of this treatise it was decided to close several plants and to reduce employment from approximately 64,000 (1974) to 50,000 (1980). (Employment in 1981 was 44,000, and a publicised plan in 1983 foresees a further reduction to less than 30,000 by 1986.)

In order to make restructuring possible, the Belgian State acquired 30 per cent of the steel industry's stock. The 1978 plan proved to be too optimistic, however; a gap *vis-á-vis* the international competitors could not possibly be closed. In 1981, a merger of the largest corporations was decided upon, with an 80 per cent government participation. The deficit financing between 1975 and 1983 (beyond the acquisition of stock) is approximately US $ 2 billion. The output of steel production is currently approximately 8 million tonnes a year.

The case of Belgium is atypical, in so far as it also reflects the deep rupture cutting across a two-nation state. In comparison, the French programme has been much more efficient, although it has left the industry in much deeper trouble than the private Germany industry.

The European Community's Davignon plan for the restructuring and scaling down of the European steel industry is valid until the end of 1985, by which time restructuring should be implemented. It is indeed the most comprehensive, thorough and painful grand-scale, international reduction and modernisation plan that has ever been decided upon. From 1 January 1986 no government subsidies or support to the European steel industry will be permitted. All member governments of the Community have to announce the relief to be given to the industry until the termination date of the Davignon plan on 30 June 1983.

The transatlantic quarrel and the steel pact of 1982

On 11 January 1982 several American steel producers, viz. United States Steel, Bethlehem Steel, Republic Steel, Inland Steel, Jones and Laughlin, National Steel and Cyclops filed petitions with the International Trade Commission of the US Department of Commerce alleging injury as a consequence of subsidies to the production of steel in Belgium, Brazil, France, Italy, Luxembourg, the Netherlands, South Africa, the United Kingdom and West Germany. On the basis of this petition, on 25 February 1982, the International Trade Commission preliminarily determined a likelihood of injury.

The US Department of Commerce announced on 10 June 1982 its 'preliminary affirmative countervailing-duty determinations concerning certain steel products' to be imported from the aforementioned countries. The US customs service was instructed 'to require cash deposit or bond on these products in the amount equal to the estimated net subsidy' on a wide range of steel products imported from those countries after 10 June 1982. A corresponding final countervailing-duty determination was declared valid by the Department of Commerce as from 24 August 1982. Then, the final injury determination was taken by the International Trade Commission on 15 October 1982 by one vote only. The duties were to become final on 21 October of the same year. However, after more than six months of negotiations between the Department of Commerce and the Commission of the European Community, the case terminated in a quota deal covering the steel qualities and the countries mentioned in the petition (plus tubular products from the Federal Republic of Germany, which previously had not been part of the claim.) The duties which comprised the preliminary determination of 10 June 1982 were to be equal to the amount of subsidies on different steel qualities awarded by the governments of the aforementioned countries, and based on the estimates by the firms filing the claim. The Customs duties to be levied would have meant virtually closing the doors to the United States for imports from the British Steel Corporation and from the major steel producers of France (Sacilor and Usinor), Italy (Italsider), Belgium (Cockerill-Sambre) and South Africa (Iscor). The duties would have ranged from 12.5 to 40.4 per cent. The second group, that would have been affected by duties ranging from 3.6 to 8.6 per cent, comprised deliveries from other firms in Belgium, Brazil and South Africa as well as the Federal Republic of Germany. A final group consisted of firms which were regarded as receiving minimal subsidies only. They were excluded because the alleged subsidy was not worth mentioning (five firms in Germany, one in Holland and one in Luxembourg).

The final determination of 24 August by the US Department of Commerce reduced some rates, but, in other cases, increased the levy of the duty rates, after investigations. The duty on British Steel, for example, was reduced from 40 to 20 per cent. Some other firms were excluded, whereas Italsider would have seen its duties drastically increased. It has been estimated by US authorities that steel imports would have been reduced by approximately 1 million tons, if the levies had become legally valid.

The quota agreement of 21 October had about the same effect by limiting steel imports from the aforementioned countries to 5.13 per cent of apparent US steel consumption. The major difference between the duty claim and the final quota arrangement is the fact that the burden of restricted exports to the United States now is distributed more equally between the affected countries, whereas, if the duty arrangement had become valid, it would have hit the heavy subsidisers but left the low or nil subsidisers unhurt.

The incident will be discussed in more detail in the section on the European Community. Its importance lies not so much in the fact that the US steel industry wanted to neutralise estimated or real subsidies awarded to firms under reconstruction in their respective home countries. (In any case, at least as far as the European Community is concerned, it was not a question of dumping: the European Community under the Davignon reconstruction plan maintains a price cartel also including export prices. No firm was deviating from the generally agreed upon prices, and the claim was rather based on the fact that the firms in question, if they had calculated on a full cost basis, would have had to quote prices different from European Community cartel prices). Neither was the problem that the'preliminary affirmative countervailing-duty determination' of 10 June was taken less than a week after the summit between the heads of major western governments at Versailles near Paris, at which the President of the United States, together with his colleagues from at least all of the European countries concerned by the decision, had agreed to adhere to the principle of free world trade amongst their countries, in order not to further the crisis in world trade.

The problem was rather that this step taken by the US steel industry's leading firms and the responses by the International Trade Commission and the US Department of Commerce with the claims of the US steel industry, implied that a further major step in a sequence of protective measures was to be taken to the advantage of the US steel industry and at the expense of trade partners in Europe, Africa and South America, following a sequence of similar activities taken since the middle of the 1960s and to be followed by new ones shortly afterwards, for example concerning alloy steel imports, imports from Japan and so on.

A particular feature of this case was a remarkable change in procedure: previously an anti-dumping and countervailing-duty investigation in response to a petition from an industry or a firm had to be carried out by two bodies, the International Trade Commission and the US Department of Commerce. According to the rules, it took a minimum of nine months. The 1982 countervailing-duty case was treated in a new and different way, disadvantaging the foreign firms as well as their governments, which were forced into government-to-government agreements under severe time pressure.

The US Government, which originally had declared its intention of staying out of the conflict, was forced into it by pressure from Congress and its powerful steel caucus, which had been able to convince the majority of the Congress members that the predicament of the American steel industry was caused by imports and that quick action was necessary because the lengthy administrative procedure of litigations under trade laws etc. would hurt the industry still more.

It should also be emphasised that the European Davignon reconstruction plan, the price agreements and the sharing of burdens of reconstruction agreements were in full compliance with the GATT (General Agreement of Tariffs and Trade) Code on Subsidies and Countervailing Duties, as agreed upon and co-signed by the US Government at the Tokyo round of multilateral trade negotiations. Further, the quarrel over steel duties or quotas fell into a period of time when US–European relations were approaching an all-time low, for several other reasons. Altogether several, certainly unrelated, steps taken or pressures applied by the US Government on its European counterparts were felt by many European governments, including the

High Commission of the European Community, to imply a drastically changed US attitude *vis-á-vis* its European partners.

The quarrel over steel duties or quotas grew to dimensions which threatened the close transatlantic (as well as transpacific) ties between the US Government and national governments abroad, as viewed in the context of 1982, and to the prospects of an obvious change to a tougher climate in the formerly rather good transatlantic and transpacific relations.

One of the aims of this book is to explain, in as objective and neutral terms as possible, the cases of the different steel industries in the United States and the major steel-producing countries of the highly industrialised world.

The plan of the book

The issues

The central issue of this volume is to describe and analyse the present steel crisis, its character, its background, but also different policies adopted, steps taken and alternatives available to overcome the crisis in the medium or long term.

In order to depict the crisis, a number of analytical approaches have been used.

To begin with, a number of statistical approaches are used to X-ray different background factors and the character of the crisis at international as well as national levels.

In a number of steps, the steel industries of major steel-producing countries are scrutinised as to their historical development in modern times, as far as it is relevant to the present situation, and its possible solutions. The emphasis is on the major agents on the international steel markets: the United States, Japan and the European Community, as well as its major steel-producing member countries: Great Britain, France and the Federal Republic of Germany.

The impact of environmental regulation upon the industry is assessed, as well as the interaction between the steel industry and its location, or the region in which it is dominant, again with particular emphasis on the possibilities to overcome the crisis.

As the innovation capacity of the industry has played an important role in the post-World War II period when it comes to changes in relative competitiveness of the steel industries of different nations, and as the innovation potential of the industry is supposed to play an important role in the possibilities of overcoming certain aspects of the crisis, the innovation syndrome is treated quite thoroughly.

The volume is concludes with a stocktake of the strategies, policies and instruments available at international, national, industrial as well as enterprise levels, in attempts to overcome the crisis.

The market potential and the market opportunities of different steel industries are assessed from a range of points of view.

In order to provide differentiated views on the major issues, experts with high international standing have been asked to contribute chapters on central issues to the volume. A differentiated view and analysis is necessary in order not only to understand the present crisis, but also to discuss and assess possible routes out of the crisis: *as little as the crisis is uniform, as little there are general, uniform routes out of the crisis* .

It is necessary to understand the different causes of the present disease. A thorough and good diagnosis is the root of effective therapy; it is a must to understand that, if the diagnosis proves that the illness has several, or many, causes, differentiated therapy measures are to be applied, in order to give efficient results.

Notes

1   Under the heading 'Protectionism, Reagan-Style: the Steel Quotas', *Fortune* concludes: 'President Reagan's decision to slap quotas and tariffs on US imports of specialty steel undercuts his lofty free trade declarations at Williamsburgh, has enraged our trading partners, and won't even do much to help the steel makers. *Fortune*, 8 Aug. 1983, p.55.

2   Europe is essentially treated as being the European Community (EC) and its member countries. Non-EC countries, although some of them are affiliated to the EC iron and steel market, like Austria, Sweden and Switzerland, are not treated despite the fact that they possess long tradition and high competence in steelmaking. As will be briefly mentioned, they are today still leaders in technology: some major inventions and innovations now in wide use have been developed there, and the countries still compete successfully with world giants as highly competent suppliers of technology.

3   This statement implies no downgrading of the influence and importance of the United Nation Organisation and its bodies, e.g. the UNIDO (United Nations Industrial Development Organisation), of the OECD (Organisation for Economic Co-operation and Development), the IISI (International Iron and Steel Institute) or GATT (General Agreement on Tariffs and Trade). They simply do not act as decision bodies, but may well be forums for discussing or even adopting certain policies,

which is a task put into the hands of e.g. the OECD Steel Commission. See, for example, Florkoski, E.S., 'Policy Responses for the World Steel Industry in the 1980s', in OECD, *Steel in the 80s'*, Paris, 1980.

4 Japan's Ministry of Agriculture and Commerce for defence purposes founded Yawata Steel Works in 1900, which in 1934 merged with Fuji to Japan Steel, but was dissolved by the supreme command of the allied powers in 1945.

The Third Reich's steel combines were given to the Federal Republic of Austria and the Federal Republic of Germany by the allied powers. In Austria it is the dominant firm, in Germany it is number six.

The British Steel Corporation was created by the Nationalisation Act of 1967. It has suffered greatly from political instability, and the original development and rationalisation plans were never therefore implemented. Political forces also prevented the implementation of the urgently needed restructuring and rationalisation programmes.

In Italy the state is the sole or majority holder in two-thirds of the integrated steel industry.

France has a long tradition of state intervention in iron and steel.

In some countries, like Belgium, the state has entered into the steel industry only very recently in order to prevent bankruptcies with consequent mass unemployment.

5 Like the 'Minette' of Lorraine also holding a high phosphorus content.

PART II
STEEL STATISTICS

2 Introduction

Long-term trends in demand

The long-term trend for iron and steel demand is heavily influenced by investment.[1] Investment may usefully be subdivided into construction investment, investment in productive machinery (including heavy transportation equipment) and private capital investment. Historical evidence proves that the build-up of industrial capacity in a country, that is its industrialisation, provides the strongest push to steel demand. All later investment moves are of weaker strength and more evenly distributed, although they are controlled by cyclical patterns of mid-term duration (for example Kusnetz cycles for industrial construction and equipment patterns and Juglar cycles for consumption patterns).

In historical perspective the industrialisation moves pushing iron and steel demand seem to last for approximately 30 years, after which period the demand continues to grow although at more modest growth rates. The strongest demand in growth in the German Reich[2] occurred between 1850 and 1880, in the United States between 1870 and 1900, and in Japan (and possibly Latin America) between 1950 and 1980. Similar patterns can also be recognised for socialist countries, which in their state planning efforts in early periods tend to emphasise raw materials and heavy industries, at the expense of suppressed and retarded private consumption.

Industrialisation growth on average shows growth figures of around 10 per cent a year. During the first growth decade steel is often imported in considerable quantities in order to build up heavy industries. This may also occur after periods of heavy destruction by wars as, for example, in Europe after 1945. After an initial period of approximately five to ten years, however, the nation or region usually becomes self-supporting on carbon steel, that is mass consumption steel, whereas imports of alloy and special steel tend to continue.

Given the above development, one can distinguish a number of regional markets with considerable internal but gradually rather marginal external steel trade:

a) Japan;
b) The North American market;
c) The Western European market;
d) The COMECON market – the market of the socialist countries;
e) Newly industrialised countries (several reasonably self-supporting markets can be distinguished, e.g. India, South America, South Korea and Taiwan); and
f) The developing countries where external demand still may be considerable but where it is mainly directed towards the closest levels above in the development hierarchy, that is, in the first place, towards the newly industrialised countries.

The higher the level of development, as, for example, measured by GNP per capita, the lower the per capita demand appears to become (see Chapters 3 and 4), even if countries enjoying high levels of development during certain periods are exporters of more or less steel-intense investment goods, particularly of industrial machinery and consumer capital goods. Obviously, however, after some time, steel-intense investment goods (like ships and transportation equipment) are often being produced in countries which gradually reach higher levels of development. Traditional shipbuilders like the United States, Great Britain, Holland and Sweden have disappeared, one after the other, from the top ranks of the shipbuilding nations. These ranks have gradually been taken over by younger industrial competitors. One can find similar developments in truck, locomotive and engine production.

Another development is that the steel intensity of investment goods seems to decline gradually: maturing products tend to get lighter and thus to use less steel.[3]

In summary, a number of factors seem to influence the demand for steel in homogenous regions (from the industrialisation point of view) to take the shape of sine curves. The steep slope of the sine curve may last for one generation or so. However, the growth seems to level off, becoming rather modest in the long term. The obviously most important factor is the apparent steel consumption's share of investment gradually becoming lower. Demand stemming from areas of high development intensity will only to a marginal extent be channelled to the highly developed countries. Thus the demand stemming from abroad is, and will remain, marginal.

It is rather unlikely that steel will recover markets it has lost to other substitute materials like aluminium and plastics. It is more likely that the substitute materials will gain at the expense of steel.

Supply

A number of structural changes can be observed when it comes to the supply side for iron and steel.

In old industrial regions the steel industry usually grew in the neighbourhood of major raw material deposits like iron ore and/or coal. In the most favoured cases, iron and coal appeared close to each other and accounted for the development of heavily industrialised regions. Access to bulk transportation at reasonable cost per tonne-mile or tonne-kilometer and low terminal cost or cost for reloading played an important role: the greatest achievement of the Ministry of Agriculture and Commerce, the predecessor of MITI of Japan, was the establishment of the Yawata Iron Works at a coastal site, near coal deposits, with access to deep-sea harbours for the imports of iron ore, which had become accessible to Japan after Japan had gained a victory in the first Sino-Japanese war. The Yawata Works were established between 1896 and 1901. Similar conditions accounted for the development of major industrial regions in the United States, around the Great Lakes, in Europe (the Ruhr District and so on) and even elsewhere.

Post-World War II development brought a continuation and gradually an accentuation of the growth of the scale at which steel could be produced in integrated steelworks. This development culminated in Japanese steelworks designed for an annual production of over 10 million tonnes (as against the average scale size of under 2 million tonnes capacity per year in the large US steel companies). This development both made necessary, and coincided in time with, fairly drastic declines in the cost for bulk transportation of ore and coal/fuel and also improved access to high-grade ore from new and often very distant locations like Africa, South America and Australia. In response to those developments, many new large-scale integrated steelworks were established at deep-sea harbour locations. Steel plants at traditional locations saw their competitive advantages dwindle, but were often in at least a temporarily favourable position to operate at much lower capital cost (accounting for approximately 25 per cent of the production costs of steel). Gradual improvement and adaptation became the defence against radical change.

At the same time as the economical scale of production for steelworks grew to gigantic levels, the capital needed to establish large-scale integrated greenfield plants rose to astronomic sums. Only the most powerful financial groups, or the state, could keep pace with such requirements. As the 1960s were a growth period of exceptional duration and intensity, the demand for steel was kept at rather high levels. World trade in steel grew from 25 million tonnes to approximately 120 million tonnes per annum in an almost perfect sinusoidal pattern, and a considerable number of independent investment decisions were taken. Thus, a large number of steel plants of considerable size were decided upon in quite a number of countries and regions, in particular in Japan and newly industrialised countries. Some such plants also materialised in Europe. A few other investment decisions were taken so late that they were never implemented.

At the same time, many older plants were modernised, or 'retrofitted', both to employ modern production technology and to adapt the old equipment to both the strengthening of environmental requirements and to skyrocketing energy prices.

As mentioned above, many investment decisions in new steel capacity were taken independently. In most cases, however, the decision base was the same or came from the same source concerning estimated demand development for steel consumption. Of course, one would not take a decision to invest 5 to 10 billion dollars in steel-producing capacity if there were not a good future market for steel.

The mere size of new capacity added per plant in most cases was unprecedented. Adding only one new plant in the beginning of the 1970s would mean the addition of capacity corresponding to 10 to 15 per cent of total world trade. So one may conclude that, through a combination of utilisation of optimistic forecasts and quite a number of independent decisions to invest in giant-size steel plants, a rather substantive overcapacity for production of carbon steel came into existence during the 1970s. Nevertheless, the 1970s still saw the best steel year ever, 1974. The later 1970s were still characterised by growth, although at a much lower rate. Only the beginning of the 1980s brought rapidly declining consumption and thus also lower production figures.

Another new trend became influential in the supply of steel: whereas some new processes (basic oxygen furnaces and, later, continuous casting) pushed the economies of scale upwards, at the low scale side, access to iron and steel scrap and to reasonably priced electricity supplies opened roads for small-scale plants with favourable locations near centres of investment growth. The so-called mini-mills gained shares of growth markets, mostly in the rather unsophisticated segments of the carbon steel market but at highly competitive prices, as their major advantage was low capital cost (1 tonne of capacity per year in a mini-mill costs approximately 100 dollars, less than one-tenth of the investment cost needed in an integrated mill).

Mini-mills can operate at low transportation cost: their raw material frequently comes from the same region to which their final products go and where they thus choose their sites. As their major raw material input, scrap steel, is already highly refined and thus is much less dependent on transport-intensive and expensive fuel, they can afford to use fuel in one of its most refined forms: electricity, often at reasonable prices (for various reasons, depending upon where they are sited). With all those advantages at their hand plus the flexibility that closeness to the market makes possible, they gained market shares at a remarkable pace. Within less than 20 years they have been able to cover one-fifth of the American market, and their growth potential is still high. At the same time, the economic production size of a mini-mill is gradually growing. Within a few years it is expected to be in the same average size range as old integrated plants are operating. This will gradually shake up a market which today is operating without due regard to capital recovery.

In summary, the tendency at the supply side of steel is that the middle-range producers, which have not been adapting to the scale economies of the modern large integrated producers but have been operating by patchwork improvement and by neglecting capital recovery, have not been able to build up resources for the re-investment which will be necessary in five to ten years' time. This may mean that the traditional medium-scale integrated sector of steel production will gradually be wiped out by competition from below and from above and by their lack of access to capital resources.

World trade of steel

World steel export is remarkably stable and static. In 1974, the peak year, 107.6 million tonnes (1978 107.0, 1979 107.9) and in 1980 105.7 million tonnes were traded.

During the same period (1974–80) the export share of world steel production was also relatively stable, hovering around 24 to 25 per cent. In periods prior to this there has been a gradual increase of the export share of world production from around 10 to 11 per cent during the first half of the 1950s, to 13 per cent during the late 1950s, 16 per cent during the early 1960s, 17 to 19 per cent during the late 1960s, 20 to 24 per cent during the early 1970s.

If analysed by product, world steel trade has also been relatively stable, except for plates (demand falling from 13 million tonnes in 1974 gradually to 8.8 million tonnes in 1980). The only other major movement has been in steel tubes and fittings (from 12.8 to approximately

15.5 over the latter half of the 1970s; with falling oil prices, pipe and tube demand fell immediately).

In a comparison of steel exporting and importing countries many amongst the top ten appear as exporters as well as importers, but not Japan (only an exporter) or Mainland China (only an importer).

The future: supply follows demand

Historical analysis demonstrates that production follows consumption, that is, steel is produced where it is consumed. The developing countries can hardly afford to buy from the highly developed industrialised countries, where the cost of production is far too high. They will rather buy from regions at approximately the same or slightly higher levels of development, which thus operate at reasonable cost levels. They will also invest in both integrated and, most likely to an increasing extent, in mini-mill capacity. Developing countries, but also highly developed countries, will remain export markets for marginal quantities of special steel or specialised steel products.

There has been a tendency to sell abroad at lower prices than at home. This does not necessarily mean dumping. It only means using marginal production capacity at marginal production cost. However, market distortions, for example anti-dumping measures, have temporarily resulted in opposite policies: steel is sold at higher prices abroad than at home. This, however, is also an effect of quantitative restrictions, which force suppliers to sell high quality, high-priced products to markets regulated by quota arrangements.

In general there is overcapacity in Europe, the USA and Japan. How much overcapacity exists is difficult to tell today since nominal figures are not necessarily realistic figures. A general tendency, however, is that whilst it is necessary to reduce steel industry capacity overall, many measures have been taken to increase capacity instead, in particular by replacing old capacity by modern high productive capacity.

A major conclusion of the statistical analysis is that there exist almost self-sufficient regions: a) the industrialised world of North America, Western Europe, and Japan, b) the socialist countries (the COMECON block), c) newly industrialised countries (South Korea, Taiwan, South Africa, Australia, Mexico, Brazil, India, and a few others), and d) the developing countries in various regions, in particular in Africa and South America but also in Asia.

In each of those regions general self-sufficiency in steel supplies matching steel demand is being established. There is a considerable inter-regional steel trade taking place which, however, has become saturated over time. Today, and most likely even in the future, only marginal quantities are being traded between the regions of either special steel or of carbon steel, in cases of temporary imbalance between supply and demand. There are also limited quantities of low-price bulk carbon steel traded between regions, but only at marginal levels.

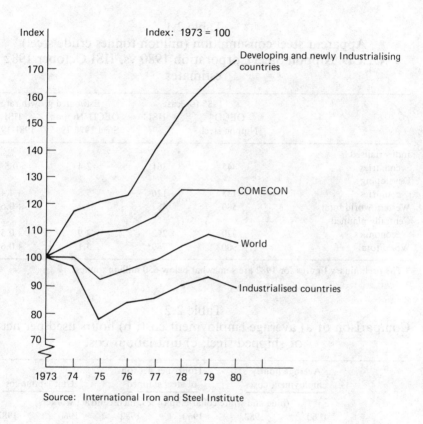

Source: International Iron and Steel Institute

Figure 2.1 Steel consumption: relative development by major regions

Comments on forecasting of apparent steel consumption

At the Paris Symposium of the Organisation for Economic Cooperation and Development in February 1980, 'Steel in the '80s', the discussion of demand and supply centred around a forecast presented by Tsutomo Kono[4] based on Nippon Steel Corporation forecasts. The forecast, which was discussed at the symposium at length but not revised, projected an apparent steel consumption of 900 million tonnes

45

of crude steel for 1985, as well as a growth rate 1978 to 1985 as shown in Table 2.1 compared with the IISI Secretariat forecast of October 1982. The October 1982 IISI forecast must be regarded as rather optimistic, even at the time it was submitted.

Table 2.1
Apparent steel consumption (million tonnes crude steel) OECD/Nippon Steel Corporation 1980 vs. IISI October 1982 estimates

| | 1985 forecast | | Estimated growth rates | |
	OECD Nippon steel	IISI	OECD/Nippon Steel 1978-1985	IISI 1981-1985
Industrialised countries	445	361	2.4	-0.5
Developing countries	135	126	5.8	+4.4
Western world total	580	487	3.2	+0.6
Centrally planned economies	320	262	2.9	+0.5
World total	900	749*	3.1	+0.6

* The preliminary figures for 1982 are somewhat below 650 million tonnes.

Table 2.2
Comparison of a) average employment cost; b) hours used per net ton of shipped steel; c) unit labour cost

| | Average hourly employment costs | | Hours per net ton of steel shipped | | Unit labour cost | |
| | (blue and white collar employees, incl. contract workers) | | | | | |
	1960	1982	1960	1982	1960	1982
US	4.12	23.00	17	9	67	207
EC	1.16	12.50	21	10	24	125
Japan	.54	10.00	51	7	28	70
Brazil	.60	2.60	n/a	20	n/a	52
South Korea	n/a	2.35	n/a	11	n/a	26

Sources: AISI, *Annual Statistical Report*, various issues and *Steel Employment News*, 11 March 1982 (white-collar employee differential estimated); Eurostat, *Wages and Incomes*, various issues; Ministry of Labour, Japan, *Monthly Survey of Labor Statistics*; NRI (a consulting firm) Tokyo; *Siderurgia Latinoamericana*, June 1981, p.7.

Mueller, H., *A Comparative Analysis of Steel Industries in Industrialized and Newly Industrializing Countries*; Conference Paper Series No. 72, Middle Tennessee State University, Murfreesboro, April 1982 (1982b).

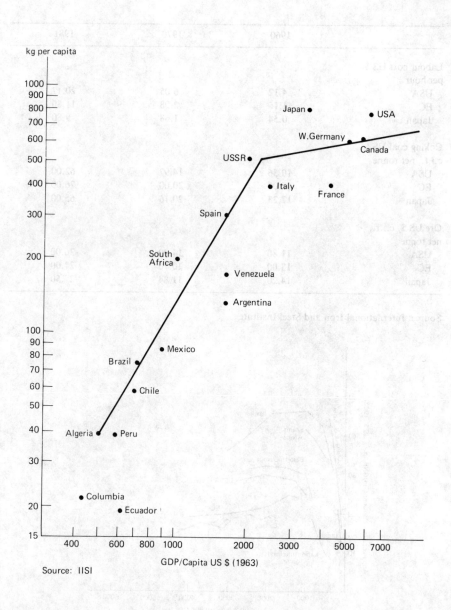

kg per capita

Japan ●
USA ●
W.Germany ●
Canada ●
USSR ●
Italy ●
France ●
Spain ●
South Africa ●
Venezuela ●
Argentina ●
Mexico ●
Brazil ●
Chile ●
Algeria ● Peru ●
Columbia ●
Ecuador ●

GDP/Capita US $ (1963)

Source: IISI

Figure 2.2 Steel consumption per capita

47

Table 2.3
An international comparison of production cost

	1960	1970	1981
Labour cost US $ per hour			
USA	4.12	6.05	20.17
EC	1.16	2.98	11.80
Japan	0.54	1.68	9.80
Coking coal US $ c.i.f., net tonne			
USA	10.56	14.97	62.00
EC	17.00	20.00	76.00
Japan	17.23	20.16	65.00
Ore US $, c.i.f., net tonne			
USA	11.80	14.90	36.00
EC	13.00	10.70	27.00
Japan	14.20	11.84	27.50

Source: International Iron and Steel Institute

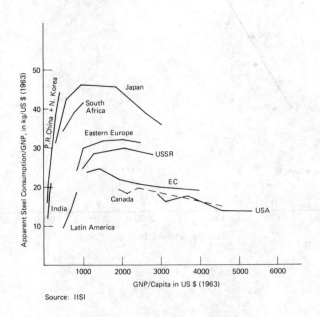

Source: IISI

Figure 2.3 Steel intensity in various countries

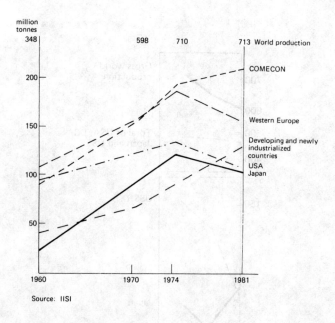

Source: IISI

Figure 2.4 Production of crude steel (million metric tonnes)

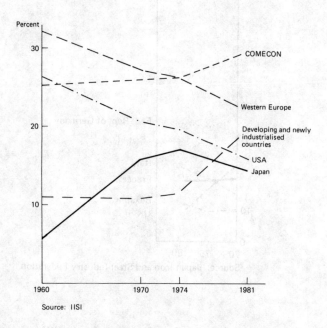

Source: IISI

Figure 2.5 Share of world raw steel production

49

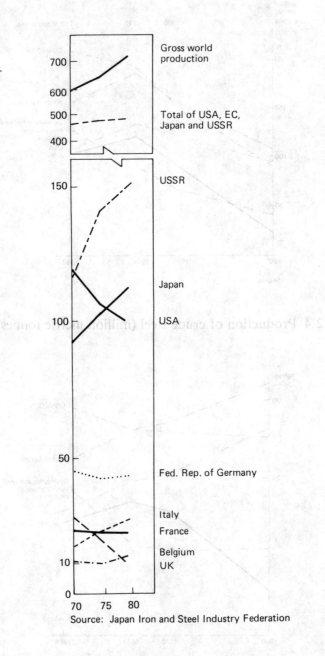

Figure 2.6 Crude steel production by country (million tonnes) 1970, 1975, 1980

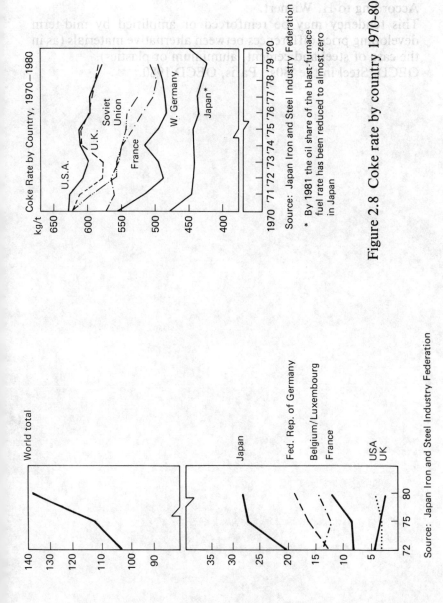

Figure 2.8 Coke rate by country 1970-80

Coke Rate by Country, 1970–1980

kg/t

U.S.A.
U.K.
Soviet Union
France
W. Germany
Japan*

650
600
550
500
450
400

1970 '71 '72 '73 '74 '75 '76 '77 '78 '79 '20

Source: Japan Iron and Steel Industry Federation

* By 1981 the oil share of the blast-furnace fuel rate has been reduced to almost zero in Japan

World total

140
130
120
110
100
90

Japan

Fed. Rep. of Germany

Belgium/Luxembourg
France

USA
UK

35
30
25
20
15
10
5

72   75   80

Source: Japan Iron and Steel Industry Federation

Figure 2.7 Steel exports of major countries (million metric tonnes) 1972, 1975, 1980

Notes

1 See Chapter 4 by H. Wienert.
2 According to H. Wienert.
3 This tendency may be reinforced or amplified by mid-term developing price differences between alternative materials (as in the case of steel and cement, aluminium or plastics).
4 OECD, Steel in the '80s , Paris, OECD 1980..

3 Crisis, adjustment and out-look of the steel industry

Carlo Maria Guerci and Sergio Treichler

The characteristics of the sector

The steel industry is an example of a mature sector. Its market growth is low, and the industry is suffering from a deep structural crisis. The main causes of these phenomena can be found in the stages of economic development of mature countries, which use less and less steel (see Figure 3.1) on account of the decreasing relative importance of the industrial sector in the GDP[1] and to the diminishing demand for steel in industry. In such a stage the more steel-intensive sectors, such as the building industry and the investments goods industry, reduce their share of total steel consumption while other sectors more susceptible to the reduction of the specific weight per unit of product, such as durable consumer goods, increase their share.

Changes in the structure of supply in the 1970s

The steel sector is highly concentrated. This oligopolistic form of supply (where the 30 leading companies produce about 60 per cent of the steel of market economy countries) has undergone a slight 'pulverisation' in Europe and Japan (see Figure 3.2), and, together with the pressure and concentration of big groups, there has been a development of smaller production units, better suited to serve new segments of the market and to adjust to its highly cyclical trend. This phenomenon is particularly evident in Italy: if compared with other EEC countries, Italy shows a recent growth of smaller-sized companies and plants (Figures 3.3 a and b). Changes in productive processes are the second important factor: the utilisation of the electric furnace is expanding in all countries and, in this case, the Italian market share –

which is higher than in other countries (see Figure 3.4) – can be explained by the higher elasticity and lower vulnerability to market fluctuations of the electric arc furnace process-based firms. After the 1975 crisis, the reorganisation of plants and the development of new capacities favoured processes which are better suited to adjust to cyclical trends of the market.[2]

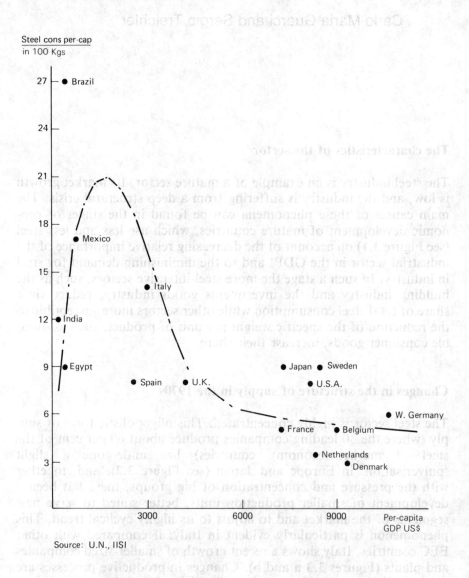

Source: U.N., IISI

Figure 3.1 1979 steel consumption and per capita GDP (1975 prices)

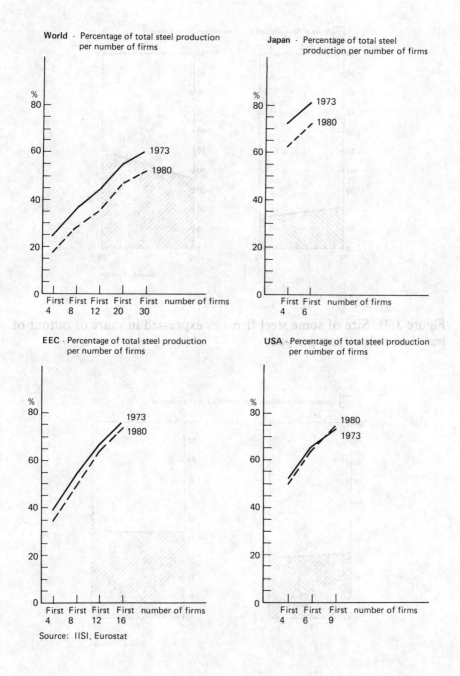

Figure 3.2 Degree of concentration in the steel industry

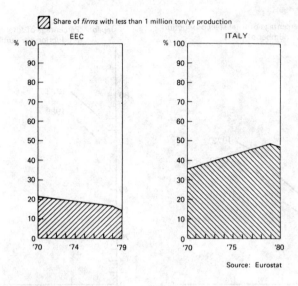

Figure 3.3a Size of some steel firms as expressed in share of output of less than million ton per year

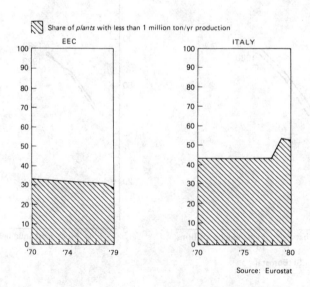

Figure 3.3b Size of steel plants

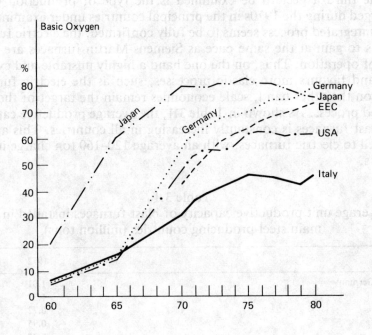

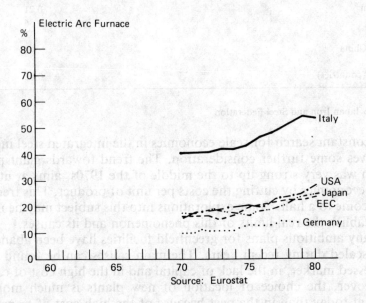

Source: Eurostat

Figure 3.4 Share of raw steel output per year according to production process

The third aspect to be examined is the type of production which emerged during the 1970s in the principal countries under examination. The integrated process seems to be fully confirmed; the electric furnace seems to gain at the same pace as Siemens-Martin furnaces are taken out of operation. Thus, on the one hand a highly unstable and cyclical demand favours more elastic processes, such as the electric furnace, but, on the other hand, scale economies remain the target of the integrated process. As shown in Table 3.1, the average productive capacity of blast furnaces is constantly increasing in all countries. This applies as well to electric furnaces, with an average 120–160 ton charge in new plants.

Table 3.1
Average unit productive capacity of blast furnaces installed in the main steel-producing countries (million tons)

	1972	1979
West Germany	1.01	1.38
Japan	1.27	1.35
USSR	0.72	1.31
Italy	0.95	1.13
UK	0.77	1.08
USA	0.97	1.00
Belgium	0.71	0.96
France	0.56	0.85
PR of China	–	0.24
EEC (7 countries)	0.82	1.12

Source: Japan Iron and Steel Federation.

The constant search for scale economies in the integrated steel industry deserves some further consideration. The trend towards giant plants, which was very strong up to the middle of the 1970s, aims at utilising scale economies by cutting the costs per unit of product. This trend has now come to a halt. Some explorations into this subject may be useful, to establish the real limits of this phenomenon and its causes.

Many ambitious plans for greenfield facilities have been abandoned or rescaled during recent years. The main causes can be found in the depressed market, in the lack of capital and in the high cost of capital. Moreover, the choice of location of new plants is much more important today than in the past because of the high cost of transportation: in many cases, locating plants with a smaller productive capacity in the vicinity of steel consumption centres appears to be more rational. The cost of a new integrated plant – with 2.5 million tons steel production capacity per year – in industrialised countries is expected to be about 4 billion dollars. Consequently, if we consider the present interest rates and the list prices of steel, the finished products would be sold with a US $ 60/ton loss. That is why, since 1973, nobody has surpassed the productive capacity established in that year by Nippon Kokan Steel at Fukuyama (16 million tons per year). Nevertheless, this does not mean that scale economies have lost their validity. Let us examine, for instance, direct reduction and electric arc furnace plants, such as the Contrecoeur-based SIDBEC-DOSCO (Canada): in 1977, the capacity of its steel plant was enlarged from 600,000 to 1,400,000 tons per year. If we consider process units instead of entire steel plants, we note the start-up of the following units, during the last few years:

Production unit	Capacity	Plant	Year
blast furnace	16m crucible diameter	Sumitomo-Cashima	1976
LD-converter	400 t	Dzershinsk (USSR)	1980
electric arc furnace	350 t	Jones & Laughlin Pittsburgh	1979

The ever-increasing capacity of units of the same type, developed over the last few years, seems to have reached its top level. This shows that the scale economy, which motivated the establishment of giant plants, retains its importance in the industry, even if the economic advantages are diminishing as plant capacity increases.

The management of complex organisations presents some difficulties, which are obviously growing to a point where they nullify gains from the purely technical effects of scale economies. These difficulties often prevent market or price leadership.

A correlation between the sizes of firm in each steel business and the total firm rentability has been calculated on a sample of European companies. Figure 3.5 shows the low correlation between scale index and rentability. It also demonstrates the presence of smaller firms enjoying a higher rentability than larger competitors'.

Trend and structural changes of demand

With regard to the changes which took place in the geographic structure of demand, the role played by the developing nations in relation to the industrialised ones is well known. Italy settles in the middle. This is due to strong growth of consumption between 1979 and 1980, contrary

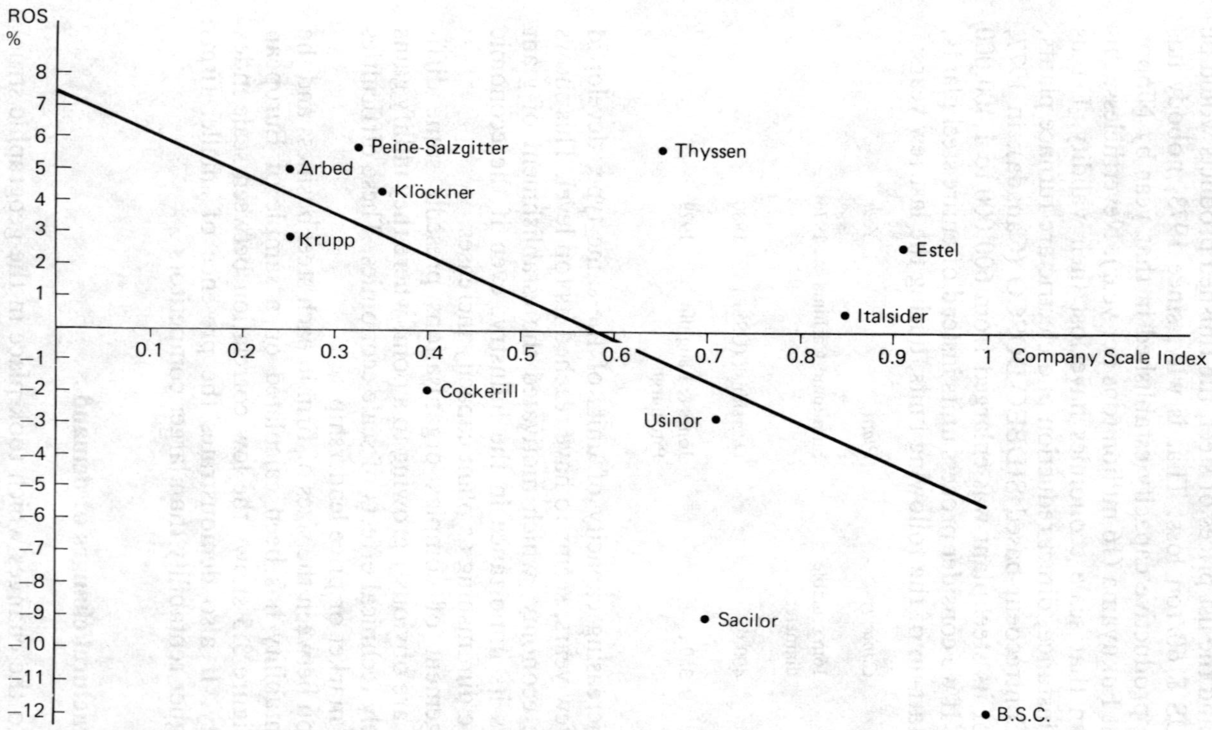

Figure 3.5 Scale index and rentability in steelmaking for the main European companies in 1979

ROS: Return on sales (i.e. net profit before taxes and interests)
COMPANY'S SCALE INDEX: Weighted average of the single product scale indexes
PRODUCT SCALE INDEX: Ratio of its volume of production to the one of the "leader"

Source: Company's Annual Reports — EEC

60

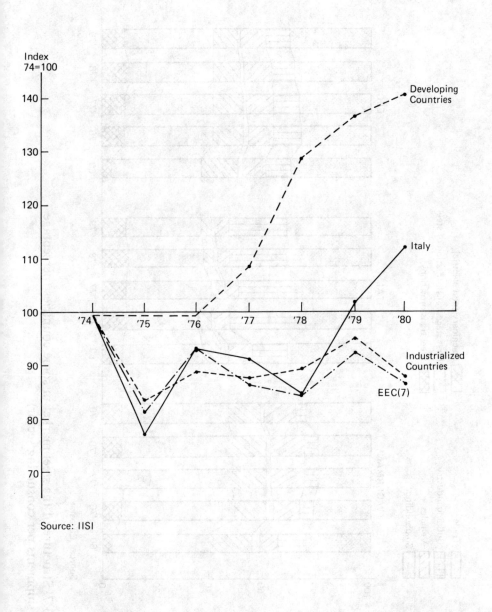

Figure 3.6 Apparent consumption of crude steel (1979 = 100)

61

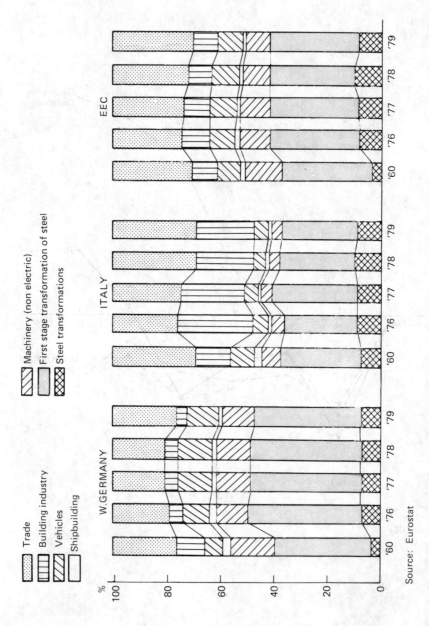

Figure 3.7 Structure of the steel market in some countries (percentage of steel shipments per consuming sector)

Source: Eurostat

62

to what happened in other countries (see Figure 3.6).

The causes of the low level of demand in the industrialised countries are to be found:

a) in the relative stability of steel final demand (see Figure 3.7), which shows that there are no new sectors of high demand;
b) in the structural crisis these countries have undergone since 1974 (see Figure 3.8), which accounts for approximately 30 per cent of the low development of consumption;
c) in the drastic reduction of specific weight per unit of product many sectors adopted during the crisis; this accounts for approximately 70 per cent of the low development of demand (see Figure 3.9).

Within a mature and stable demand, however, deep changes took place according to a diversified pattern, which corresponded to the different growth rates of the various steel products. There are many managerial implications for the steel industry. It will have to face a difficult market in the 1980s, since steel can no longer be regarded as a simple undifferentiated commodity, but as a range of products for specific utilisation segments, where technical assistance and distribution gain a key role in purchases.

Location factors in the steel industry

The steel industry is highly capital-intensive. On the one hand, the low rate of innovation permits depreciation on a 15–20 year basis. On the other hand, the ever-increasing amount of money needed to start a steel plant has drastically reduced the opportunities for developing countries.

As shown by means of a standard profit and loss account of an integrated steel plant (Figure 3.10), labour cost accounts for the major part. Table 3.2 shows the difference existing between wages in industrialised countries. It is estimated that the investment cost share is about US $ 2,200 per ton of steel in the developing countries and about US $1,600 in industrialised ones.

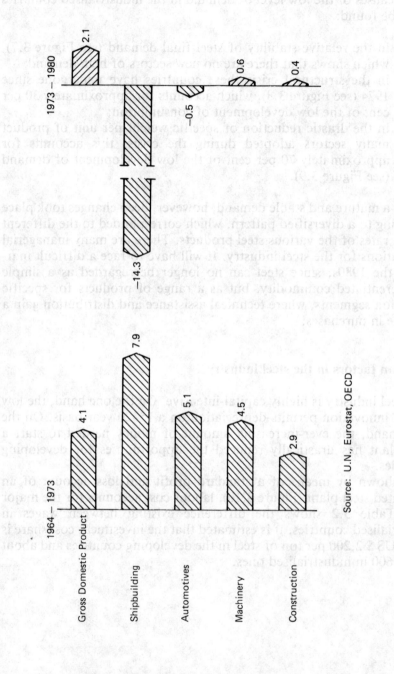

Figure 3.8 Trends of steel using sectors in EEC member countries

Source: U.N., Eurostat, OECD

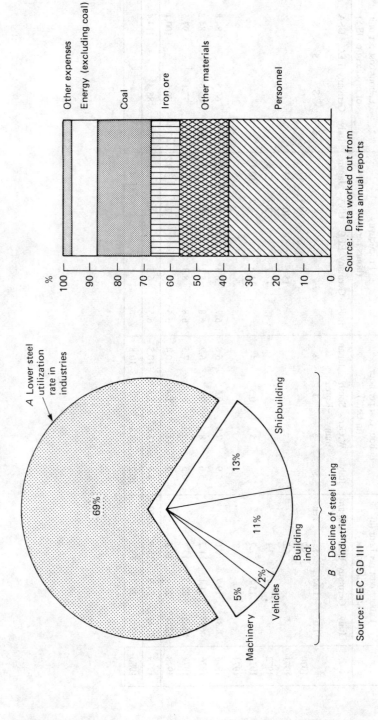

Figure 3.9 Estimated causes of the decline of steel demand within the EEC

Figure 3.10 Structure of standard industrial costs of an integrated steel plant within the EEC 1979

Table 3.2

Labour cost, productivity and labour cost per unit of product in the various steel industries

Year	Labour cost per hour (in national currency)				Labour cost per hour (in US $)				Productivity (hours per ton)					Labour cost per unit of product (in US $)				
	Italy Lit	Germany DM	USA $	Japan Yen	Italy $	W.Germany $	South Korea $	Japan $	Italy	Germany	EEC	USA	Japan	Italy $	Germany $	EEC $	USA $	Japan $
1970			5.7							13.8		10.4	13.7					
1971			6.2							15		10.5	13.7					
1972			7.1							12.7		9.7	12.8					
1973		15.2	7.7			5.7	0.5	4		11.6		9.2	10.1					
1974																		
1975	3881	19.7	10.6		5.9	7.9	0.5	5.8	6.7	8.3		8	6.2	39.5	65.5		84.8	35.9
1976	4697	21.1	11.7		5.6	8	0.7	6.2	6.8	8.2		7.2	5.9	38	65.6		84.2	36.5
1977	5585	22.3	13		6.3	9.3	1	7.5	6.8	8.3		7.5	6.1	42.8	77.1		97.5	45.7
1978	6231	23	14.3		7.3	11.3	1.4	10.4	6.5	7		7	5.9	47.4	79.1		100.1	61.3
1979	7343	24.9	15.9	2303	8.8	13.6	1.6	10.5	6.5	6.5	6	7.2	5.1	57.2	88.4		114.4	53.5
1980	8741	26.3	18.5		10.2	14.4		10.8					4.9					

Table 3.3 shows that developing countries absorb a very high share of the total investments in steel, related to the still low quantity of steel they produce.

Table 3.3
Share of a) production of steel and b) investment in the steel industry in 24 countries, 1980

	Production	Investment in Steel industry
EEC (7)	27.1	19.1
Other Western countries (5)	5.7	6.1
Canada	3.7	3.0
USA	23.6	20.1
Latin America	5.8	20.7
Japan	25.9	17.2
South Africa and Oceania	3.9	2.9
India and Republic of South Korea	4.3	10.9
	100	100

Source: IISI

The causes of the crisis and actions taken to overcome it

The causes of the crisis

The steel sector crisis was caused in the major industrialised countries mainly by an imbalance between demand and supply (Figure 3.11). Sluggish demand at the beginning of the 1970s was met by an increasing productive capacity, due to investment decisions taken under the influence of very optimistic demand forecasts. The subsequent low rate of utilisation of plants (Figure 3.12) caused:

a) a necessary phase of rationalisation/reduction of variable costs;
b) the creation of integrated areas protected by means of industrial politics, aimed at preserving the steel industry from 'price wars', at maintaining high production levels, keeping down the cost of units produced even if this policy drove many firms into heavy deficits.

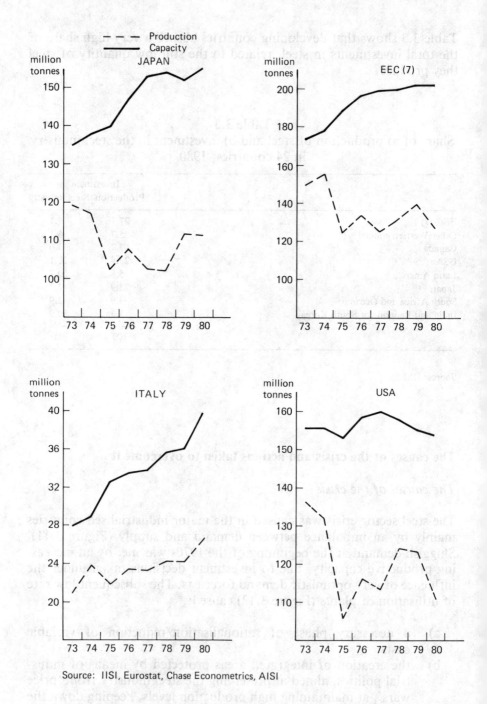

Figure 3.11 Trends of production and capacity of the main steel producers (million tonnes of crude steel)

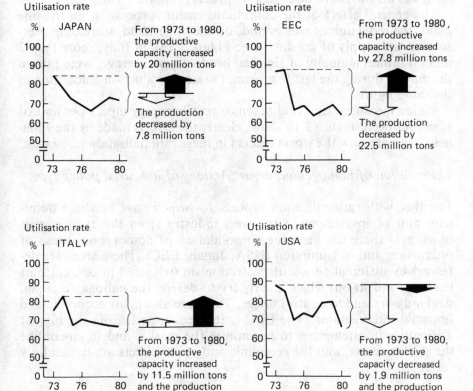

Figure 3.12 Utilisation rates of productive capacities in the various steel industries

Rationalisation

All the main countries are now reducing their investments (Figure 3.13), thus showing the maturity reached by the sector. Substantive financial resources are devoted to rationalisation measures at the existing plants and to the improvement of production yields, with a particular focus on the continuous casting process (Figure 3.14).

The general effect of the continuous casting process is, on the one hand, the reduction of wastes and, on the other hand, increasing pressure on the supply of products (see Figure 3.15): in Italy, more than 2 million tonnes annually of finished products, on average, were put on the market during the last few years, the steel production amounting to the same figure.

Major measures are being taken to reduce energy inputs per unit of steel produced. Figure 3.16 shows clearly the efforts made by the Japanese, who are now the world leaders in these rationalisation processes.[3]

The creation of homogenous areas. Actions of industrial policy type

Together with rationalisation processes – which have been the dominant aim of investment in the steel industry from the energy crisis onwards – there has been the consolidation of homogenous areas of production and consumption (USA, Japan, EEC). These areas are defended by different forms of protectionism (triggered prices, customs barriers, and so on) which clearly try to defend the national/regional steel industry and to control prices. There are also many economic and financial interventions. The EEC is the area experiencing the biggest difficulties in attempting to harmonise the market and to streamline the steel industry, and the economic and social effects are particularly severe.

Changes in international trade

Steel's share of world trade is approximately 6 per cent. This trade developed in conformity with the volume of world trade in manufactured products, but has slowed down during the last few years (see Figure 3.17). To an increasing degree various countries tend to produce steel for internal consumption. This is the first result of a careful analysis of international trade.

The internationalisation of the steel sector is now slowing down (Figure 3.18), with big differences in the various regions/countries and *an ever-increasing presence of cyclical exports, as a consequence of declining internal demand in different economies*. This is clear for Italy (where in 1978 exports reached about 50 per cent of production). The Japanese position, however, deserves some further consideration.

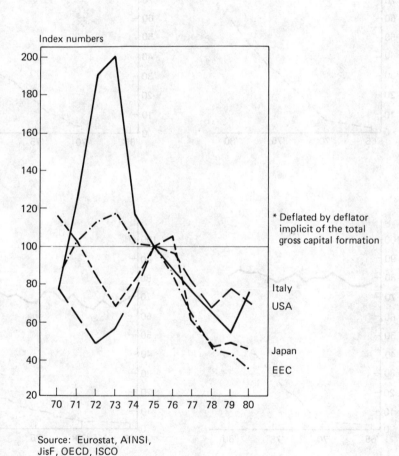

Figure 3.13 Relative development of real investment in the steel industry* (1975 = 100)

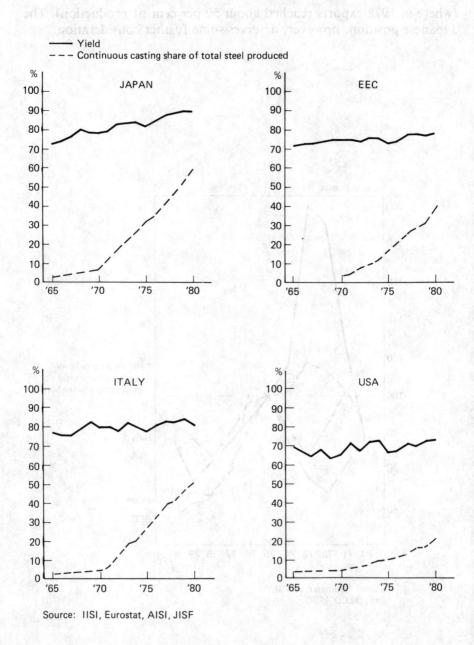

Source: IISI, Eurostat, AISI, JISF

Figure 3.14 Trend of production yield (relation between finished products and raw steel) and of the quantity of steel obtained by continuous casting

72

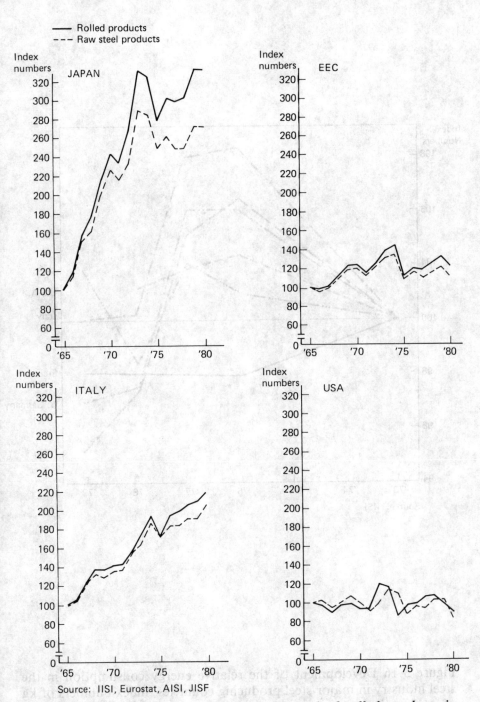

Source: IISI, Eurostat, AISI, JISF

Figure 3.15 Trend of raw steel production and of rolled products in the various steel industries

73

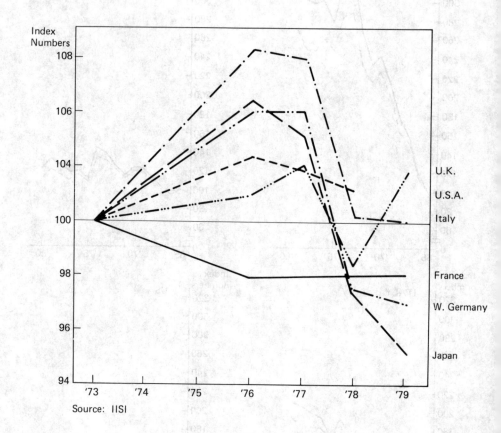

Figure 3.16 Development of the relative energy consumption in the steel industry in major steel producing countries (index numbers of kg of coal equivalent per ton of produced steel) 1973 = 100 (average for each individual country)

74

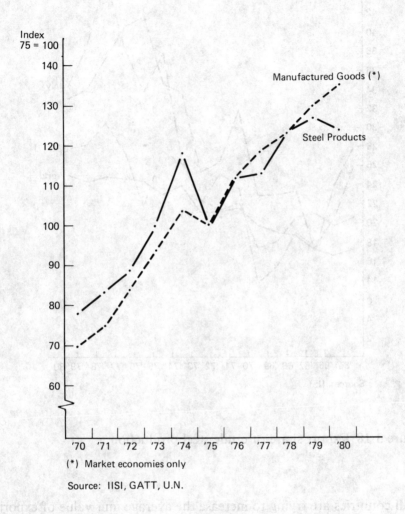

Index
75 = 100

Manufactured Goods (*)

Steel Products

'70 '71 '72 '73 '74 '75 '76 '77 '78 '79 '80

(*) Market economies only

Source: IISI, GATT, U.N.

Figure 3.17 World trade size (1975 = 100)

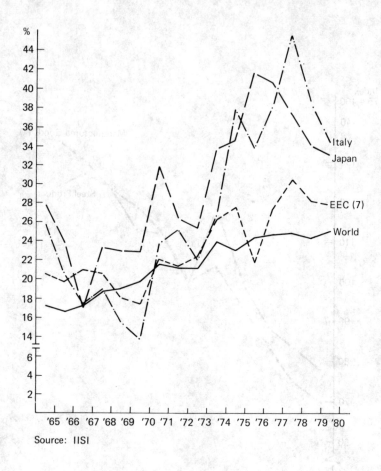

%

44
42
40
38
36
34
32
30
28
26
24
22
20
18
16
14
6
4
2

'65 '66 '67 '68 '69 '70 '71 '72 '73 '74 '75 '76 '77 '78 '79 '80

Italy
Japan
EEC (7)
World

Source: IISI

Figure 3.18 Export share of crude steel production

All countries are trying to increase the average unit value of exports rather than the quantities (see Figure 3.19). Since 1976 Japan has been losing market shares in international trade (see Figure 3.20). This trend corresponds to an increase in indirect trade. The international strategy of the EEC (and Italian) steel industry seems less clear. In the EEC, the export share continued to decline while there has been a stronger penetration of the internal market, implying a growth of trade between the member countries (Figure 3.21). The growth of trade within the EEC has not been met by a clear production specialisation: the specialisation indices calculated gave extremely homogeneous values for all the member countries and lack statistical significance.

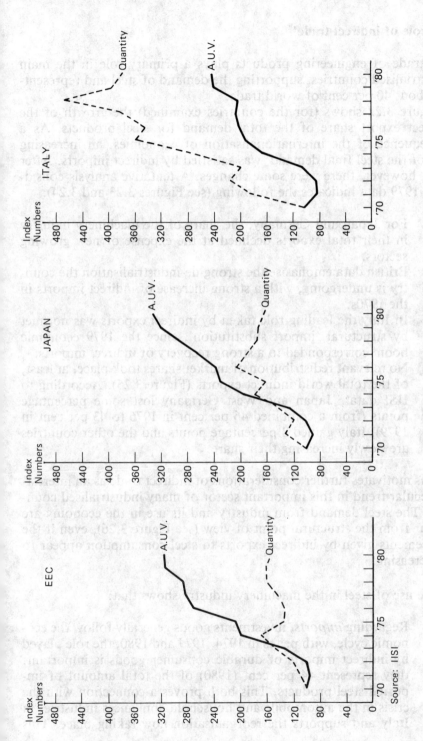

Figure 3.19 Comparison between exported steel quantities and export average unit values (AUV) (1970 = 100)

Source: IISI

The role of indirect trade[4]

The trade of engineering products plays a primary role in the main industrialised countries, supporting the demand of steel and representing about 40 per cent of world trade.

Figure 3.22 shows (for the countries examined) the growth of the indirect export share of the total demand for steel products. As a consequence of the internationalisation of economies, an increasing part of the steel final demand[5] was satisfied by indirect imports. After 1976, however, there were some changes. A tentative analysis considering 1979 data indicates the following (see Figures 3.23 and 3.24):

1   For Japan and Germany, the share of the machinery industry in their total exports declined at the expense of new growing sectors.

2   British data emphasise the strong de-industrialisation the country is undergoing, with a strong increase of indirect imports in the 1970s.

3   In Italy the leading role taken by indirect exports was not met by structural 'import substitution', since the 1979 economic boom corresponded to a strong recovery of indirect imports.

4   No relevant redistribution of market shares took place, at least, of the total world indirect exports (Figure 3.25). According to IISI data[6], Japan and West Germany lost some percentage points (from a combined 46 per cent in 1976 to 43 per cent in 1979), Italy gained 3 percentage points and the other countries are slowly increasing their share.

This motivates further considerations of indirect trade as an interesting secular trend in this important sector of many industrialised countries. The steel demand from industry and its use in the economy are similar from the structural point of view (see Figure 3.26), even if the inducements given by indirect exports to steel consumption appear to be increasing.

The use of steel in the machinery industry shows that:

1   Regarding *imports*, investments goods generally follow the economic cycle, with peaks in 1974, 1979 and 1980; the role played by indirect imports of durable consumer goods is important: they represent 44 per cent (1980) of the total amount of imported steel products. This both proves a connection with the crisis in the automobile and household appliances industries in Italy and supports the reorganisation now taking place. This

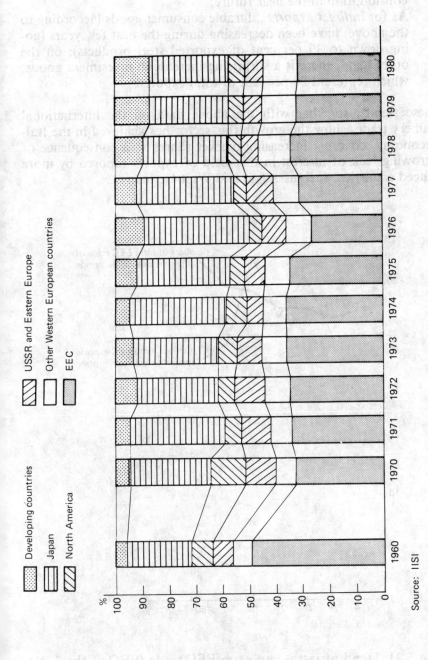

Figure 3.20 Percentage market shares of total steel exports (volume) (intra-EEC and intra-COMECON trade excluded)

Source: IISI

79

reorganisation should induce a stronger growth of steel consumption in the near future.

2   As for *indirect exports* , durable consumer goods (according to the above) have been decreasing during the past few years (going down to 30 per cent of exported steel products); on the other hand, there is a surprising increase in investment goods, which represent 20 per cent of total exports.

These trends, together with the above data on the international machinery trade, show the growth this sector has achieved in the Italian economy, covering increasing market shares as a consequence of the growing lack of interest in this field of exports exposed by more advanced countries, such as Japan and West Germany.

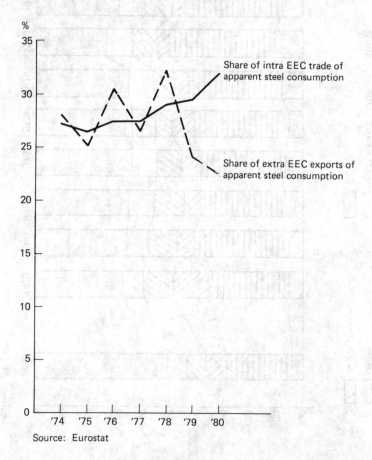

Source: Eurostat

Figure 3.21 Trend of intra- and extra-EEC trade (EEC of the 7, i.e. without Denmark, Eire, Greece)

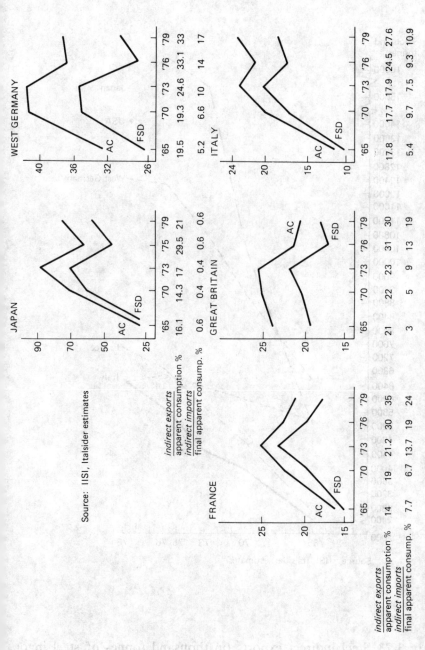

Source: IISI, Italsider estimates

Figure 3.22 Trend of apparent consumption (AC) and of final steel demand (FSD) in the various countries (million tonnes of raw steel)

81

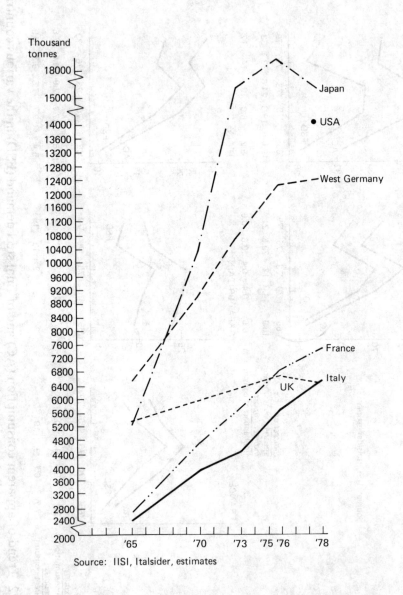

Figure 3.23 Steel indirect exports (in thousand tonnes of steel ingot equivalents)

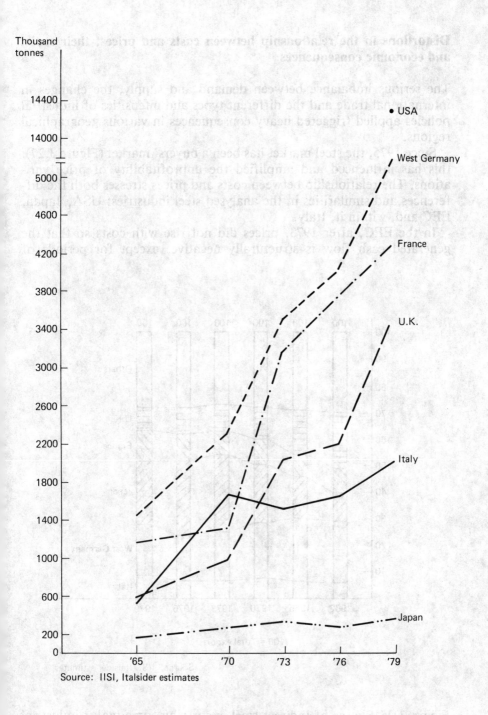

Thousand tonnes

- 14400
- 14000
- 5000
- 4600
- 4200
- 3800
- 3400
- 3000
- 2600
- 2200
- 1800
- 1400
- 1000
- 600
- 200
- 0

• USA

West Germany

France

U.K.

Italy

Japan

'65 '70 '73 '76 '79

Source: IISI, Italsider estimates

Figure 3.24 Steel indirect imports (in raw steel ingot equivalents)

83

Distortions in the relationship between costs and prices: their social and economic consequences

The serious imbalance between demand and supply, the changes in international trade and the different types and intensities of industrial policies applied triggered heavy consequences in various geographical regions.

Since 1975, the steel market has been a buyers' market (Figure 3.27): this has influenced and amplified the unprofitability of price variations. The relationship between costs and prices stresses both the differences and similarities in the analysed steel industries: USA, Japan, EEC and, within it, Italy.

In the EEC, after 1973, prices did not rise with costs so that the generated cash flow is structurally negative, except for periods of

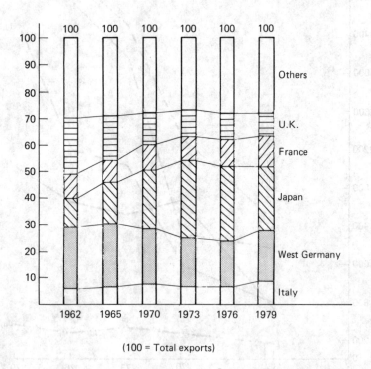

(100 = Total exports)

Source: IISI, Italsider estimates

Figure 3.25 Shares of indirect steel exports by some major industrial countries

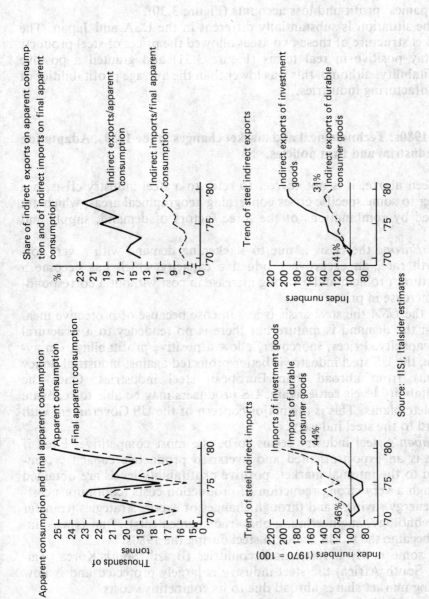

Figure 3.26 Structure of final steel demand in Italy

Source: IISI, Italsider estimates

85

particular strain of demand (Figure 3.28). The consequences were dramatic: heavy employment losses (Figure 3.29) and heavy losses in companies' profit and loss accounts (Figure 3.30).

The situation is substantially different in the USA and Japan. The market structure of these two areas allowed the prices of steel products to stay positive in real terms (Figure 3.31) and granted a positive profitability, although this was lower than the average profitabilities of manufacturing industries.

The 1980s: Technological and market changes in the 1980s. Adaptation of industrial and firm policies.

As seen above, it is not correct to refer to a steel industry crisis, but rather to some specific crises concerning geographical areas, which are caused by combinations of the three factors of demand, supply and cost.

In *Europe* the crisis is due to slackening demand with a very low growth rate, an excessive productive capacity which has become a structural problem and a strong increase in cost without a corresponding increase in prices.

In the *USA* the steel crisis is less intense because of protective measures: the demand is mature but there is no tendency to a structural overcapacity. Prices, moreover, allow a positive profitability. On average, the US steel industry is better protected against industrial policy actions from abroad than European steel industries. Once the profitability levels settle again, US producers may be able to renovate obsolete plants. This is the major concern of the US Government with regard to the steel industry.

Japan's steel industry seems to be the most competitive. Even if there is an export-oriented and oversized productive capacity, compared to the internal market, positive profitability levels are obtained through a very strong reduction of production costs (continuous casting, energy savings) and through changes of export strategies (transfer of technology, engineering, global trade agreements). That is why Japan became the world leader in steel during the 1980s.

In some newly industrialised countries (Brazil, South Korea, Taiwan, South Africa) the steel industry is largely protected and is now gaining market shares abroad due to its competitive costs.

Against this background, it is useful to carry out a more detailed analysis of the EEC area, where a more harmonised industrial policy is needed to overcome the crisis.

An important problem concerns EEC steel industries' competitiveness compared with the other areas. The European steel industry can

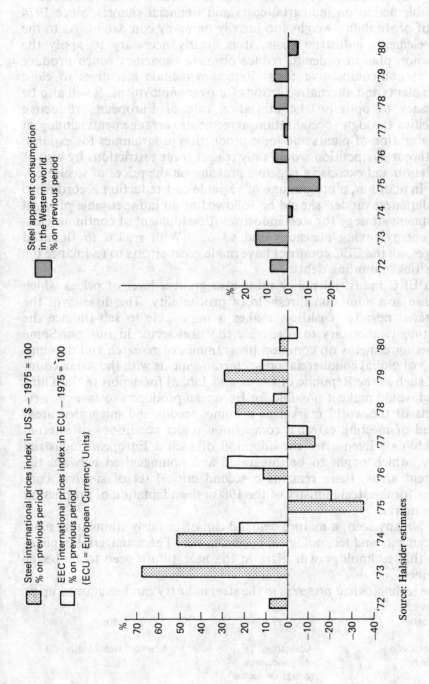

Figure 3.27 Comparison between the trend of prices and of demand in the steel industry

Source: Halsider estimates

Steel apparent consumption
in the Western world
% on previous period

Steel international prices index in US $ – 1975 = 100
% on previous period

EEC international prices index in ECU – 1975 = 100
% on previous period
(ECU = European Currency Units)

87

survive only through a reduction of its production costs. This involves a double action on industrial costs and financial charges: since 1974 loss of profitability weighs too heavily on many companies. As to the first element, industrial costs, it is clearly necessary to apply the Davignon plan in order to reduce obsolete capacities which produce steel at non-competitive costs. Remedies include incentives to close down plants and alternative actions for overemployment. It will also be necessary to optimise the utilisation rate of European productive capacities through specialisation agreements, arrangements aiming at the saturation of plants and joint production programmes for exports. Cut-throat competition would only trigger trade restrictions by way of retaliation and exercise a negative pressure on the prices of steel products. In addition, a programme of variable cost reduction according to the 'Japanese model' should be followed as an indispensable physical investment strategy for steel industries (development of continous casting, energy saving measures and so on). With regard to financial charges, all the EEC countries have made great efforts to re-finance the industries' mounting debts.

An EEC industrial policy must also grant a level of prices which stabilise at a minimum threshold of profitability. The disaster of the generated negative cashflow makes it impossible to self-finance the investments necessary to reactivate the steel sector in Europe. Some further agreements on common programmes of research and development,[7] of global commercial policy (arrangements with the Asian countries, such as the Republic of China and Japan) for export to the Third World would make it possible for European producers to face the new aspects of the world market as a homogeneous and integrated area, instead of meeting external competition under conditions of internal squabblings. Even after the adoption of such a European industrial policy, which ought to be integrated and homogenised between the different areas, there remains a second critical set of survival conditions for the steel industry of the 1980s: the adaptation of production to the market.

As already seen, a mature demand can offer fairly stimulating market segments and technological opportunities. The managerial implications that technology will offer in the near future need to be better examined.

The technological progress in the steel industry can be summed up as follows:[8]

1960s	1970s	1980s
Mass production technology	Adaptation. Temporary measures (energy problems, rationalisation)	Advanced technologies

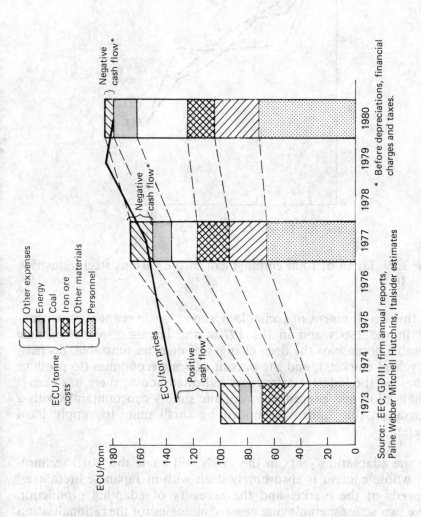

Figure 3.28 Trend of the structure of costs and of prices in the EEC per average of sold ton of steel (in European Currency Units, ECU)

Source: EEC, GDIII, firm annual reports, Paine Webber Mitchell Hutchins, Italsider estimates

* Before depreciations, financial charges and taxes.

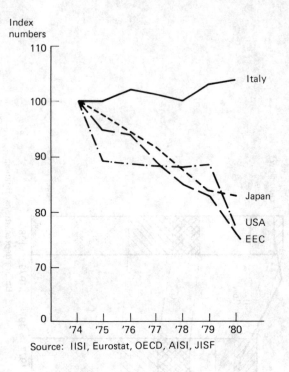

Source: IISI, Eurostat, OECD, AISI, JISF

Figure 3.29 Trend of total employment in the various steel industries (1974 = 100)

As this development of technology shows, scale economies consolidated in the 1960s and in the 1970s; the large-size plants became vulnerable because of the deep changes of demand (unstable, decreasing, cyclical markets); and the pursuit of scale economies (to produce volume) led the industry astray. As indicated above, there has been a rationalisation and concentration of the supply concomitantly with a pulverisation (big units surrounded by small units to supply local markets).

After the adaptation years in the 1980s will come the 'soft' technologies, whose content is absorbingly dealt with in Japanese literature.[9] The needs of the market and the necessity of adapting production involve two actions: employing new technologies for the rationalisation of production costs, but also offering new products as well as new uses of existing products; and employment of appropriate non-sophisticated technologies, which are suitable for new or cyclical markets (with the role, for example, of transferring know-how to developing countries).

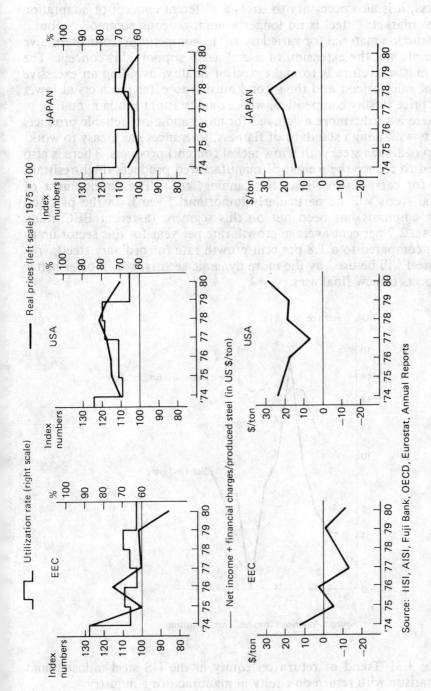

Figure 3.30 Trend of real prices, of utilisation rates of production capacities and of profitability of the various steel industries

Besides the technological evolution and its relevant opportunities for success, it is also necessary to stress a different concept of adaptation to the market:[10] steel is no longer a homogeneous commodity, but a construction material for varied uses. The competition from alternative materials and the extension of users' needs support this concept. The first must therefore is to make steel of quality, avoiding an excessive rise of sales prices; and the second must is to offer products at lower sales price to stay competitive, where quality is not fundamental.

There is furthermore a demand for more and more reliable products (sheets with a high standard of flatness, tolerances etc.), easy to work, cheap (stainless steel with a low nickel content) products. There is also a need to reduce the weight of manufactured products (high-resistant steel for cars, tin-free steel for canning, etc.). The development of special steels will be particularly important. Even if in the past too much emphasis has been put on this segment (a recent EEC study shows a 2.7 per cent average growth rate per year for this sector up to 1985, compared to a 1.8 per cent growth rate for ordinary steel), special steel will be used by the more dynamic sectors, which offer better prospects of new final uses.

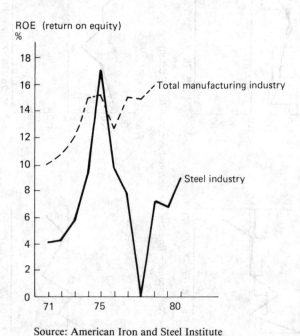

Source: American Iron and Steel Institute

Figure 3.31 Trend of return on equity in the US steel industry and comparison with return on equity in manufacturing industry

Conclusions

This chapter has tried to prove that:

1   There is an international crisis of steel demand (maturity of the sector, scarce innovation, alternative materials), but the structural crisis is mainly a European phenomenon.
2   The threats stemming from the crisis of the steel sector can become opportunities, if steel is no longer considered by management as an undifferentiated commodity, but as a building material for different uses.

The implications for managerial and industrial policy are as follows:

1   There is a necessity to harmonise the steel industry policies between the different countries to protect the phases of consolidation and reactivation of national/regional industries.
2   There is a necessity for a homogeneous policy supporting the industry's reorganisation which will save only companies which are capable of rationalising their productive processes to diminish costs.
3   Change-oriented policies should be employed.

Notes

1   See Hosoki, S. and Kono, T. ' *Japanese steel industry and its rate of development* ', NSC paper, 1978.
2   OTA *Technology and steel industry competitiveness* US Dept of Commerce, Washington DC, June 1980.
3   Nippon Steel Corporation, *How Japan's Steel Industry has improved its Productivity* , 1981.
4   Imports and exports of manufactured products containing steel.
5   Steel final demand = apparent consumption less indirect export plus indirect imports.
6   IISI data consider 85 per cent of world trade of steel products.
7   See, on this point, the proposals of the EC Commissariat au Plan.
8   UNIDO, *Picture for 1985 of World Iron and Steel Industry* , Vienna, 1980.
9   See Hosoki, S., Kono, T. and Imai, K., *Iron and Steel* , Japanese Economic Studies, 1975.
10 Hamermesh, R.G. and Silk, S.B., 'How to compete in stagnant industries', *Harvard Business Review* , vol 57 no. 5, 1979, pp.161–68.

4 On the development of steel consumption in highly industrialised countries: The case of the United States of America

Helmut Wienert

Aims of the chapter

All forecasts dealing with the future development of steel consumption in the highly industrialised countries come to the conclusion that growth will continue although at a more or less reduced rate. Eight years have passed since steel consumption reached its last peak, so it does not seem completely unjustified to call these forecasts doubtful.

The aim of this chapter is to find out whether such doubts are justified for the US steel market.

The USA was selected for the following reasons:

1 It is the country with the highest level of development.
2 It has the biggest domestic market.
3 Its development since World War II has been the least affected by disturbing influences.

First of all, we select a suitable explanatory series to explain the development of apparent steel consumption (ASC). Steel consumption is then divided into a *specific* component and into a *volume* component, and their developments are then analysed. Finally, the chapter gives some forecast considerations as a conclusion.

Development of the steel intensity of physical investment

Steel is a basic material in industry and is used in the production of various goods in the steel processing industries. The overall steel consumption will therefore change if the steel input per article changes, and/or the production volume of any one article changes. However, since an overall and complete assessment of such changes is not possible we always have to use more or less aggregated values.[1]

For this purpose we can use either production figures from the steel processing industries or overall economic demand aggregates. The following steel-intensive production series were tested:

a) production of durable goods (PDG);
b) production of automotive products (PAP);
c) production of production equipment (PEQ);
d) production of metal products (PMP).

The following steel-intensive demand aggregates were tested:

a) gross domestic product (GDP);
b) gross private investments (GPI);
c) construction spending (EST).

The explanatory power of the above series was determined by means of linear regressions on apparent steel consumption (ASC) for the period 1950 to 1980. We used the regression functions

$$(A) \; ASC = a + b \, x, \text{ and}$$

$$(B) \; ASC/x = a + b \, t,$$

where x is the explanatory series.

GPI has the highest explanatory value for ASC. This aggregate does not only cover production equipment and construction, that is the main sectors of steel consumption, but also includes changes in inventories, that means a congenial cyclical component[2] most suitable for explaining the *apparent* steel consumption (Figure 4.1). The results can be seen from Table 4.1.

Table 4.1
Testing for explanation of apparent steel consumption and regression results

Regression of ASC on	Approach	Parameter		Test criteria				
				T-values				
		a	b	t_a	t_b	R^2	V	DW
PDG	A	43.8	0.51	10.1	11.2	0.81	8.9	1.3
	B	143.6	−2.27	46.0	−13.4	0.86	7.9	1.3
PAP	A	48.4	0.41	12.6	11.5	0.82	8.7	1.2
	B	134.0	−2.34	40.2	−12.9	0.85	9.4	1.5
PEQ	A	48.4	0.49	10.3	9.4	0.75	10.3	1.3
	B	162.9	−2.93	25.1	−8.3	0.70	15.2	1.0
PMP	A	35.1	0.59	6.7	11.9	0.80	8.2	1.1
	B	121.2	−1.34	46.1	−9.3	0.75	7.2	1.3
GDP	A	39.4	0.05	6.7	8.9	0.73	10.7	1.3
	B	122.4	−0.15	34.3	−7.9	0.68	9.9	1.6
GPI	A	38.6	0.35	9.8	13.6	0.86	7.6	1.3
	B	82.3	−1.09	51.6	−12.5	0.84	6.7	1.8
EST	A	17.8	0.69	2.5	10.2	0.78	9.7	1.3
	B	96.9	−0.56	24.6	−2.6	0.19	12.2	1.1

V = coefficient of variation, in per cent.

Regression function B of GPI may be interpreted as the trend in steel intensity (SI) of investments, so that

$$ASC = SI \times GPI.$$

As b is negative, SI drops in terms of time. This decline can easily be interpreted with a view to economy: technical progress is time-dependent, and, as a rule, has such steel-saving effects as constructive improvements, improved steel qualities which reduce the material cost, and substitution of material and product innovation, which reduces the importance of steel-intensive investments.

It is of crucial importance for the future development of ASC whether the decline of SI speeds up in terms of time. From the estimated linear trend we obtain, for example, a decline by 1.4 per cent in 1960, which will have increased to 2.0 per cent by 1980.

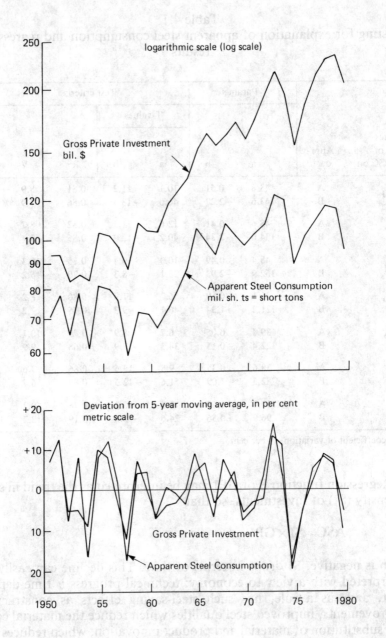

Figure 4.1
Steel consumption and investment, United States of America, 1950-1980

Estimating SI with a function that implies constant relative declines gives just as good a fit as with the linear formula.

The question of whether average steel consumption per unit of investment accelerates or decreases at a constant rate cannot be answered in this way. The reason for this may be that the tendencies were contrary within the given research period. When we break down the period we shall have to take the cyclical developments into account. If industrial production[3] is to be used as an economic indicator, the given observation period may roughly be divided into the following decades: 1950–60; 1960–70; 1970–80. The first decade covers about three cycles, the second decade only one 'super-cycle', and the third one covers two cycles.[4]

If we use semi-logarithmic regression functions to estimate the development of steel intensity of investments within these decades, we obtain the following average annual growth rates:[5] -1.9 per cent in 1950–60; -0.8 per cent in 1960–70; -2.3 per cent in 1970–80. As we can see from the growth rates, there was a slow-down in decline during the 1960s but an accelerated decline in the 1970s.

Fluctuations in the decline in steel intensity may be traced back to changes in investment structures. For reasons of simplicity, we take *physical investments* and *inventory build-up* together to form A, and *residential investments* and *industrial construction investments* to form B. We thus obtain two components to which one trend each (specific consumption) can be allocated by a fourfold regression. The regression equation reads:

$$ASC = (1.10 - 0.0169 \cdot t)A + (0.63 - 0.0081 \cdot t)B$$
$$(5.0) \quad (1.9) \qquad\qquad (3.8) \quad (1.0)$$
$$R^2 = 0.93$$
$$DW = 1.65$$
$$V = 5.57$$

The average rate varies despite the fact that the decline in specific consumption continuously speeds up both for physical investment as well as for construction investment. The average annual decline adds up to: -1.8 per cent for 1950–60; -1.5 per cent for 1960–70; -2.0 per cent for 1970–80.

The reason why the growth rates vary is that the share of physical investment has increased. If there had been no changes in investment structure, the decline in steel intensity of investments would have speeded up[6] from -1.5 per cent in 1950–60 to -1.9 per cent in 1960–70, and to -2.4 per cent in 1970–80.

The steel intensity decline components derived from the equations are shown in Table 4.2.

Table 4.2
Steel intensity change factors

		of which due to					
		change in the weight of			specific consumption of		
	$\triangle$SI	A	B	A+B	a	b	a+b
	total	in kgs per # US $ 1,000					
1950–1960	– 140	– 57	+ 32	– 25	– 66	– 49	– 115
1960–1970	– 94	+ 68	– 40	+ 28	– 79	– 43	– 122
1970–1980	– 109	+ 50	– 31	+ 19	– 90	– 38	– 128

The effects caused by structural changes can be eliminated if the overall steel consumption is related to the PEQ investment. The regression function will then be:

$$SI_A = SV/A = 2.40 - 0.0461 \cdot t$$
$$(65.6) \quad (23.1)$$

$$R^2 = 0.95$$
$$DW = 2.69$$
$$V = 5.98$$

The decline in steel intensity thus moves continuously from – 2 per cent in 1950 to – 4.5 per cent in 1980 (Figure 4.2). Estimating the decades with semi-logarithmic trend we obtain the following mean annual decline rates: – 2.9 per cent in 1950–60; – 3.1 per cent in 1960–70; – 4.7 per cent in 1970–80.

The share of investment in production capacity increased and that made the decline in steel intensity of investments slow down during the 1960s, due to the fact that steel consumption per dollar of equipment is higher than that per dollar of construction. This share cannot go up infinitely so that in the long run the tendency will develop into a progressive decline in average steel intensity of investments.

Development of investment volume and steel consumption

The volume of GPI grew at an average annual rate of 3.6 per cent in the period 1950–80. Within the individual decades the following rates were reached: 1.6 per cent in 1950–60; 5.3 per cent in 1960–70; 2.6 per cent in 1970–80.

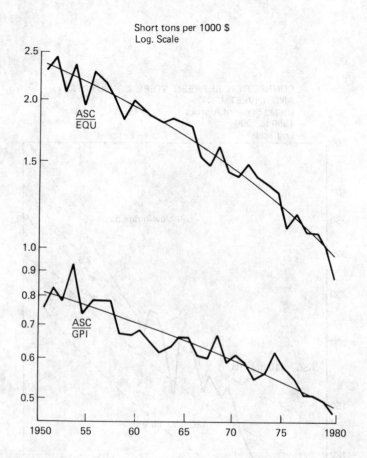

Figure 4.2
Steel intensity of investment United States of America 1950-1980

That shows that investment activities were particularly strong in the 1960s. At the same time steel intensity dropped less drastically during that time so that there was an enormous increase in steel consumption with an average annual rate of + 4.5 per cent. The 1950s, however, had brought about an annual decline of − 0.3 per cent, and in the 1970s only a slight annual growth of 0.2 per cent was reached.

Thus one could classify the 1950s and the 1970s as periods of stagnation, and the 1960s as a period of growth.

If we look more closely at the remarkable development in the 1960s it is clear that investment and steel consumption had been growing by 50 per cent, that is at an annual rate of about 10 per cent from the low point in the cycle up to 1966 or 1965 respectively without the occur-

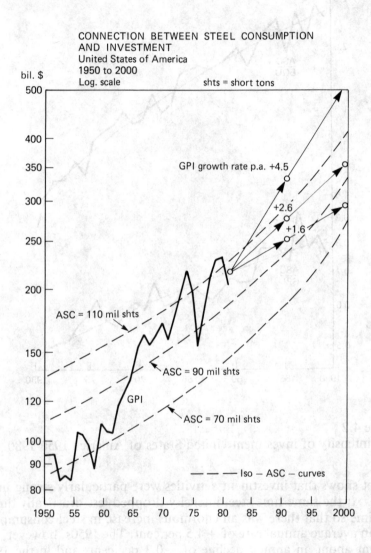

Figure 4.3
Connection between steel consumption and investment United States of America 1950-2000

rence of a similarly rapid decline in subsequent years.[7]

The time-related concentration of investments forced steel consumption up because the time-dependent slowdown of steel intensity was more than compensated for by the increase in volume.

The correlation is illustrated by a diagram: using the estimated steel intensity function quoted above and transforming it, we obtain the equation:

$$GPI = ASC / (0.84 - 0.012\ t)$$

Lines of equal steel consumption and development of investments as related to time may then be entered into a diagram with GPI and t coordinates. The steel consumption (ASC) increases only if the investment curve rises higher than that of the iso-steel consumption curve (see Figure 4.3).

Forecast considerations

From the findings of our study obtained so far we can draw some conclusions with regard to the future development of steel consumption.

Firstly, it can be expected that the steel intensity of investments will continue to decrease. Following the regression function

$$SI = 0.82 - 0.0109 \cdot t$$

the steel intensity of investments will drop from 0.482 t per $ 1,000 in the year 1980 to 0.373 t per $ 1,000 in the year 1990, which equals an average annual decline of 2.5 per cent. Assuming that steel consumption does not change, investments would have to increase at an average annual rate of 2.6 per cent.

If the former 'trend' continues, steel consumption in the USA is likely to keep growing, as investments grew by an annual average rate of about 3.5 per cent from 1950 to 1980. Assuming that this growth rate will be maintained up to 1990, the steel consumption would be 113 million short tons compared with 104 million short tons in 1980 (trend figures) (102 and 94 million tonnes respectively.)

If there were reason to expect the 1980s to experience a rapid growth similar to that of the 1960s, steel consumption would go up to even 125 million short tons by 1990 (113 million tonnes).

The constellation of the 1960s, which was most conducive for growth, was, however, an 'exception' already during the period 1950–80 and for the inflation-ridden 1980s it would be rather an extreme assumption. If we therefore take the lower growth tendency of the 1970s, future investments will only increase by 2.6 per cent, and the 1990 steel consumption of the USA would in this case not exceed that of 1980.

The question raised at the beginning concerning the long-term trend in steel consumption in the highly industrialised countries cannot be definitely answered for the USA. On the one hand, there is a long-term and stable trend towards a decreasing steel intensity but, on the other hand, intense investment phases in the future might also over-compensate this tendency, despite the fact that it is becoming increasingly difficult for the highly industrialised countries to exceed the 'critical rate' that tends to increase.

At any rate, the estimated figures are certainly considerably lower than those of other forecasts that have been available so far.

The American Iron and Steel Institute (AISI) issued a forecast quoting a figure of 138 million shts for the year 1990.[8] That is 13 million short tons or 10 per cent more than indicated in our study for an extremely keen investment activity. If our study were based on the lower growth limits[9] quoted in the AISI study, then, following the development of steel intensity described in the present study, we would obtain a decline in steel consumption to 92 million short tons (83 million tonnes).

Notes

1   If we relate overall steel consumption of a country to an explanatory series we obtain, however, an average consumption which is not only influenced by the technical input situation, but also by structural changes within the explanatory series employed.

2   Investments prove to be a good explanatory indicator for steel consumption, not only for the United States but also for almost all countries. See also IISI, *Causes of the Mid-1970s Recession in Steel Demand*, Brussels, 1980, p.155f.

3   See US Department of Commerce, *Business Conditions Digest*.

4   1st cycle: end of 1949 to beginning of 1954,
     2nd cycle: beginning of 1954 to the end of 1958,
     3rd cycle: early 1958 to the end of 1960,
     4th cycle: early 1961 to the end of 1970,
     5th cycle: end of 1970 to early 1975,
     6th cycle: early 1975 to mid-1980.
     See also Zarnowitz, V., Moore, G.H., 'The Recession and Recovery of 1973–1976', p.508 in *Explorations in Economic Research* , vol.4, 1977, p.471ff.

5 Taking the cyclical figure for 1960 of the period 1950–60 and the cyclical figure of 1970 for the period 1970–80 for the period from 1960 to 70 we obtain a mean annual growth rate of -1.2 per cent instead of 0.7 per cent for the period 1960–70.

6 If a stands for the specific steel consumption for equipment, and b for construction, we obtain:

$$ASC = a \cdot A + b \cdot B/GPI$$

$$SI = a \cdot \frac{A}{GPI} + b \cdot \frac{B}{GPI}$$

The formula to calculate this is:

$$SI_t - SI_{t-1} = a_t \cdot \left(\frac{A}{GPI}\right)_t - a_{t-1} \cdot \left(\frac{A}{GPI}\right)_{t-1} +$$

$$b_t \left(\frac{B}{GPI}\right)_t - b_{t-1} \left(\frac{B}{GPI}\right)_{t-1}$$

If we extend it and transform it we obtain:

$$\triangle SI = \left(\frac{A_t}{GPI_t} - \frac{A_{t-1}}{GPI_{t-1}}\right) \cdot a_{t-1} + \left(\frac{B_t}{GPI_t} - \frac{B_{t-1}}{GPI_{t-1}}\right) \cdot b_{t-1}$$

$$+ (a_t - a_{t-1})\frac{A_t}{GPI_t} + (b_t - b_{t-1}) \cdot \frac{B_t}{GPI_t}$$

7 This period of growth was caused by an expanded money volume. With stable labour/unit costs and moderate price increases the result was a profit-financed strong acceleration process which received additional incentives from the public road construction and school construction schemes.

8 AISI, *Steel at the Crossroads — The American Steel Industry in the* 1980s, Washington, DC, January 1980, p.29

9 'Lowest reasonable' growth of GNP, assuming a constant investment rate; one may use that rate for investment as well.

PART III
THE MAJOR STEEL-PRODUCING HIGHLY INDUSTRIALISED COUNTRIES

5 Introduction

As demonstrated in the introductory and statistics parts, the steel industry crisis in different countries is by no means uniform. Historical, political and other factors have shaped the industry structures in different ways. They also have a bearing on the understanding of the crisis and its consequences, and the ways and means used in different countries by different industries and firms to cope with it. A proper understanding of the differentiated character of the crisis is necessary in order to be able to consider measures to be taken not only to minimise its consequences but also to get out of it.

The description aims at providing a basis for the reader's understanding of how the crisis developed, the consequences it has, and how national governments (in the European case the European Community as well), the industry and its firms have reacted to, and tried to overcome, the crisis.

The most important agents in the crisis are the United States, Japan and Europe, between which rather serious tensions have developed, here briefly labelled as the transatlantic and transpacific quarrels over steel.

In the European theatre the most important actors have been selected as examples: the European Community, the steel industries of the Federal Republic of Germany, of France and of Great Britain, since they reacted in most different ways to the crisis, but also since they also have to pull in the same direction because of their membership of the European Community.

In order to illuminate the problems of the US steel industry in as thorough a way as possible, two expert views are presented: Robert Crandall takes the market economy point of view, whereas Joel Hirschhorn advocates a US steel industry policy.

6 The European Steel Industry

The European Coal and Steel Community

The European Coal and Steel Community (ECSC) is a super-national organisation aiming at the establishment of a common market for coal and steel. The basic agreement (to be effective from mid-1952) was reached in 1951 between the governments of Belgium, the Federal Republic of Germany, France, Italy, Luxemburg and The Netherlands. The initiative was taken by the former French Secretary of Foreign Affairs, Robert Schuman, (the Schuman Plan) and was part of the rethinking of French–German relations (which had seen a series of wars between the two nations).

The union had three aims:

1   *Politically* the plan aimed at replacing the allied control power over the German Ruhr district by a system aiming, in the long term, at a political union of Europe (at least major parts of it).
2   From the point of view of *economic structural policy* the union aimed at a rationalisation of the split-up markets by replacing them by a common market for coal and steel.
3   The third, *social policy* aim was to harmonise and improve working and living conditions for employed persons in the industries.

In 1967 the union became part of and was replaced by the *European Community* (EC), which took over its functions.

The economic consequences of the union were the abolition of customs and other trade regulations, with repercussions even outside its immediate realm (for example the liberalisation of transportation regulations). The union resulted in a considerable reorientation of intra-European trade flows in steel. The consequences of the union were, however, considerably smaller than originally envisaged because the decline of the coalmining industry (due to the availability of cheap oil) came to occupy the authorities. The achievement of the political aim, the establishment of the European Community, loosened the ties and weakened the power of the coal and steel union.

The Community's role with regard to the steel industry diminished in favour of (at best rudimentarily coordinated) national policies of widely differing natures. The Community's internal role during the 1960s and the first half of the 1970s can be characterised as 'policing', that is scrutinising 'improper' practices. Only when the crisis came between 1974 and 1975 did the role of the EC shift quite drastically, now becoming a combination of assistance and monitoring in crisis management. At the same time the external role of the Community (which during certain periods was the only visible one) increased in importance, opposing trade restrictions imposed in particular by the United States (for example the Voluntary Restraint Agreement of 1968 and its sequels) and exercising – some – influence on Japan in order to shelter Europe against the massive flow of Japanese steel pouring into the European market because of newly established overcapacity and US import regulations.[1] This external role also had to be coordinated with the thus strengthening internal role of the Community and its executive body, the Commission.

The most important measures taken by the Commission in reacting to the present steel crisis have been

a) The establishment of a restructuring and price cartel – the Davignon plan (1977, valid to 1985);
b) Negotiating, internally with member countries, a quota deal (autumn 1982) for bulk steel (plus pipe and tube) with the US government;
c) Doing the same in 1983 for speciality and alloy steel.

The Davignon plan

The Davignon plan caters for a capacity reduction between 1977 and 1985, at the same time avoiding cut-throat pricing which would imply throwing the burden of consequent bankruptcy and unemployment at the feet of the public. (One of the underlying assumptions was that the victims of competition would mainly be found amongst the fairly efficient private firms, whereas the government-owned or operated plants would be able to survive, because of their greater chances of obtaining financial relief to weather the crisis.) The plan also comprises a programme to take care of the burden from unemployment caused as a consequence of the restructuring efforts.

The price cartel, the third leg of the tripod, was to render the necessary cash flows to realise restructuring and social relief plans.

A basic prerequisite of the Davignon plan were, and are, corresponding national programmes to implement the ideas and to provide the means agreed upon. The implementation of the plan, however, ultimately requires participation and discipline from the individual firms – private or public – in the EC member countries' steel industries.

The overall objective of the Davignon plan is 'to make the steel industry sufficiently productive at reduced production capacity, to allow it to continue to exist in a world, where trade remains open and free, a situation which maintains the Community as a net exporter'.[2] 'The Commission considered that a basic restructuring policy could effectively be implemented only if there was real consensus among industry, trade unions, and users within the Community, as well as agreement with trading partners outside the Community.' 'The Community takes the firm view that enterprises are primarily responsible for designing and implementing restructuring programmes.'

The restructuring plan (which was voluntary from the beginning in 1977 but became mandatory, with regard to output quotas and future capacities, in 1980) is based on a number of assumptions:

1   A basic condition is a strict cooperation between the EC and national governments on all parts of the plan (the 'technical', political, financial, social and timing parts of it).
2   The plan aims, within a time frame of eight years (1977–85), at
     a)   cuts in productive capacity
     b)   primary concentration on the EC market, as the most profitable one in the long run
     c)   re-establishment of the competitiveness of the industry as well as
     d)   financial restructuring of the remaining enterprises.

The implementation of the plan calls for

- new negotiations aimed at agreements upon self-constraints in the production and trading of steel products;

- organisation of a functioning EC steel market, to be specified in prices and quantities for principal product groups; and

- a temporary neutralisation of competitive advantages in the hands of the most able enterprises in favour of those who have lost their competitiveness because of insufficient restructuring or modernisation efforts.

The EC Commission presents a quarterly list containing market quota and prices per product group, country and enterprise. The lists are negotiated jointly between the Commission and representatives of national governments, producers (EUROFER, a cartel of the leading firms in Europe[3]), users, dealers and unions. The prices are set on the basis of the most efficient producer's production cost, in order both to induce rationalisation and restructuring, and to avoid harm to the users.

The EC Commission has established policies and procedures for information to, and negotiation and agreements with, the most important external partners (for example the US Department of Commerce, the US steel industry, MITI of Japan, Japanese steel industry etc.). Such agreements are necessary in order to establish an action space and time for the restructuring efforts within the EC steel industry.

The EC is controlling the subsidies, rendered in various forms and for different purposes, to the steel industry and to individual enterprises or product groups within the member countries. It has developed a subvention code based upon the GATT code. Under pressure from inside (in particular Germany) and outside (mainly the US steel industry and the US Government) the code has been tightened, most recently during the autumn of 1981. The total amount spent on subsidies between 1975 and the end of 1983 has been estimated at more than US $37 billion.

Subsidisation – no matter for what purposes or how important in the individual case – does not only imply a sometimes massive distortion of the market mechanisms. (During the summer and autumn of 1982, the US Department of Commerce claimed that certain products are subsidised by up to 40 per cent of the price quoted. This is an extreme, of course.) In fact, it also means a renationalisation of the steel industry, thus withdrawing it from the jointly agreed principles and rules of the EC. If the EC were to tolerate this, it would dig its own grave.

A third consequence of subsidisation is that investments are heavily distorted. Between 1977 and 1981, the German steel industry, which gets no government support, spent on investments in productive capacity per ton of raw steel production US $14, the Belgian industry US $18.5, the French industry US $17.5, the British US $26, and the Italian steel industry US $23 approximately.

The consequences are both a short-term (steel prices) and long-term (production characteristics and cost) distortion, disregard of the jointly

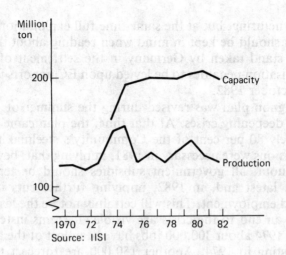

Figure 6.1
Steel-producing capacity and apparent production in the European
Communities (the Nine). Million metric tonnes raw steel

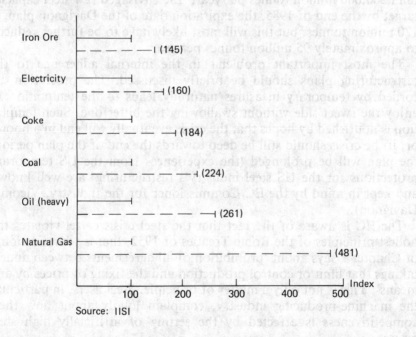

Figure 6.2
Material and energy input prices in the production of European raw
steel, 1974(= 100) and 1981 (market prices) (1974 is after the first oil
price shock)

agreed restructuring, but at the same time full exploitation of the price cartel. This should be kept in mind when reading about the seemingly very harsh stand taken by Germany in the settlement of the quarrel over subsidisation penalties to be levied upon EC imports to the United States in October 1982.

The Davignon plan was revised during the summer of 1980, in the light of the deepening crises. At that time, the plan came to comprise approximately 80 per cent of the Community's steelmaking capacity. Further revisions were necessary in 1981, implying that, beyond further restricted quota, all government subsidies should be terminated by 1985 at the latest and, in 1982, implying further cuts in long-term capacity and employment. This will certainly not be the last revision as no signs of an end to the crisis are visible. It seems instead to accelerate. Since 1974 about 300,000 jobs have been lost of the approximate 800,000 existing in 1974. Another 150,000 are forecast to disappear between 1983 and 1987. Capacity utilisation has fallen to about 50 per cent of the capacity still in operation.

Against the presently available steelmaking capacity of slightly less than 200,000 million tonnes per year,[4] the envisaged raw steel capacity target by the end of 1985, the expiration date of the Davignon plan, is 120 million tonnes, but this will most likely have to be further reduced to approximately 95 million tonnes per year.

The most important problems in the internal adherence to the restructuring plans should be briefly discussed. The protection afforded by temporary measures naturally leads to the temptation to enjoy the sweet side without swallowing the bitter one. Such temptation is nourished by hopes that the crisis eventually will end in a boom, or, if the crisis should still be deep towards the end of the plan period, the plan will be prolonged (the experiences from the US temporary protections for the US steel industry's restructuring[5] are well known and kept in mind by the EC Commissioner for the Industry, Vicomte Davignon).

The EC is aware of the fact that the steel crisis cartel violates the holiest principles of the Rome Treaties of 1952, that is what Section 1 of Chapter 1 says about 'prohibition of all agreements between undertakings that limit or control production and the fixing of prices by any means'. This is not only a matter of principle. Steel users, in particular the machine-producing industry, complain loudly about how their competitiveness is affected by the setting of artificially high steel prices.

The progress of restructuring is clearly too slow, however, not only because of the reluctance on the part of victims, but also because of the progressive worsening of the crisis. The capacity utilisation target of 85 per cent is far from being reached yet, and about one-fifth of raw steel

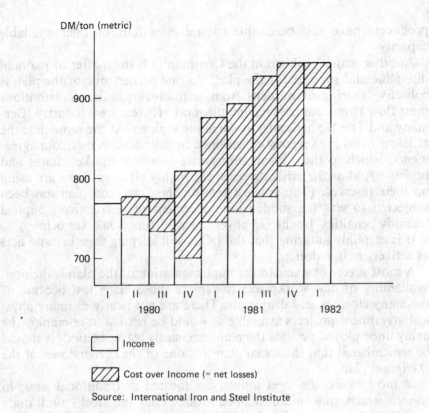

DM/ton (metric)

Income

Cost over Income (= net losses)

Source: International Iron and Steel Institute

Figure 6.3
Cost for and income from rolled common and quality steel, European Community, 1980, 1982 DM/tonnes

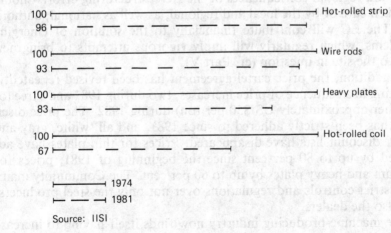

Source: IISI

Figure 6.4
European Communities steel export prices 1974 (= 100) and 1981

producers have not been able to use even half of their available capacity.

Another major problem in the Community is the matter of national discipline and support for the plan. As one cornerstone of the plan is collective sharing of burdens from restructuring, huge contributions must flow from countries having the most efficient steel industry (Germany and The Netherlands) to the ailing regions. At the same time the efficient firms are severely constrained in their actions by quota agreements, which to them mean substantive losses of market shares and profits.[6] And as the crisis deepens, the highly efficient firms are using up their reserves. (The severest tension the Davignon plan has been subjected to was the sudden announcement of retroactively applied 'subsidy penalties' by the US government in June 1982, see below.)

It is certainly amazing that the EC is still keeping together and acts as actively as it is doing.

A most severe obstacle to the implementation of the plan is the non-availability of new workplaces to replace those being lost because of the many closures and dismissals. There are not nearly as many physical investment projects available as would be needed to re-employ the many unemployed persons the plan necessarily leaves behind. It should be remembered that the social plan was one of the cornerstones of the Davignon plan.

In most cases, the steel industry is located at traditional sites, in regions which grew and developed around coal and steel – until quite recently safe growth industries. Within a few years, prosperous regions (with often high average incomes) have turned into regions of stagnation and decline. The social plan is based on the assumption that the local and regional consequences of the EC restructuring efforts should be taken care of by the local and regional, as well as national, authorities. The EC will contribute financially to the solution of emerging problems, which regularly will imply vigorous attempts to bring new jobs to the site in question (cf. Part V).

In addition, the price cartel agreement has been revised repeatedly, resulting in a sequence of price increases, two during 1981 and three (of together approximately US $50 per ton) during 1982. The price discipline has been strictly adhered to since 1981, and all 'white, gray and black' discount lists have disappeared. Prices for thin plates have advanced by up to 80 per cent since the beginning of 1981, prices for medium and heavy plates by up to 60 per cent. The Community maintains strict controls and regulations over not only the steel producers, but also the dealers.

The machine-producing industry now finds itself having to increase its prices by approximately 20 per cent just because of the increases in the prices of the steel used. The reactions from the machine industry have been manifold, ranging from material substitution wherever possible, to imports, either of semi-produced parts (not subjected to the controls) from non-EC sources or by purchase of steel from non-EC controlled producers (for example the mini-mills of Brescia, Italy, the so-called Bresciani) or 'second-grade' declared imports from Brazil. (As a consequence, the EC Commission withdrew a decision of 1956 implying that second-grade equalled downgraded quality production and imports were to be exempted from the monthly registration and reporting procedure. The pertinent decision taken in July 1982 is to be applied retroactively as from 1 January 1982.)

The 'quota deal'

The US steel industry has been successful in lobbying for protection against imports ever since foreign competitors started to appear on the US domestic market after World War II (see Chapter 11). After the Voluntary Restraint Agreements (VRA, 1968–74) the Trigger Price Mechanism (TPM) was introduced (from 1975 onwards).

Early in June 1982 the heads of government from the most important industrialised countries held a summit meeting at Versailles. The principal aim was to confirm the credo in free, unrestricted trade among the western economies, a plea against neo-mercantilism. The summit was meant to be a demonstration against trends towards trade restrictions in response to the deepening economic world crisis.

Only a few days after the meeting, the US Secretary of Commerce, M. Baldridge, telephoned the EC Industry Commissioner, Vicomte Davignon, at 4 a.m. to tell him that the US Government would introduce measures to protect the US steel industry against steel being offered in the United States at dumping prices. The Vicomte was able to tell the US Secretary of Trade that the subsidies overtly, and thus undoubtedly, given to the steel industries of Great Britain, Belgium and France were well within the framework of the subvention code of GATT. The Vicomte and his colleague, the EC Commissioner of Foreign Affairs, immediately departed for Washington DC to talk to US politicians and administrators. They returned with the strong feeling of having met an attitude of understanding and insight into the European steel situation. They had also offered a deal similar to the old Voluntary Restraint Agreement.

One week after President Reagan had co-signed the statement of Versailles against 'protectionist pressures', his own government decided to introduce such pressures and measures against European steel firms. From the day of its announcement in early June 1982, European suppliers would have to deposit an amount equal to the subsidies received and thus not contained in the invoiced amount for deliveries made after 10 May 1982.

The subvention penalty would hit British, French, Italian and Belgian firms particularly hard, whereas Dutch and German firms, which never had turned to the state for help, would be hit by low excises. *In toto* , however, imports of 2 million tons of steel from Europe would be hit by a 'preliminary subvention excise' amounting to US $ 3 billion.

The European Community decided not to fire back from the hips, in order not only to avoid an escalation which, besides jeopardising the orderly implementation of the Davignon plan, would hit other industries as well, but also to avoid an open transatlantic trade war which would be a disaster in many respects. There were also hopes that the US side would understand that the European share of US steel imports – less than 7 per cent – could not imply any severe damage to the US steel industry as was being claimed by the industry itself.

The summer of 1982 was hot weatherwise, and as far as US–EC trade was concerned. Under strong pressure from the US Government, Brussels had to accept a reduction of the European market share from 6.5 per cent to 5.7 per cent. But the US steel industry felt that this was not enough.

There were severe tensions within the EC as well. In particular German firms and the Bonn Government refused to be punished for rigorously keeping to the laws of the market. They would suffer not only from the new quota restrictions but also from the unrestricted import pressure upon the German market from the British, French, Belgian and Italian firms, which effectively would be cut off from selling to the US market. In fact the US measures did play a role in stranding advanced plans for an intra-German merger between two of the better performers on the market, thus jeopardising a reconstruction, which now will cost a much higher price. It is therefore not difficult to understand the hard German feeling *vis-á-vis* the US Government as well as *vis-á-vis* the EC steel policy.

It is surprising indeed that the German Government, very reluctantly, but nevertheless in the end did swallow the bitter pill of loyalty to the EC and to the USA. (It was, it should be remembered, also a very hot summer in other respects, for example, the Siberian pipeline, some important Nato-related decisions and, last but not least, an unusually deep internal political crisis in the Federal Republic itself. And, finally, who does not remember the Falklands and the Lebanon war?)

One should also recall that the American attitude during the negotiations was fairly stiff. The timing, immediately after the Versailles summit, was a surprise, to use an understatement.[7] The time constraints set by the US Government – settlement within eight weeks during the short European summer – made the Europeans feel blackmailed. Nevertheless, an agreement between the EC Commission and the US Government was reached (and also officially blessed by President Reagan) within these time constraints, despite the fact that Europeans had been keeping within the boundaries of the GATT subsidy regulation (Article VI (3)).

The deal foresaw a quota system for eleven steel products. Within these groups, total European exports to the US market would be limited to 5.754 per cent, a cut of 9 per cent against the existing quota agreement. The deal would also have implied a subsidy penalty of about 2 per cent to be levied upon the non-subsidised US imports from Germany. The deal did comprise stainless steel but not special steel and tubes. No claims to that extent had been raised by the US steel industry in the first round. (A few weeks earlier, the US Government had rejected an EC offer to reduce the steel exports of the 'subsidisers' within the EC – Belgium, France, Great Britain and Italy – by 10 per cent, leaving the quota from the non-subsidisers unchanged. It should be remembered that the quarrel started with complaints about the subsidisers.)

However, the above deal was rejected by the US steel industry. What a surprise! No, no surprise if one knew that the US steel industry's capacity was utilised to only 43 per cent during the summer of 1982 and that no improvements were envisaged. And as the US steel industry only exports negligible quantities, it had no reason whatsoever to help to avoid an escalating transatlantic trade war.

An agreement was finally signed on 21 October 1982. The EC will undertake to limit the EC import share of eleven products to 5.75 per cent of the US market insted of 6.4 per cent against the valid quota deal. The quota for each of the eleven product groups varies, however. Included in the deal are now also alloy steel, and tubes. The agreement is labelled 'arrangement', since it violates established fair conduct in international free trade. The last hard fights were about tube imports (which had grown rapidly because of the new US oil and gas projects), of which German Mannesmann held 70 per cent of the EC imports, equivalent to 9 per cent of the US market for high quality tubes.

The final tube arrangement catered for a market share of 5.9 per cent equal to the average of 1979–81. Strictly speaking the tube deal is not part of the 'arrangement' but is covered by an 'exchange of letters' between the US Secretary of Commerce and the EC Commissioner of Industry. If tube imports should exceed 5.9 per cent, this will have no consequences for the 'arrangement'. Most likely the tube market will decline anyhow as a consequence of the faltering of OPEC.

As the Germans both had the largest share of European imports and could not (with one exception, to be discussed in some detail later, see Chapter 7) be blamed for subsidising, they had taken the toughest attitude amongst the EC partners – and won a little at least. Beyond the separate tube deal they gained an additional share of 17,000 tonnes per year, but by the 'arrangement' they effectively lost 500,000 tonnes per year.

The US claim for subsidisation penalties was withdrawn. The 'arrangement' is valid until 31 December 1985, the expiration date of the Davignon plan. 'What would the EC be without Vicomte Davignon!' commented the EC Ambassador to Washington DC.

Summary

The European Communities are an assembly of nations which individually maintain sovereignty over their national economies. Decisions at EC Commission level all have to be negotiated with the national governments. National governments in turn have to negotiate with individual firms or groups of firms. Thus, as evident from both cases of EC activities in the steel sector, the individual state and firms are the ultimate decision points. There, the decisions are taken which establish the degrees of efficiency of each single EC programme or decision. As the examples have demonstrated as well, the steel industries of the individual member countries are organised in idiosyncratic ways (more on this in Chapters 7 to 9 on some individual EC member countries). In order to understand the policies applied, but also to build up an understanding of problems and of possible solutions to them, it is thus necessary to briefly report on the steel industries of the most important member countries at least.

Notes

1   The Japanese, being used to, and sensitive to, the activities of MITI, were greatly disappointed by the impotence of the European Commission to act as an entity on the occasion of the oil price increases dictated by OPEC. The EC members acted or reacted on their own, completely disregarding both the EC and the joint interest of the Community members. This experience has certainly both downgraded the importance of the EC in Japanese eyes and influenced the Japanese attitudes taken when having to negotiate with the Community after the mid-1970s.

2   OECD, *Steel in the '80s* , Paris, 1980, statement by Vicomte Etienne Davignon, p.150. The European steel industry is very trade-intensive. The intra-EC trade is most important, but dur-

ing 'normal' years approximately 30 per cent of production is exported to non-EC countries.

3   As the smaller producers, amongst them the new very vivid and profitable mini-mills, felt bypassed and neglected by EUROFER, in 1981 they formed an association of their own, EISPA (European Independent Steel Producers Association, 68 members). The inaugural president was the legendary German steel entrepreneur, Dr Willi Korf.

4   All estimates of capacity are today of low reliability. In many cases the mills could not be put on stream again. A reason to blow up the figures is the capacity reduction premiums: the larger the (nominal) reduction, the higher the premium. A reasonable estimate is that of 200,000 million tonnes per year; only approximately 150,000 million tonnes are 'real' (1983).

5   But also the disappointing experience of the European fibremaker crisis cartel, which after five years ended in October 1982. It at best saved companies from bankruptcy when the crisis was most severe. The aimed at capacity reduction has never been achieved.

6   Klöckner of Germany would be permitted to run its brand-new hot sheet plant at a capacity utilisation of only 27 per cent. It constantly neglects the orders from Brussels (as many other European producers do as well) and refuses to pay the fines (accrued to, by January 1983, approximately US $ 60 million). As to be expected, the firms, being in severe financial difficulties, find it hard to keep to the rules. It is far more comfortable to violate the requested discipline than to go bankrupt. And which bank would today lend the violator the sums needed to pay the fines?

7   The weeks after the annual summit of 1983 – at Williamsburg again devoted to solemn declarations *pro* free trade, against protectionism, brought an almost identical playback of the steel débâcle of 1982, only this time the US intervened in the field of speciality steel imports.

7 The Federal Republic of Germany

The Federal Republic of Germany reconstructed its steel industry after the heavy damages during World War II. It ranks traditionally amongst the world's foremost steel producers, during the last 15 years occupying the third rank after Japan and the USA. Unlike the USA, Germany is a great importer and exporter of steel and regularly makes a fairly large export surplus.

Table 7.1
Steel production (million tonnes raw steel per annum) and employment (in thousands)

	1966	1974	1981	1982	1983 estimate
Raw steel production	35.3	53.2	41.6	35.9	32–35
Employment	378	344	269	256*	

* One in two steelworkers were working part-time only during winter 1982–83

Thyssen (reducing its raw steel output from 20 million tonnes in 1973 to 10.2 in 1982), Krupp and also Mannesmann have over the years developed into conglomerates. Steel's share of Thyssen's total turnover is approximately 30 per cent; for Krupp the figure is nearly 25 per cent. Mannesmann's main business has traditionally been steel tubes, for which it produces its own raw steel. Mannesmann is today also an important producer of sophisticated machinery. Thyssen is active in heavy physical investment goods as well as in industrial constructions and machinery. The three firms have extended into the service sector as well. This differentiation has helped them to remain profitable.

Table 7.2
The most important steelmakers in the Federal Republic of Germany
(1982 production of raw steel in million metric tonnes)

Thyssen	11.1	Peine-Salzgitter[b]	4.1
Krupp	4.9	Röchling-Burbach[c]	2.9
Klöckner	4.8	Korf[d]	1.1
Estel Hoesch[a]	4.7	Others	3.6
Mannesmann	4.5		

[a]   Estel-Hoesch is one of the very few multinational firms in steel (some Japanese giants and 'little' German Korf own subsidiary plants in the United States in the midi and mini-mill range), coming out of a merger between German Hoesch and Dutch Hoogovens. The aim of the merger was concentration of steelmaking in a deep-sea harbour location near Rotterdam. The merger is under dissolution.

[b]   Peine-Salzgitter is the only state-owned steel combine in Germany, for historical reasons. It is managed on a strict business basis and is one of the very few state-owned steel firms with a respectable profit record over the years.

[c]   Operating in the troubled Saar region, see below.

[d]   In the mini-mill range. Korf in January 1983 appealed at court for an official financial reconstruction settlement.

During 1980–82, Thyssen lost approximately US $ 160 million on its steel sector, but the corporation was able to cover most of it from gains on other business sectors. In 1981–82 the steel sector broke even. (Thyssen's US daughter Budd Co. suffered severe losses, US $ 80 million in 1980–81, US $ 160 million in 1981–82, mainly because of the slackening demand from the US automobile industry.)

Krupp (employment 1981 82,000) today makes only 25 per cent of its turnover in steel, and about 16 per cent in industrial construction and turnkey projects. Machinery accounts for approximately 12 per cent of the turnover, services and trade for approximately 30 per cent. Krupp is the largest in Germany in special and alloy steel, which account for about one-third of its steel output, the second largest (approximately one-third of Krupp's output) in alloys being Thyssen. The company is losing heavily on its steel sector (1982 approximately US $ 80 million, the 1983 forecast being as high as US $ 160 million) and has been selling off some of its holdings of land and physical property to cover the losses in steelmaking.

Other firms have suffered more and all have lost on steelmaking: from 1975 to 1982 most likely more than US $ 6 billion.

The German rule has been that the firms have to look after their own business. Klöckner is a remarkable example of such a restructuring effort. There has, however, been one major exception from the principle, the case of the Saar (see below).

There was one often quoted example or exception before, in 1967. Germany was suffering the first severe recession since the war, and one

126

of its well-known steelmakers was in severe trouble because of under-capitalisation: Krupp. The former weapon forgery of the Reich had suffered severely from the allied powers' segregation legislation (*Entflechtungsgesetze*) and was virtually stripped of capital. Its solidity (equity's share of the balance sum) was only 16.6 per cent. In fact it reached 20 per cent only ten years later, despite respectable profits. (Only Mannesmann has – for a long time – had an equity basis of above 30 per cent. Krupp is today with its approximately 20 per cent a good number two amongst the German steel-based firms.) As Krupp, whose exports since the post-war reconstruction always have been higher than the sales to the domestic market, had been engaging itself in large projects in the COMECON area, it found itself in a position where, because of the recession, it could not meet the cash flow required.

Krupp asked the state for a guarantee for a loan, in order to be able to fulfil its contracts in the East. The guarantee was granted under quite stiff conditions – short duration, until 31 December 1968 – and with cooperation from the banks. The banks finally granted altogether US $ 120 million. At the same time, however, they requested the conversion of the Krupp family firm, which it had been since 1811, into a corporation with limited liability under public control, thus also the establishment of a board of directors with seats for the banks, and a reorganisation. What several wars had not been able to shake up had come into being: the power of the Krupp family over its firm was broken.

The amount guaranteed by the state was never utilised to more than 30 per cent. One should remember that it was never intended as a subsidy, but only as a safety signature for the banks. The taxpayer did not lose a penny. As a matter of fact, it was a very profitable deal for the state, as it rendered good tax earnings over the years and also fairly stable employment.

Estel-Hoesch, in 1981 holding the fourth rank in Germany in raw steel production, was the result of a Dutch–German steel merger, contemplated in 1969 and implemented in 1972 when the future of steel was very bright. The aim of the marriage between Koninklijke Nederlandsche Hoogovens en Staalfabrieken NV (with production facilities at Ijmuiden on the coast near Amsterdam in the Netherlands) and Hoesch AG of Dortmund (in the eastern off-Rhine part of the Ruhr District) to Estel NV was to move their steel production to Rotterdam (one of Europe's largest deep-sea harbours) with a final capacity of 10 million tonnes of steel per year.

With the advent of the new steel crisis Estel put their advanced blueprints back into the drawer. Instead, the works at Ijmuiden were modernised and enlarged, but little was done at the German site. The new firm essentially came to function as an accounting umbrella under which differences in profits and losses could be settled. The older German factory went into red figures as the crisis progressed.

In 1982 it was finally decided to break up the marriage (Hoesch had found a more attractive partner to cooperate or marry with – Krupp). The settlement of the joint assets and liabilities turned out to be difficult – 'it is always easier to make an omelette out of two eggs than two eggs out of an omelette', commented Mr Hooglandt, the Chairman of Hoogovens. In this case, Hoesch has to carry two-thirds of the total sum (approximately US $ 200 million) at stake because of the divorce.

The temporary marriage has left a vacuum in the Ruhr (see Chapters 20 and 21): the workers as well as the regional government are demanding a replacement for the rundown Hoesch plant. There are plans to build a new steel plant in the Dortmund area, at an estimated cost of US $ 800 million, and these have been speeded up, as the investment ought to be made before the Davignon plan expires at the end of 1985 in order to stand a chance of being blessed with public support, which, as things stand now, will not be permitted after 1985. Hoesch has applied for approximately US $ 280 million investment support from the authorities. But will there be a market for such a new plant? And, even if it should happen to get the so much desired and needed investment bonus from the government, will Hoesch be able to raise the capital needed? Of course, it will need state guarantees for the loans to be taken. And now it matters that Hoogovens has a close to US $ 200 million claim against Hoesch because of the 'divorce' settlement. More worrisome still is that Krupp (which in the meantime has a 50 per cent surplus raw steel capacity) will, within a year or so, have its brand new electro-steel plant at Bochum, only ten miles west of Dortmund, ready to go on stream. Is there a need to invest in more new capacity? Is Hoesch perhaps planning for the possibility that its new bridegroom will change his mind at the last minute? Hoesch's boss, Rohwedder, seemed to indicate this kind of thinking when informing the Bonn Government of 'conditional investment plans' amounting to US $ 1.25 billion during early autumn 1982.

As already briefly mentioned, Hoesch had a new marriage in mind when abandoning Estel's bedside. The new wedding is to be part of a twin wedding: Krupp is aiming at two beds, one with Hoesch (approved by the shareholders of both corporations in February 1982), to form Ruhrstahl AG for highly efficient bulk steel production. Krupp is, however, not only active in the coal steel sector but, as already mentioned, also in special and alloy steel, where Thyssen is second largest. Thus, the Krupp–Thyssen marriage in special and alloy steel is under way. (This was obviously a shock for Hoesch, who thought that the alloy sector would be part of their deal with Krupp.) Depending upon the sequence of marriages, Krupp might not even be the bridegroom of both marriages. Most likely, Ruhrstahl AG will be replacing it in the alloy steel merger with Thyssen (Deutsche Edelstahlwerke AG, to be owned 50–50 by Krupp and Thyssen).

As the future prospects in the alloy and special steel sector seem to be a little brighter than in coal steel,[1] Ruhrstahl must be a much less attractive new home for Hoesch than it appeared when special steel was still part of the dowry. Perhaps Krupp feels that Hoesch is not a really first-class bride, so it is better to put the pet baby into another basket (the alloy steel merger would result in a concentration of 90 per cent of German alloy steel production in one company, employing approximately 24,000 with an estimated turnover of US $ 2.5 billion a year). One consequence of the marriages is clear to the unions: employment in steel is dwindling rapidly. Another 6,000–10,000 jobs would soon disappear.

The Federal Government would have liked to see its own Peine-Salzgitter become part of the new grouping. The unstripped (of alloys) Ruhrstahl would have been an ideal new home for the only government-owned steel firm.[2] Peine-Salzgitter is situated near meagre ore deposits some 150 miles north-east of the Ruhr District, a much poorer location still than the Ruhr (of which the eastern part already is much worse off than the western one along the river Rhine), neither of them accessible via deep waterways for bulk transports without reloading if modern and economic bulk tonnage is to be employed. The off-the-main-road location is costing Peine-Salzgitter a lot, approximately US $ 25 to 30 million a year. Nevertheless, Peine-Salzgitter is one of the most efficient government-owned enterprises in Europe. It has had to meet the cold winds of competition without a shelter. Its most efficient and dominant part of the product portfolio is in heavy profile steel.

The Saar – a troubled region

The exception from the German rule not to subsidise the steel industry is the traditional coal and steel district of Saar, neighbouring the heavily crisis-stricken French Lorraine district. It first suffered through the decline of the coalmining industry, and, as a matter of fact, many miners found new jobs in the growing and prospering steel industry during the coal crisis. The restructuring in the Saar has been cited as the exemplary showcase of a successful joint effort by federal and regional government and industry in saving a region from severe decline.

In 1977–78 the Saar steel industry faced severe problems. The regional government of the Saar and the Federal Government in Bonn contributed to a dual restructuring programme:

1   To make possible financial and technical restructuring of the region's steel industry, the two governments jointly guaranteed bank and debenture loans in the range of approximately US $ 1 billion.
2   A second programme aims at restructuring the region, which will lose 10,000 jobs between 1977 and 1984. Attempts will be made to attract new industries and workplaces to the troubled region.

The case of the Saar deserves some attention, not only because it is a departure from the otherwise tough attitudes exposed by both German steelmakers and authorities.

A short historical review of the Saar's fate is called for. The Saarland is one of the states making up the Federal Republic of Germany. After World War I, the Treaty of Versailles stipulated that the Saar – German populated border-country between France and Germany – should be administered for 15 years by the League of Nations. Then a plebiscite should be held about the Saar's political future. The Saar mines would become French property. In 1925 the Saar was integrated into French customs territory. The plebiscite resulted in a 90 per cent vote for Germany and the Reich bought the mines back from their French owners.

After 1945 the French Government re-established the status of 1925–35, with only slight changes. The Saar mines rendered approximately one-third of French coal production.

The Saar became an associate member of the European Parliament (1950) and of the European Community for Steel and Coal (1951), in which it was represented by the French Government. Pro-German parties were not permitted there. After a plebiscite (1955) the Saar was integrated into the Federal Republic of Germany.

For a proper understanding of the present steel crisis it is essential to remember the many shifts and uncertainties about the political and economic future of the Saar (not only during this century).

Despite the drastic decline of coalmining (oil was much easier to handle and had become very cheap) at the beginning of the 1970s, 45 per cent of jobs were still in coal and steel. Jobs lost in mining often were replaced by jobs in steelmaking and refining.

In 1970 half of the value of the Saar's production was in raw and rolled steel. If one adds the steel construction and machinery industries, the steel and steel-dependent share was more than two-thirds. The Saar was a region in drastic and thorough structural change after it had become part of Germany again. From the steel-making point of view the Saar is a disadvantaged location: it has access only to low tonnage waterways; it depends on ore (and oil) imports to the region, but has access to local coal (although this is scarcely regarded as an advantage any more). After the economic integration of the Saar into the Federal Republic of Germany, the steel industry, consisting of small to medium-size rather modern, often family-owned plants, has been under continuous reconstruction. For the time being, after a major reconstruction in 1978, the steel producers of the Saar are owned by French holdings (Dillinger, employment about 10,000) and Luxemburg ones (ARBED-owned Röchling-Burbach with daughter Neunkircher Eisenwerk, employment approximately 21,000).

Table 7.3
Employment (in thousands) in the Saar mining and steel sectors

Year	Mining	Steel	Together
1950	66	28	94.5
1960	56	42	98
1970	27	39	66
1980	24	31 (of which ARBED 21)	55*
1987 (forecast)		25 (of which 15.2 by ARBED, Rogesa and ZKS)†	

* Saar unemployment 1982 41,000
† a jointly owned coking plant

The reconstruction of 1978, when ARBED increased its holdings, thereby opening possibilities of concentrating some major processes in economically reasonably sized plants, resulted in substantive cuts in raw steel-producing and steel-forming capacities. The reconstruction, requiring a total physical investment of US $ 1.25 billion and compensation to the laid-off personnel of approximately US $ 200 million, was supported by the Federal German and the Saar State Governments, by means of debenture loans and guarantees for bank loans amounting to US $ 850 million (which in December 1982, to avoid bankruptcy, had to be topped up by a further US $ 25 million, against securities in ARBED's total holdings in Germany). It is also accompanied by wage cuts ranging between 10 and 30 per cent and temporarily withheld payments of 50 per cent of a month's wages. The workers are losing between US $ 80 and 320 a month.

A further step towards reconstruction, made necessary by continued drastic slackening of the market, is under implementation: ARBED Saarstahl (the financially reconstructed holdings of ARBED in the Saar region) and Dillinger will co-operate 50–50 on a concentration of their raw steel (Rogesa) and coking (ZKS) capacities. The consequent physical investment for the years 1982–88 amounts to another estimated US $ 1.25 billion. Total steelmaking employment by 1987 is estimated to be less than 25,000, provided ARBED Saarstahl can survive.

ARBED's annual report for 1982 shows debts amounting to approximately US $ 1 billion (the turnover being approximately US $ 790 million. The burden of interest alone is in the range of US $ 95 million (or about 13 per cent of the 1982 turnover).

During the early autumn of 1983, ARBED (then employing 17,200) had to tell the banks and the Saar and Bonn Governments that bankruptcy was unavoidable, unless both governments were prepared to sign guarantees for new loans to be taken. Also 5,100 employed (essentially all those over 50 years old) would have to be put on an early retirement scheme. ARBED offered them 82 per cent of their current earnings, and requested the unions to abstain from pay increases for the active workforce for the next few years to come. All management personnel would experience cuts of 25 per cent of their present earnings for 1983– 85.

The unions, after some weeks of desperate resistance, finally accepted the deal – otherwise all 17,000 would be losing their jobs. The settlement of November 1983 means public subsidies during 1983 and 1984 of about US $ 65 million, the extension of the pay-off period and exemption from interest on those loans guaranteed by the state (since 1978 approximately US $ 420 million of such loans have been accrued).

In addition, ARBED's banks are not going to press for payment of interest on and settlement of loans. By those means they improve ARBED's liquidity by approximately US $ 140 million over the next few years.

Again, ARBED was rescued at the last minute from seemingly unavoidable bankruptcy. Nobody knows, however, if the US $ 1 billion of support at the taxpayers' expense will actually suffice to secure ARBED's survival.

'Little steel'

There is also 'little steel' in Germany. Dillinger Hütte has been mentioned amongst the traditional steelmakers but there are a few others which, in justice, should have been mentioned.

The most interesting development has, however, taken place in the mini-mill range. The outstanding entrepreneur is Dr Willy Korf, who until very recently has run profitable product lines in a way that has become established as the prototype of the mini-mill concept. (In contrast to the northern Italian 'Bresciani', his plants are not favoured by artificially cheap electricity supplies.) Korf has also gone multinational by entering the US production market, holding 70 per cent of Korf's Industries Inc., Charlotte, North Carolina, the remaining 30 per cent, the same as of the German holding, being in the hands of the Sheikdom of Kuwait. Korf Industries Inc. is the mother of two producing daughters, the Georgetown Steel Corporation, Georgetown, South Carolina (which also produces wire in the Andrew's Wire Division) and the Georgetown Texas Steel Corporation, Beaumont, Texas.

Korf has lately suffered severe blows – who would not in such a chaotic 'market' as the steel market. He has had to learn bitter lessons from partners as well.

The most remarkable case so far, Nordferro, started as a superb idea, a typical Willy Korf brainchild. Dependent as mini-mills are on scrap prices, he ventured into direct reduction, a process rendering a high degree (approximately 95 per cent) ferrous input into steelmaking, without using blast furnaces.

The partners who met in the project seemed to be the absolutely ideal couple. Korf steel is not only a strong supporter of the iron-sponge method, but even produces and sells worldwide turnkey direct reduction plants.

The idea of having a coastal plant was first taken up with Estel NV, but Dutch natural gas proved to be too expensive. However, the Norwegian State Oil Corporation-held Sydvaranger, which has a pipeline terminal from the North Sea Ekofisk Field at Emden, Germany, would provide not only natural gas at a good price, but also ore from new north Norwegian mines. Still better, Sydvaranger would contribute 75 per cent (Korf 25 per cent) of the equity. And as the coastal region is a notorious crisis and unemployment strip, the couple would only need to put less than 12 per cent cash into the plant, the remainder being public job creation subsidies (14 per cent) and bank and EC loans, for which the Federal and State Governments would guarantee slightly less than two-thirds. This meant that the couple's own risk for the loans taken would be 27 per cent.

The plant went into operation in the spring of 1981. By that time, however, the price for Norwegian gas had increased by 250 per cent, so that the unit price was now 40 per cent above the maximum economically tolerable price.

There was more trouble to come: not only did scrap prices fall to lows no-one could remember (and scrap prices determine the price for iron-sponge) but, so far, not one single ton of ore has been shipped from the north Norwegian mines. The Royal Treasurer of Norway does not want to put any more Kroner into operations of which no-one knows if they ever will become profitable.

Nordferro went into bankruptcy during the summer of 1982.

Korf's steelmaking operations are for the first time in the red because he has been growing very fast and his operations carry the burden of the last years' high interest rates. The man who could and would sell, at good profits, at prices considerably lower than big steel could offer during the last years, found himself outcompeted by heavily subsidised French (state) steel, selling at prices ranging US $ 120 and more below direct production cost. No wonder that early in January 1983 Willy Korf had to file a petition for official financial reconstruction not only of his holding corporation, but also of his producing corporations. Will this be the end of Willy Korf, who most likely would have made his way certainly not into big steel but rather into large midi-steel, based on electricity and scrap/direct reduction? Most likely the very efficient steel factories in Hamburg and in southwest Germany will survive.

Like the big steel firms, the Korf group has avoided dependency on steel alone but comprises technology and turnkey production project divisions/corporations as well. The group's turnover in Germany is approximately US $400 million, of which US $ 100 million come from engineering products. Turnover worldwide is about US $ 1 billion. Employment is 3,400 in Germany, approximately 10,000 worldwide. The steel production of all Korf factories is between 2.5 and 3.00 million tons of raw steel and 2.3 million tons of rolled products, all figures per annum.

Co-determination – Legislation for steel and associated industries

German mining and steel-producing corporations with an employment of 1,000 and more are subjected to legislation (1951, 1956, 1965, 1967) which aims to give the employees an equal number of seats on the board of supervisors to the employers. Usually there are 11 board members: a chairman elected by the other 10 members; five employees' representatives, of which a minimum of one blue-collar and one white-collar is employed in the firm; and three other members who are nominated by the unions and represent the firm's employees in wage/salary negotiations etc. The other five members are nominated by the firm or the owners.

Further, there is a 'work director', who is an equal rights member of the directorate of the firm, with the task of looking after the employees' social interests. He cannot be put into or out of office against the majority vote of the employees' board members.

Even firms in which employment in mining or steelmaking has declined, so that it only represents a minor share of the firm's employment or turnover, are still held to maintain the rules for the two industries. (In other industries board representation is prescribed as well, but never in the parity rule. During 1982, Mannesmann planned to re-organise in a way which was understood by the unions and the then Labour-dominated government as a means to 'escape' the law. Against the resistance of the employees' board representatives the re-organisation could not be implemented.)

There are other rules for employee co-determination as well, valid for steel corporations in the same way as for firms belonging to other industries and lines of business. Such generally valid rules are not further elaborated upon here.

The future of 'big steel'

The future of German 'big steel' has been in the melting pot during 1983. The 'crisis' is forcing the industry to take long-term solutions earlier, quicker and in a more drastic mode than expected. This crisis is essentially imposed from the outside: the German steel industry, despite its inland location, has had to meet the shifts in the market without protection and subsidies. Thus high flexibility, efficiency and productivity were the only means of competition at hand, and they have been used, together with structural adaptation and change.

The 1970s and, in particular, the early 1980s brought heavy distortions in the markets in the shapes of subsidies and regulation – for example protectionist measures but also the EC quota system. The size of the subsidies to the steel industries of Great Britain, France and Italy, ranging between 20 and 40 per cent, was laid open during the steel quarrel of the summer of 1982 between Brussels and Washington.

We have seen above that the conglomerates, Thyssen, Krupp and Mannesmann, who diversified into other business long before the first signs of the present deep steel crisis became visible, have all been losing heavily on their steel sectors, but, because of their diversification, were able to survive by covering deficits in their steel business with their profits from other sectors of their portfolio. But all big steel is not in such a lucky position. Hoesch, Klöckner, Peine-Salzgitter, each of them highly efficient in their specialities, have lost their financial reserve cushion, without having access to profits from other business.

The European super-guardian of free trade, the head of the Federal German Department of Economics and Trade, Otto Count Lambsdorff, called the steel industry (Thyssen, Krupp, Hoesch, Peine-Salzgitter, Klöckner, but also ARBED Saarstahl and Korf) to a crisis meeting in November 1982. As a result of the meeting the industry agreed to ask a group of independent experts to produce some alter-

135

native proposals for a thorough restructuring of German big steel. Neither the Federal or State Governments nor the unions became involved in this mission, which presented its proposals in January 1983.

The experts' mission was to review the situation and future of German steel, in order to submit proposals for measures to be taken to 'stabilise the market' and 'to cooperate beyond the borderlines of presently existing enterprises'. The treatise does not have the status of a government paper. It contains records of meetings between the experts and the representatives of the above enterprises, during which consensus was reached on market measures, but disagreement persisted about options to cooperate or to merge. The experts emphasise that 'decisions about measures to be taken are the exclusive prerogative of the responsible bodies of the concerned corporations'.

The treatise in brief concludes the following:

1   *Market proposals* . The experts propose an 'ultimate trench' for German bulk steel, which should be defended against dumping and other destructive measures from abroad. The experts assume and claim the agreed-upon EC production quotas to be maintained, as planned, to the end of 1985. Temporary agreements should be reached with EC partners and other countries regularly selling on the German market, in order to 'stabilise the German market' whilst restructuring the industry.

2   *Cooperation proposals* . The experts propose a re-grouping of the bulk industry into two sectoral groups named 'Rhine' (Thyssen and Krupp) and 'Ruhr' (Hoesch,[3] Peine-Salzgitter and Klöckner-Maxhütte). As an immediate measure the experts propose the establishment of two temporary marketing organisations for flat products and heavy profiles respectively. Special products, wire and concrete steel would stay outside. Krupp, Hoesch, and Peine-Salzgitter should, as already being planned, drop their production lines in the last-mentioned ranges to improve capacity utilisation for other producers. These should cooperate in two marketing corporations, 'Southwest' (ARBED Saarstahl, Maxhütte and Korf-Baden) and 'Northwest' (Thyssen and Korf-Hamburg). In the course of the restructuring efforts an intra-German switch of production quotas, within the frame of the EC agreement, should take place.

The financial means needed to implement the restructuring are estimated to range between DM 2 and 3 billion (US $ 800 million to 1.25 billion). An EC estimate ranges in the neighbourhood of US $ 2.5 billion. As steel industry enterprises today are bare of resources, the public would have to put up the money needed to finance the restructuring efforts. A major (not specified) proportion should be granted as conditional subsidies. Interest-free loans should be used wherever possible (the support should be allotted according to the participants' 1981 raw steel production) and subsidies should be granted immediately, in some urgent cases like Klöckner, which is in deep trouble, having lost much substance in the course of its major restructuring efforts (see above). Now, in summer 1983, 30,000 jobs at Klöckner are in acute danger.

The *implementation* of this first comprehensive restructuring outline for the entire German bulk steel industry is contingent on decisions to be taken by many interested parties: the firms and their shareholders, trade unions, state governments, the Federal Government, banks, but also the EC Commission. Time at disposal is scarce – and turbulent. Subsidies proposals in all EC member countries had, under the present Davignon plan, to be submitted to the Commission after March 1983. The Commission then undertook the most delicate task to decide which subsidies were 'legal' and which were not by the end of June 1983. This part of the Davignon plan aimed at constraining the subsidisation run.

The Commission will not be inflexible in the case of Germany, where subsidies up to now have been rare exceptions. The fact that the German steel industry still is, and intends to remain, in private hands will also be taken into consideration, as many more decision points are involved than would be the case in a mainly state-owned industry.

A major question not answered in the treatise is the number of jobs to be given up or to be maintained. It was simply impossible to even estimate the employment consequences within the two months available for the experts to produce their proposal. This, however, means that the unions and the concerned state governments will not be willing to cooperate until they know the 'whole truth'.

A few points should be made about the viability of the proposal, which is radical but still rests upon the basic assumption of a largely market economy. (Monopolies in singular lines of business or production have been strictly avoided. Two competitors are foreseen to remain in all major production lines.) The big question is, however, whether even such a radical restructuring and consequent rationalisation will be able to generate enough gains for German steel to survive in a market where subsidies effectively distort all regular competition.

State governments have had no influence on the treatise. Some of them feel heavily disadvantaged and are trying to influence the outcome in order to minimise regional losses in employment and industrial structure. In particular the expected imbalance between the strong Rhine steel corporation and the much weaker and even poorer Ruhr

137

group is causing trouble. Shareholders, management, the unions and concerned state governments feel uneasy about the Ruhr group, and this is not only because of Klöckner's obvious and acute problems. In the meantime, Thyssen and Krupp are taking steps to implement the Rhine merger, which, after the formation of Deutsche Edelstahl AG, will mean that both Thyssen and Krupp in the future will only have joint holdings in steel and, as the Rhine group will request only a minor share, if any at all, of the restructuring subsidy, the external influence upon the formation of the Rhine group is low.

Early in November 1983, whilst the life saving attempts in ARBED's favour were in their most critical phase, the last parts of the reorganisation proposal were torn into pieces. Its seemingly strongest part, the Krupp–Thyssen merger, was rejected by the two parties. Whereas the Bonn Government was prepared to contribute approximately US $ 175 million to this part of the total solution, Krupp and Thyssen claimed a total of US $ 415 million (half of which was a loan to Krupp) to cover a reasonable fraction of the structural adjustments to be undertaken.

It had been claimed that the Krupp–Thyssen merger would have implied savings of about US $ 50 per tonne of steel produced jointly. Both Krupp and Thyssen have announced drastic cuts to be undertaken at very short notice. (Thyssen will cut its raw steel capacity by one-third to only 11 million tonnes per year, most likely implying lay-offs around 8,000, whilst Krupp will dismiss 4,000 in 1984 and 1985.) Amongst the measures to be taken is a cooperative venture between Krupp and Thyssen in the forgeworks sector yielding close to 4,000 jobs at a turnover of US $ 200 million per year. A loss of another 5–6,000 jobs seems, however, unavoidable.

In concluding the account of this attempt to reorganise German 'big steel' it is of interest to note that the preparedness of the Bonn Government to contribute over US $ 1 billion (approximately 415 million to new investment and close to US $ 600 million to the scrapping of capacity and to 'social cost' to cover some of the consequences of increased unemployment due to the plan's realisation) did not suffice to move the industry even one step towards a 'grand solution'. Everyone claims to be prepared to find his own way out of the dark. In the meantime the demand for steel on the German market is being covered by growing imports – in 1983 50 per cent of the total supply of bulk steel came from abroad.

Summary

In summary there is reason to believe that German steel will not only survive with big and small plants, but has a fair chance of being one of the very few survivors in Europe able to produce at a profit. Its strength in technology, its maintenance of a fairly strict market and customer orientation, and tough management are the main factors of competitiveness that will count.

The unions have played, and are playing, an important role. In retrospect, their high wage policy has forced the industry to stay at the forefront of technology and to produce products for which customers worldwide still pay the cost price. For years now, the unions – also in their capacity as board members because of the unique status they are granted – have had to watch their membership dwindle, and there is no end yet to the steep slope.

Government has – not only because of the pressure from industry at large – been careful not to enter the subsidisation spiral. The Saar programme, which was designed and implemented with the EC Commission's consensus and permission, where a dual-structured, economically and historically very specific coal and steel region has been hit within only two decades by two deep crises, may be regarded as the exception which proves the rule. Germany (like Holland) is not without subsidisation cases, but, as the recent US accusations against the European Community have shown, it is still a most modest subsidiser. Subsidies are strictly targeted towards modernisation and capacity reduction and creation (but hardly maintenance) of jobs in other lines of business. They are also given for research into new products and processes.

Notes

1. There is no exact borderline between bulk and alloy/special steel. Modern mills make it possible to enter the more exclusive markets. If the bulk market remains dull, which is more likely than not, there will appear many new competitors, in particular Japanese ones, in the speciality market.

2. A heritage from the Third Reich's quick and efficient build-up of steel capacity before and during World War II – Reichswerke AG für Erzbergbau und Eisenhütten 'Herman Göring' – founded in 1937. After the war four-fifths of the plant was removed as a war damage payment (a second part of the group were 'Reichswerke Alpine Montanbetriebe', after the war to become the core of the Austrian government-owned steel industry).

3. This would mean that the Ruhrstahl AG merger of Hoesch and Krupp would not take place.

8 France

French steel production accounts for 3 per cent of the world's total production, giving France the rank number 7 (after the Soviet Union, the United States of America, Japan, Germany, China and Italy). The 1983 deficits for Usinor and Sacilor are estimated to range around US $ 500 million *each*. French output is estimated to stay around 17 million tonnes in 1983.

Table 8.1
Production of raw steel (million tonnes) and employment
(in thousands)

	1974	1981	Change	1982
Production of raw steel	27.0	21.2	-21.5%	18.4
Employment	156.6	97.2	-38.3%	

About 40 per cent of the steel produced is exported; in 1980 over 9 (1982 7.5) million tons, imports approximately 7.5 (1982 7.5) million tons.

The old steel centres are Lorraine in the east, the north (bordering Belgium) and central France. As a result of the 'planification' the coastal sites at Dunkirk and Fos,[1] near Marseilles, were developed with great efforts. French steel had undergone a drastic expansion and concentration process already during the 1960s, when raw steel production grew from 4.5 to 27 million tons between 1946 and 1974. The new capacities were ready to be put into production when the crisis came. This resulted in another concentration into a few enterprises so that since 1978 Usinor and Sacilor have produced 80 per cent of French raw steel. French steelmaking is today in the hands of the state. Creusot-

Loire, an integrated steel/machinery-steel construction combine went bankrupt in 1984. Vallourec is the dominant tube producer.

The reduction of jobs has been dramatic during the last ten years. Since 1974, 42 per cent of the jobs in steel have been lost, most of them in the Lorraine (the Saar is the immediate eastern neighbour of the Lorraine; see Chapter 7). Despite its rather low average age, the industry suffers from a rather moderate productivity. This fact, in combination with the high debts and the consequent high cost of capital, has hit French steelmaking rather hard and has made necessary financial reconstructions, for which only government was able to raise the capital. The final socialisation of the industry by the Mitterand Government was therefore a formality at best. The reconstruction plan of 1977–78[2] foresaw both a closing of old plants and modernisation and, consequently, a reduction of jobs by 20,000. Since then, the target has grown to over 60,000.

The results of the reconstruction during the late 1970s gave good progress towards higher productivity, at a rather high price again. Some of the steelworks today come close to Japanese and German productivity levels and have made remarkable achievements such as record low coke consumption per ton steel produced of low grade ore, almost 60 per cent (1982) continuous cast steel (equal to the share in Germany), growing shares of refined products, and technological advances in converters.

The level of subsidisation, for example as published during the US–EC quarrel about steel subsidies during 1982, for the French steel industry is rather high in some product categories in the opinion of the US steel industry: the average subsidy for Sacilor is 30 per cent, for Usinor 20 per cent of the full cost of production.

After heavy losses in 1981 and 1982 (approximately US $ 650 million in each of the two years or 38 per cent of the total deficit of French government-owned industry, with the annual cost to the French taxpayer of maintaining the steel industry presently exceeding US $ 1.5 billion p.a.) to be covered by state loans as permitted by the EC, still another reconstruction programme taking place between 1983 and 1986 has been decided upon by the French Government. It again gives Usinor and Sacilor new capital (about US $ 2.5 billion) for modernisation. The plan foresees break-even for French steel after 1986, at a capacity of 24 million tons of raw steel and a loss of another 12,000 jobs (estimated employment in 1986 being 78,000). Several EC member countries are objecting to the planned capacity target of 24 million tons per year as being far above reality and beyond the common European targets.

At the beginning of 1982 the production of alloy steel was reorganised. It is now in the hands of Sacilor.

Notes

1. These are joint ventures of Sacilor and Usinor. Both are basic oxygen process-based continuous casting plants.
2. Stoffaes, C., Gaddoneix, P., 'Steel and the State in France', *Annalen der Gemeinwirtschaft* , no.4, 1980, pp.404–21.

9 Great Britain

British steel production reached an all-time high, 28.3 million tons of raw steel, in 1970, 1973 being another peak year at 27 million tons. During the strike year of 1980, only 8.4 million tons were produced and 1981, a 'normal' crisis year, rendered 13.2 million tons. The British steel industry was hit earlier and more severely than any other European steel industry by a sequence of crises.[1] Imports had been rising continuously since 1970, at the expense of British Steel[2]. The principal causes of the first crisis were slow technological progress,[3] low productivity[4] and failures to meet average quality standards. The British Steel Corporation was almost exclusively active in bulk products, thus lacking entries into the special and alloy steel sectors almost entirely.

Private steel in Britain holding approximately one-third of the steel-producing capacity, but accounting for only one-sixth of raw steel production, was able to defend its market shares much better, mainly because of a more suitable product portfolio.

Approximately 50 per cent of domestic apparent consumption is provided by British Steel, 22–25 per cent by private steel producers. Imports have been growing from less than 20 per cent to close to 30 per cent between 1974 and 1981.

Since 1975–76 BSC has been running at a deficit. Between 1975 and 1980 employment was reduced by 60,000. Nevertheless, neither the productivity targets nor the planned break-even could be achieved. A further reduction of 80,000 jobs during 1980–83 is cutting BSC employment to approximately 90,000 and the new – spring 1983 – employment target is 75,000 jobs. Despite this drastic reduction in employment, BSC is still running at large deficits. Capacity is being reduced to 14 million tonnes and many measures have been taken, perhaps rather late, to improve the quality of products and of customer services.

A major problem of BSC has been the location of its plants in monostructured regions, where steel is the only and thus entirely dominant industry. Although similar problems exist in many countries (in particular in Belgium and France), the British situation seems to be very difficult as it has been very hard to attract other industries to the troubled regions. Thus it has been politically impossible (even during the extremely tough Conservative government under Mrs Thatcher) to

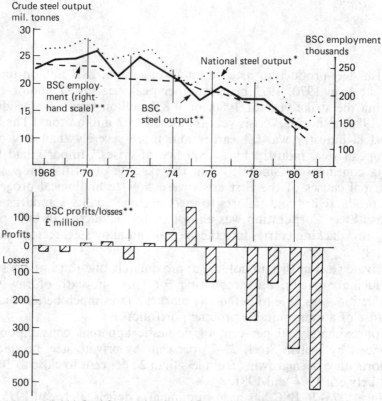

Source: British Steel Corporation Annual Report; Annual Abstract of Statistics

* Calendar years.
** Financial years ending on 31 March of the years shown. Profits/losses after depreciation but before taxation, long-term interest and extraordinary items.

Figure 9.1 British Steel Corporation — selected indicators, 1968-81

close big plants which, because of their inefficiency, should have been closed. In December 1982, the British Government decided to prohibit the closure of any of the remaining five large plants, despite ever-increasing deficits.

During 1980 the present government went to hire Mr Ian McGregor, Scottish-born US citizen, for the position of BSC board chairman. The contract with Mr McGregor, covering a three-year period ending mid-1983, was a remarkable one, as it stipulated certain success criteria to be met by Mr McGregor. Despite Mr McGregor's attempts to reduce the cost burden of BSC (for example by means of drastic reduction of employment in the industry), BSC at the time Mr. McGregor's contract expired, was losing £8 million a week, against £7 million when he took over. The steel crisis has been worsening considerably, without any doubt, and government has, under the impact of growing political opposition to the very drastic cuts in steel employment, had to put brakes on the supermanager's Procrustean drive. A quite severe conflict is looming up at judgement day, that is when government and Mr McGregor will have to agree upon why the performance targets of his contract have not been met.

The British taxpayers put approximately £7 billion into BSC between 1975 and 1982. Still, BSC has been losing market shares in a declining market, through foreign imports. Whereas in 1970 BSC held 70 per cent of the domestic market (private steel producers 25 per cent, imports 5.5 per cent), in 1976–77 BSC's share had fallen to 55 per cent (19 per cent imports) of an apparent consumption of approximately 16 million tons. It further declined to 50 per cent (28 per cent imports) of a 13 million ton market in 1980–81. During 1982 BSC's share of the domestic market slumped to 43 per cent. Exports are negligible. Whereas Britain ranked number 5 amongst the steel-producing countries in 1967, in 1981 it ranks number 9, after Poland and before Czechoslovakia.

Notes

1   For a more detailed description, in particular of the restructuring efforts taken during recent years, see Chapter 10.

2   The British Steel Corporation was established in 1967 by a merger and consequent nationalisation of the 14 largest raw steel prod ucers. Only firms producing less than 500,000 tons per year remained in private hands.

3   An indicator of technological progress is the introduction of the so-called basic oxygen process. Its share of the production of raw steel was

	1964	1968
UK	10.5%	23.9%
USA	12.2%	37.1%
Japan	44.2%	73.3%

4   In 1967 the output of crude steel per manhour in the UK was one-third of that of Germany or the USA, half of that of Japan. Employment cost per unit of crude steel output in the UK in the same year (incidentally being equal to the US cost) was about three times higher than that of Germany (and Japan), according to A. Cockerill, in collaboration with A. Silberston, *The Steel Industry: International Comparison of Industrial Structure and Performance* , Cambridge, 1974.

10 Restructuring British Steel

Anthony Cockerill and Sara Cole

Introduction

The United Kingdom (UK) in common with most other major steelmaking nations has experienced a prolonged recession spanning most of the past decade. Steel consumption has shown an unrelieved downward trend, and the increasing penetration of imports together with the steady level of exports has caused steel production to decline more rapidly. The fall in activity has been especially severe in the last two years. The historic peak of output was in 1970 when 28.3 million tonnes of crude steel were produced, followed by a smaller peak of 26.6 million tonnes in 1973. By 1978, however, output had fallen to 20.3 million tonnes and collapsed to 11.3 million tonnes in 1980, largely as a result of a strike that halted virtually all production in the first quarter of the year. Output in 1981 is estimated to have been about 14 million tonnes.

The reduction in demand has been partly due to the fall in the rate of growth of overall output and partly also to longer-term structural changes that have taken place in the composition of the economy. Manufacturing fell as a share of gross domestic product from 33 per cent in 1970 to 29 per cent in 1978.[2] Within this, the decline in the activity of the metal-using industries has been particularly pronounced. The British Steel Corporation had amassed losses of £3.8 billion between its formation in 1967 and the end of the financial year 1980–81, and many of the independent steel companies that remain in the private sector of the industry had begun to make losses by the end of the 1970s. The declining financial performance of the industry is due to the fall in demand, to persistently rising costs of energy, employment, finance and raw materials, and to weakening prices. A surge in investment in the mid-1970s coincided with the fall in consumption to pro-

149

duce substantial overcapacity in the industry.

The purpose of the recent reorganisation and financial reconstruction of the British Steel Corporation (BSC) has been the latest step in a series of attempts to come to terms with these problems. The initiatives also envisaged increased linkages between the public and the private sectors of the industry, and these have been reinforced recently by specific aid to the private sector. The corporate plan, published in December 1980, is a convenient basis on which to examine the restructuring of the British Steel Corporation (BSC) and its future prospects.

BSC dominates the iron and steel industry in the UK. It was formed in 1967 by the nationalisation of the thirteen largest iron and steel-producing companies. The purposes of nationalisation included the modernisation and improved efficiency of the industry as well as the intention to bring it more firmly within the ambit of public policymaking.

The Corporation accounts for virtually all of the iron manufacture in the UK, about four-fifths of crude steel production, and approximately 70 per cent of the production of all finished steel. Capacity in the Corporation has been reduced by the closure of many plants in the period since its formation. It now operates five major integrated works together with about 20 other steelmaking, rolling and finishing works. It makes a full range of flat and long products and accounts for about one-half of the total production of special steels.

The independent sector of the industry comprises more than 100 firms, of which only some 30 produce crude steel, the remainder undertaking a variety of rolling and finishing processes, taking their supplies of semi-finished steel from domestic sources including BSC as well as from abroad. Two-thirds of the private sector's output of between 3 and 4 million tonnes a year is of common grades of steel, produced in mini-works using electric arc furnaces and continuous casting machines. The remaining output is of special steels coming mostly from four main groups located in Sheffield.

The corporate plan

In this section the main operating aspects of BSC's corporate plan are reviewed.

Market share

BSC's home market share has been falling for the past decade. In 1970 it was 70 per cent but had declined to 54 per cent by 1978–79. As a

150

result of the strike in the first three months of 1980 and keen pricing within the European Communities (EC), market share fell still further during the year (Table 10.1)

Table 10.1
Share of supplies of finished steel to UK consumers

	1970	1978-79	1979-80	1980-81[a]
BSC	70.4	54	48	50
Imports	5.5	20	25	28
Private steel companies	24.1	26	27	22

Note: [a] estimated

Source: *BSC Annual Report and Accounts* 1980-81; *BSC Annual Statistics for the Corporation;* Iron and Steel Statistics Bureau, *Annual Statistics* (various years).

Increases in import penetration have accounted for most of BSC's loss of market share with the private sector maintaining its overall share of a much-reduced market.

A central feature of the corporate plan was to regain the pre-strike market share of 54 per cent. BSC is endeavouring to achieve this by competitive marketing and pricing and by improving quality and consistent delivery. The main losses of market share occurred in products for which BSC has traditionally enjoyed dominance of supply; that is, plates, tinplate, tubes, sections and certain other types of commercial and special steels.

Capacity

The intention of the corporate plan was to reduce BSC's overall manned capacity to 14.4 million tonnes a year. Capacity has been progressively reduced in recent years as plants have been closed in the face of falling demand and output in an effort to maintain operating rates in the remaining steelworks. A large proportion of the Corporation's capacity is concentrated on five sites: sections and commercial steels are produced at Scunthorpe in Lincolnshire and Lackenby on Teesside; strip products are manufactured at Ravenscraig in Scotland, and Port Talbot and Llanwern in South Wales. The remainder is contained in a number of smaller plants producing special steels, tubes and other products. The corporate plan proposed the closure wholly or in part of a number of steelmaking and associated works in addition to those previously announced.[3] As a result, utilisation of manned steelmaking capacity was to have increased from 64 per cent in October 1980 to well over 90 per cent in 1981-82.[4]

Exports

An increase in export tonnage was another main feature of the corporate plan. In recent years BSC's exports and its share of world trade in steel have fallen. This has been the result of the recession in the international steel market coupled with a strategic decision to withdraw from the supply of certain export products on the grounds of unprofitability. Currently BSC's exports stand at less than 3 per cent of world trade in steel. Under the corporate plan BSC intended to raise export tonnages to approximately 2.7 million tonnes a year, an increase over recent levels but below the peaks of the past.

Investments

Capital investment in BSC is being restricted to the completion of projects in progress and to essential new items. Required levels for the next five years are estimated at £200–250 million at 1980 prices. Most of the £200 million investment earmarked for 1981–82 was for projects which had already been approved. Three major new projects were included so that detailed design and preparation for future completion could begin. The three projects will increase continuous casting capacity and improve liquid steel and finished product quality by the introduction of vacuum degassing and refurbishment of the Port Talbot hot strip mill. Further work on these schemes will depend on the targets of the corporate plan being achieved.

Prices

Prices for steel in the UK fell in 1980–81 and the continuing overcapacity in the industry internationally suggests that competitive pressures on prices and profit margins will continue, particularly within the EC. In the second half of 1980, the situation with regard to prices deteriorated rapidly when EC prices dropped back well below 1978–79 levels. This was made worse in the UK by the appreciation of the sterling exchange rate which reduced the price of imported steel still further. Increased UK costs and relative cheapening of ECSC imports forced BSC to reduce prices in order to remain competitive. The European Commission's manifest crisis measures halted further price erosion and the corporate plan anticipated 'a modest recovery of price'.

Employment and productivity

Employment in BSC has reduced steadily over recent years and at the end of March 1981 stood at 121,000.[5] The corporate plan called for a further reduction of approximately 17,000–20,000 during 1981–82. The fall in employment has been brought about by the decline in demand and efforts to improve productivity. The plan lists four methods of reducing personnel: eliminating tasks no longer considered necessary; improving working practices; improving plant availability, and cutting absenteeism and unnecessary overtime. The Corporation intends that the majority of personnel reductions should come from improved effectiveness rather than plant closures. The intention of the corporate plan was that by March 1982 BSC productivity levels would equal those currently prevailing in the rest of the EC (8.5 manhours per tonne of liquid steel).

Financial reconstruction

A counterpart of the physical reorganisation of the Corporation was the need to revise its financial structure. BSC made losses in 10 out of its 14 accounting periods between 1967–68 and 1980–81, by which time the accumulated revenue deficit amounted to £3,804 million.[6] The financial year 1980–81 saw the record loss to date of £668 million. A considerable proportion of the long-term funds provided to the Corporation since its formation has been supplied in the form of interest-bearing debt, with the consequence that interest charges (£183 million in 1980-81) have made an important contribution to the losses. Furthermore, with the fall in the Corporation's output and the reduced prospects for future sales and profits, the balance sheet overvalued the assets, both because a reduction in capacity was necessary and because of the reduced earning power of the facilities remaining in use.

Under the Iron and Steel Act 1975, BSC's medium and long-term finance is authorised by the Secretary of State for Industry with the approval of the Treasury. At the end of March 1981, this consisted of four types:

1. *New Capital* . This amounted to £2,988 million and has been provided since 1978 under Section 18(1) of the 1975 Act. The statute empowers the Secretary of State to supply to the Corpo ration 'such sums as he sees fit'. The funds are supplied as irredeemable state equity; there is an obligation for the Corp oration either to propose a dividend each year on the funds, or to satisfy the Secretary of State as to why no dividend should be paid. The Secretary of State retains a default power in cer tain circumstances to direct the Corporation to pay such divi dend as he specifies.

2. *Public Dividend Capital* (PDC). This was the form in which equity funds were subscribed to the Corporation between 1969 and 1978. Dividends were expected to be paid on these funds at the same rate as if they had been supplied as interest-bearing debt, taking one year with another. In the event, dividends were paid in two accounting periods only. The amount of PDC outstanding to the Corporation in March 1981 was £1,824 million.

3. *Long-term borrowings from the National Loans Funds* (NLF). The NLF is the Treasury account with the Bank of England through which the majority of government borrowing and domestic lending is carried out. Borrowings by the nationalised industries from this fund are at a fixed rate of interest for the term of each loan. The sums outstanding from this source stood at £509 million at the end of March 1981.

4. *Overseas borrowing.* Subject to government approval, BSC may borrow funds from abroad, and during the second half of the 1970s this was the major source of long-term loans. The amount outstanding at 31 March 1981 was £616 million.

In addition, BSC may also raise *short-term funds* and these have been an important source of finance at times. At the end of March 1981 they amounted to £326 million.

The reconstruction of the Corporation's finances is proceeding in stages. First, £1,141 million of assets were written out of the accounts between April 1979 and March 1981. Second, the Iron and Steel Act 1981 extinguished entirely the sums outstanding from the NLF and reduced by £3,000 million the amount of New Capital. Total capital was thus reduced by £3,510 million, and the revenue deficit almost eradicated. Table 10.2 shows the effect on the Corporation's balance sheet at 28 March 1981 of the reconstruction, had the legislation been enacted by that time. Third, the Act also contains provision for the New Capital to be reduced by further amounts not exceeding £1,000 million in aggregate by the end of 1982. Fourth, the foreign loans currently outstanding will be progressively repaid, either as they fall due, or earlier. These funds will be replaced with New Capital. Most of the repayment will take place after 1981–82, and the loans are expected to be extinguished by 1985–86.

This operation does not require the net injection of new funds into the Corporation. The revenue deficit represents sums that are irretrievably lost and that have been supplied in the form of capital subscriptions, some of which are interest-bearing. The reduction in the value of the fixed assets will reflect more accurately their potential future earning power. The eradication of NLF funds and the replacement of foreign loans with New Capital will progressively reduce the interest burden from long-term funds until, on present policies, it is extinguished completely.

Financial control

The financial reconstruction was accompanied by some changes to the process of financial control by government. These will increase the degree of government influence over the Corporation in some respects, and reduce it in others. Government exercises financial control and influence over the Corporation through the Department of Industry and the Treasury in a number of ways. Long-term funds are authorised by the Industry Secretary, who also has statutory powers of approval over major investment projects and the corporate plan. There are limits to the borrowing powers for long-term funds that can be raised only with the approval of Parliament. Since 1977, in line with other public enterprises the Corporation's access to external finance has been governed by an annual external financing limit (EFL), set administratively by the Treasury.

Table 10.2
British Steel Corporation financial reconstruction: impact on balance sheet at March 1981

	Pre-reconstruction £'m	Capital reconstruction £'m	Post reconstruction £'m	Post reconstruction %
Fixed assets, investments and long-term debtors	1,870		1,870	79.8
Working capital	800		800	34.0
Short-term borrowings	(320)		(320)	−13.8
	2.350		2,350	100.0
Capital	4,820	(3,000)	1,820	77.4
Reserves (deficit)	(3,630)	3,510	(120)	-5.1
	1,190		1,700	
Long-term borrowing	1,140	(510)	630	26.8
Minority interest	20		20	0.9
	2,350		2,350	100.0

Source: HC336, *Effects of BSC's Corporate Plan*, Vol. 1, HMSO London, 1980–81, p. xxii.

The latter is the most important of the control devices, in the short term at least. External financing limits for nationalised industries form part of the public expenditure planning total and thus influence the public sector borrowing requirement (PSBR). Government takes account of the projected cash needs of BSC each year in setting the EFL, but it also has in mind its own targets for the Corporation's financial performance and for the level of public expenditure. The funds approved under the EFL are to meet expenditure on fixed assets; increases in working capital; operating losses; interest payments, and extraordinary expenses such as closure costs and severance payments. The Corporation's EFL for 1981–82 was set at £730 million. The Treasury does not subdivide the aggregate of the EFL between the various expenditure heads. The result is that no distinction is made between the Corporation's operating and investment cash flows, or between these and the extraordinary expenses. Higher than projected expenditures under some heads (as a result of recession, for example) have to be offset by reductions under others. In extreme circumstances, where trading conditions are much worse than expected, or the financing limit unrealistically low, government may find it necessary to raise the EFL during the course of the year. The result is that the EFL is a very crude and indiscriminate means of financial control, but there are no prospects of any major changes in its operation in the immediate future.

Government also needs to keep in touch with changes in short-term trading conditions and to ensure that the appropriate managerial decisions and actions are taken. In view of this, the Department of Industry has extended its short-term monitoring system. A comprehensive range of commercial and financial indicators, analysing performance by at least 11 businesses within BSC, are now reviewed every month and discussed with the Corporation. As a result there is now 'a slightly greater... area of monitoring (of) management than there has ever been before' by the Department.[7]

Whilst in some areas government scrutiny remains close, in others control is being relaxed. BSC is being encouraged to form jointly-owned subsidiaries with private-sector companies in activities other than bulk steelmaking in efforts to reduce excess capacity and rationalise production, thereby increasing efficiency and profitability. The first of these ventures was the establishment in 1981 of Allied Steel and Wire Ltd, a company owned jointly with GKN Ltd. Other similar projects are being considered, and further developments may be expected. The intention was that BSC would subscribe 50 per cent of the capital of the subsidiary companies, the other half being provided by the private sector interests. One effect of this would be that external funds supplied to the new companies would fall outside the public expenditure total. The partners would provide the initial capital to set up the new companies but would not have any financial obligations beyond that. In addition to the efforts to form joint subsidiaries, the management structure of BSC has recently been reorganised. There are now 18 separately identified businesses, based mainly upon products. If the financial performance of all or some of these can be improved, there will be scope for private capital to be introduced into them, or for them to be transferred entirely to private ownership. Such a process of 'privatisation' would be very much in line with the policies of the present Government.

Progress

The corporate plan was described as optimistic in its targets both by the chairman of BSC and by the Department of Industry.[8] This optimism extends to the targets set for output, revenue and cost levels. Four aspects in particular are discussed here – sales volume, prices, exchange rates and costs and productivity.

Sales volume

The plan called for the Corporation to regain 54 per cent of the UK market for steel in 1981–82, equivalent to the share held in the period before the 1980 strike, and towards the end of 1981 it became clear that considerable progress had been made towards this target. In the longer term, however, the prospects for the Corporation will depend not only on market share, but also on the level of steel demand, both at home and abroad.

It is possible to project the annual level of BSC's supplies to the UK market and the margin of capacity available for export until 1985 on the basis of estimates of home steel consumption made available to the Industry and Trade Committee.[9] The results are shown in Table 10.3. The Corporation's present manned capacity is 14.4 million tonnes of liquid steel a year, and it is assumed that its market share is held constant at 54 per cent for each year from 1981 onwards. Column 2 gives the projected level of apparent domestic steel consumption in terms of crude steel equivalent in the UK. Demand increases slowly but steadily until 1984 after which it shows a marked fall. Over the period, however, the level of consumption remains on average well below the levels of the two previous decades. The assumption of increasing overall demand until 1985 together with the expectation that BSC will regain its former share of the market during 1981 implies that the Corporation's supplies to the home market will increase in tonnage

157

terms until 1984 (column 4). This is in marked contrast to actual experience since 1976, during which time BSC's output has fallen progressively. Sales by BSC of semi-finished steel to re-rollers and finishers in the private sector of the industry are a significant part of total sales volume, and column 5 gives a projection of the annual levels of these. They remain fairly steady at approximately one million tonnes a year. Column 7 shows the amount of capacity available to supply export markets. This is considerable, and an expansion of world demand coupled with an aggressive pricing and marketing strategy by the Corporation will be required if it is to be filled.

Table 10.3

Projections of UK steel demand and BSC output including export margin, 1980–85 (crude steel equivalent)

Year	Apparent consumption[a] m tonnes	BSC market %	BSC output for UK final consumption m tonnes	BSC sales to private sector[b] m tonnes	BSC total output for UK consumption m tonnes	Margin for export[c] m tonnes
1980	17	48.5	8.2	0.9	9.1	3.9
1981	15.5	54	8.4	0.8	9.2	3.8
1982	16.25	54	8.8	0.9	9.7	3.3
1983	17.25	54	9.3	0.9	10.2	2.8
1984	18	54	9.7	1.0	10.7	2.3
1985	16.25	54	8.8	0.9	9.7	3.3

Notes:

a   HC 336, *Effects of BSC's Corporation Plan*, Vol.2, 1980–81, p.13.

b   Based on ratio of sales of semi-finished products to total home deliveries in 1979.

c   Assuming 90 per cent utilisation of manned capacity of 14.4 million tonnes

Prices

In common with other steel producers, BSC's cash flow and profitability are very sensitive to the level of prices. Steel prices throughout the markets of the EC have been depressed since the mid-1970s. The corporate plan looked for a modest increase in UK prices during 1981–82 despite the fact that, recently, prices in the home market have been above those in other Community markets.

The pricing policy currently used by the Corporation is a keenly competitive one designed to regain market share, but a general increase in prices is necessary to improve the relationship with costs. With effect from 1 July 1981 Commissioner Davignon reviewed the package of measures imposing mandatory quotas on the production of steel products representing more than 70 per cent of the total output of products defined within the Treaty of Paris. These controls replaced the similar measures that were introduced in 1979, the hope being that they would raise Community steel prices by 15-20 per cent. An increase in BSC revenue from higher prices depends on the success of these measures, as left to itself the market would show little sign of improvement. There is still 'too much steel chasing too few markets'[10] which accounts for the collapse of prices; improvements will occur only when there is either an expansion of the market or a contraction in capacity.

BSC prices weakened in the first quarter of 1981 after a slight strengthening at the end of the previous year but overall were rather above those prevailing in the EEC, resulting in import competition. Prices in the first half of 1981-82 were on average 3 per cent lower than had been expected in the plan, owing to the steel price weakness in Europe, despite efforts by the European Community to strengthen prices generally. Because of the higher prices prevailing in the UK market BSC was unable to increase prices to cover rising costs without jeopardising home market share.

Following efforts by Eurofer and the European Commission there were significant increases in European prices in the last few months of 1981. January 1982 saw a further 12 per cent price increase. Export prices have recovered somewhat from the very low levels of the latter part of 1980 but they are still below home prices. Increases in prices are vital if revenues and cash flows are to be improved but, in seeking rises, Community producers must bear in mind the impact on their market share targets, the anticipated degree of cooperation from their competitors, and the risks of a surge in supplies to the European market following any reductions in the level of imports in the United States.

Exchange rate

The high rate of exchange of the pound sterling against most other major countries from 1979 damaged BSC's export sales and increased the competitiveness of imported steel in the home market. More importantly, it reduced the demand for the output of the major steel-using industries. The fall in the exchange rate that has taken place recently therefore will bring some benefit to BSC, both directly by increasing its competitiveness, and indirectly by providing some relief to its major customers. However, it must be noted that the fall in the exchange rate so far has taken place principally against the dollar, the rates with the major European currencies have changed little. Thus the pressure of European competition has not abated very much, whilst the strengthening of the dollar has raised the sterling cost of imports of raw materials (including iron ore) and of oil.

159

Costs and productivity

The most important cost items in steelmaking in the UK are materials and employment, which together account for almost three-quarters of the total. Of the remainder, energy costs and financial charges are the most significant. In recent years the overall trend of costs has been firmly upwards at a time when there has been little change in the level of prices. The rising cost of imported raw materials was alleviated somewhat by the rise in the sterling exchange rate but the UK industry suffers a penalty estimated to be approximately £3 per tonne of liquid steel as a consequence of energy prices that are higher than elsewhere in Europe.[11]

Steelmakers can do little directly to influence most of their costs, but they can try to contain employment charges, which form almost one-third of the total. They can do this by limiting increases in the level of earnings and by raising productivity. Unit labour costs are a function of hourly wage rates and the number of manhours required per tonne of steel output (the reciprocal of output per worker, or productivity). In the past, both hourly wage rates and productivity have been below the average levels in the other major European steel-producing nations[12] Wage rates have not been sufficiently low, however, to offset entirely the productivity differential, with the consequence that unit labour costs have been high in relation to some other European producers, although the fall in the sterling exchange rate somewhat helped to improve the competitiveness of UK steel.

The extent of the productivity gap and the relationship with unit employment costs for BSC and the steel industries of France and Germany are shown in Table 10.4. A major objective of the restructuring of BSC and of the corporate plan was to raise productivity to the levels currently achieved in other European countries – that is, to reduce labour input per tonne from approximately 12 to 8.5 manhours.

The Corporation's present strategy is to reduce capacity and employment to bring them into line with the lower level of prospective demand and to raise productivity. This has been done by the closure of some entire works and parts of others, so that the reduced level of production can be concentrated upon the largest and most modern units. Demanning, accompanied by efforts to increase the flexibility of workforces, is an integral part of this, and the corporate plan called for the loss of at least 20,000 jobs during the year. This is a continuation of a trend that has been in operation for some time; between September 1979 and the end of 1980, for example, total manpower fell by 55,500, a reduction of 30 per cent.

The indications so far are that some success is being achieved in the drive to improve productivity. This is most noticeable at Port Talbot

Table 10.4
Comparisons of productivity and employment costs in the European steel industry

	Manhours per ton	Emp. cost per ton £*
BSC		
1978	15.1	44.8
1979	13.6	50.4
1981 (P)†	11.8	52.0‡
Post-corporate plan forecast	8.5	
France		
1978	10.1	53.4
1979	9.3	52.0
1980 (P)	8.5	48.2
Germany		
1978	7.9	51.6
1979	7.0	49.3
1980 (P)	7.2	49.0

Notes:

*   Costs are shown in £ sterling and are based on the average rate for the relevant year.

†   BSC figures are based on the first quarter of 1981 because 1980 figures were distorted by the strike and short-time working.

‡   The reduction in employment costs per tonne when expressed in sterling terms of 9.7 per cent in France and 5 per cent in Germany while BSC employment costs per tonne have risen by 16 per cent is very much influenced by the fluctuation in exchange rates and relative rates of inflation.

(P)   = Provisional

and Llanwern, the two major works in South Wales. In addition to a sharp reduction in the number of jobs, many restrictive work practices have ended, including demarcation, under which certain tasks are reserved solely for certain categories of employee. At the same time, efforts to raise productivity elsewhere are being less successful. This is so at Teesside, where a deep-water harbour, sintering lines, coke ovens and a giant blast furnace have been built adjacent to the Lackenby steelworks. In addition to technical problems involving coke and iron production, the workforce has been reluctant to abandon long-standing manning levels and work practices, the standards for which have traditionally been set by the shipbuilding industry, another major employer in the region.[13]

Postscript

In the event, the main objectives of the survival plan for 1981–82 were achieved. BSC reduced its loss to £358 million (before extraordinary items) as compared with the deficit of £668 million the previous year. Cash requirements were contained within the external financing limit. Employment fell by 17,200 to reach 103,700 by the end of the year, assisting the steady increase in productivity.[14]

Against this background, there were hopes that the annual loss could be halved to £180 million in 1982–83 and that trading could be at a breakeven rate during 1983. Major uncertainties still remained, however. Community exports to the American market were threatened by anti-dumping and countervailing duty suits entered by US steelmakers, alleging unfair competition from subsidised foreign producers. It was vital for the European Commission to be successful in negotiating effective production quotas for Community producers in order to achieve an increase in the general level of prices, while at the same time avoiding sharp increases in imports from third countries.

BSC's overall trading conditions have deteriorated in recent months and it is now clear that the intended improvement in financial and operating performance will not be achieved, at least in the time span initially envisaged. The European Commission reached a voluntary agreement with the American authorities in October limiting most classes of steel imports from the Community to 5.5 per cent of the US market, but it is not yet clear what this represents in tonnage terms or how it is to be shared between the various steel producers.

Nevertheless, it seems likely that the export sales of most producers will be reduced considerably. The depressed state of the world market for steel means that these sales cannot be made up elsewhere and that there will be increased pressure on other markets as a result of the steel diverted from the American market. The UK market has remained depressed throughout 1982 and there are no signs of an upturn despite the rise in sales of manufactures which has been met in the main by imported goods. BSC's market share has been eroded by a surge in imports from countries outside the Community.

After a period of improved trading earlier in the year, BSC's losses were running at £7½ million a week in September 1982. The losses for 1982-3 were much greater than anticipated at the start of the year, approaching the same order as in 1981-2 and averaging £8 million a week during the summer of 1983. It is not yet clear whether the low level of UK steel consumption represents a permanent reduction in demand or whether it is rather more transitory. Future policy towards the industry will depend upon the results of this analysis, and the need for further reductions in capacity and employment, involving the closure in part or in total of some of the remaining large-scale integrated works, remains a serious possibility.

Notes

1   This is a thoroughly revised and updated version of a contribution to the Atlantic Economy Association Conference, London School of Economics, 20 August 1981. Thanks are due to E. Denham, J.F. Pickering and J.C. Siddons for helpful comments. Output, capacity and sales volumes are in metric tonnes (1 metric tonne = 1.1023 US tons = 0.9842 long tons).

2   Central Statistical Office, *National Income and Expenditure 1980* , HMSO, 1981.

3   As a result of these additional reductions in capacity, Normanby Park Works, Appleby-Frodingham No. 1 rod mill, Lackenby bar mill, Distington ingot mould foundry and Ebbw Vale 4-stand cold mill have been closed. Templeborough works will be run as a mini-plant, coking capacity will be reduced by closing ovens at Shotton and Hartlepool, Orgreave and/or Brookhouse will continue until alternative fuel arrangements are made for works in the Sheffield area, and Velindre (tinplate) will operate on a restricted basis.

4   HC 366, *Effects of BSC's Corporate Plan* , Fourth Report from the Industry and Trade Committee (2 vols), vol.2, HMSO, London, 1981–82, p.6.

5   British Steel Corporation, *Annual Report and Accounts* 1980–81.

6   Ibid.

7   HC 366q. 510, p.101.

8   Ibid., q.2, p.19; House of Commons *Debates* , Vol.999, 24 February 1981, col. 745.

9   HC 366, vol.2, p.13.

10 HC 366, vol.2, q.248, p.56.

11 HC 61, *Minutes of Evidence, British Steel Corporation* , House of Commons Industry and Trade Committee, HMSO, London, 25 November 1981.

12 A. Cockerill, *The Steel Industry* , Cambridge University Press, Cambridge, London, 1974, p.32.

13 *Financial Times* , 28 July 1981.

14 HC 237, *Financial Statement and Budget Report 1982–83* , HMSO, London, 1981–82, p.24.

11 The steel industry of the United States of America

Historical overview

The US steel industry[1] has, unlike its European counterpart, never been devastated by war. It has enjoyed periods of almost world monopoly, and, when subjected to competition from abroad, has obtained good access to a very powerful lobby, the so-called 'US steel caucus'. To trace periods of tough international competition, one has to go back a century. When growing in a boom market, the fight for market shares was often cut-throat and, like other young US industries, the steel industry was aggressive and highly innovative. As resources – except labour – and markets were abundant, the American way of running industries was invented: management, technology and marketing – a new factor combination which was developed to full bloom. But management was a bottleneck factor and it was systematically greenhoused by some big financial groups.

During a few years at the turn of the century, two concentration moves were undertaken, one by the Carnegies, Morgans and Moores (which, however, failed to stabilise the market) and the other the formation of the US Steel Corporation, bringing two-thirds of US steelmaking capacity together. From this point in time, competition was replaced by agreements and cooperation resembling a duopoly situation, although there was one dominant firm on the one side and a whole range of firms on the other, who followed the leader, even through tough times.[2]

The deep crisis of the 1930s, when only one-fifth of the total capacity available in the steel industry was used, did shake up the cartel (the National Industrial Recovery Act, NIRA, of 16 June 1933, F.D. Roosevelt's first 'New Deal'). When during the second 'New Deal' in 1935–36 some large industrial concentrations were broken up and

NIRA was declared by the Supreme Court to be incompatible with the US constitution, the steel industry was left untouched despite the fact that the industry had maintained its cartel and even extended it through agreements with the European International Steel Cartel.

As the United States Steel Corporation applied a tolerant pricing policy, a number of efficient medium-size steel companies were able or allowed to increase their market share: in 1980, US Steel Co. held only one-fifth of the nation's raw steel-producing capacity against two-thirds at its foundation in 1901. The big decline in US Steel Co.'s market share occurred between 1902 and 1938, in which year it held 33 per cent or only half of its start-up share of the US market. In 1980 the top four held 50 per cent and the top eight 74 per cent of the market.

World War II, the devastation of the European steel industry, the post-war European Recovery Plan (ERP), the Korean war, pushed the US steel industry to new heights. Much new capacity was added to the industry (albeit, with pre-war technology) and melting capacity increased by almost two-thirds. Two dozen new blast furnaces and ten hot strip mills were added.

To summarise the development for a moment, the industry had been working under a strong cartel since the beginning of the century – the short episode at the beginning of the 1930s lasted only two to three years and was soon forgotten again. The immediate post-World War II period saw the US steel industry in spectacular growth, under conditions of a maintained domestic cartel and, internationally, of monopoly or at least dominance in wide ranges of the steel product portfolio, in particular those qualities needed for the post-war reconstruction of the European and Japanese industry, including the steel industries. It is important to remember that the industry was able to mobilise and maintain a lobby, which not only sheltered it against anti-trust but, later, also was prepared to, successfully, fight for protection in crisis situations.

The post-World War II investment boom could have given the US a technologically superior steel industry. Obviously, however, the continuous lack of pressure from competition on the one hand, and scarcity of resources on the other, had now turned into a context which would lead to a centennial decline of vitality and competitiveness. Strangely enough, the capacity installed during and after the war did not keep the US in the forefront of technological progress. Many of the heavy investments of the 1950s were in conventional technology, which was ripe for modernisation only some ten years later. Furthermore, practically all of the investments went into traditional, coal and/or ore deposit-orientated, steel regions instead of to the growth markets or to cheap coastal locations favoured by the Europeans and the Japanese when building up their new steel-producing capacity.

As the US steel industry had access to cheap energy inputs, it did not need to bother about energy saving or high yield per unit of ore and fuel technology. And, again, there was virtually no price competition.

The conservative policy so generally embraced by the industry planted the seeds for the decline of first its international, and later, even domestic competitiveness. We will now turn to an analysis of what factors accounted for the not only rather mediocre performance of the industry, when seen in an international perspective, but also the gradual emergence of needs for more and stronger protective measures. All of them were originally aimed at providing temporary relief from international competition, to give the industry an opportunity to restructure itself, but all essentially left the industry in a gradually worsening shape.

Changes in the 1950s as either causes or early indicators of present problems

A number of thorough changes came at the end of the 1950s.

1   The competitive advantage of the United States' steel industry consisting in access to low-cost (c.i.f.)[3] high quality raw material (which previously made it possible to compensate for high relative labour cost) disappeared as such. Raw material became available at even lower c.i.f. prices at coastal locations elsewhere in the world.

2   Modern high yield and quality rendering, highly productive facilities became operative in Europe, in newly industrialised countries and, most importantly, in Japan. This not only meant a sudden loss of market shares abroad, but also drastically increasing import competition on the domestic US market, particularly in some major growth regions where c.i.f. prices of US produced steel soon rose far above c.i.f. prices for imported steel. This development not only implied a decline in domestic market shares to the domestic steel-producing industry, but also a deterioration of the steel-using industries' (automobiles, machinery, tools and so on) competitive power. Those industries both lost markets and tried to substitute steel with other materials. The growth rate of apparent steel consumption in the United States consequently levelled off: a growth rate of slightly above 2.2 per cent during the 1960s was followed by a growth rate of only 1.6 per cent in the years 1969–79. During the same period, the growth rate in the European Community of the Nine was 23 per cent and in Japan approximately 19 per cent.

From the mid-1950s, world steel exports started to rise sharply. In 1955 there were approximately 25 million US tons,[4] in 1960 approximately 35 million and in 1970 approximately 150 million tons traded worldwide. The US share of the world steel trade during the entire period hovered around 3 to 7 million US tons. Thus, the US steel industry had lost its international competitiveness very early and never regained it.

An important conclusion from the above development is that steel has become an increasingly internationally traded commodity which is available today from many more sources and markets than before. The dominant situation of the US steel industry in the world market has gone for ever.

In the 1960s and 1970s steel prices rose in constant US dollars (1971) by 30 per cent, cement and aluminium prices nil, whereas plastic prices decreased by 40 per cent. These price trends certainly have had an impact and will continue to have one on attempts to replace steel by other materials wherever feasible.

3   Imports to the United States essentially started in 1959 because of the 119-day strike. In 1968 imports covered 17 per cent of apparent steel consumption in the United States; during the 1970s the average figure was 15 per cent and in 1981 approximately 20 million tons or 19 per cent of apparent consumption were imported, of which 6.5 million tons were from the European Community. This accounted for approximately one-third of total imports; the Community's share has been continuously declining from an average of 50 per cent during the 1960s. From that time onwards, Japan has been increasing its market shares; later also Taiwan, Brazil and Canada have been gaining.

4   Despite the often heard claims that too little cash is available for investment in modern high productivity, integrated, production capacity, the industry invested quite vigorously in the late 1960s and mid-1970s (between 1965 and 1979 the US steel industry outspent its European and Japanese rivals per ton of capacity added or replaced), however seemingly without adding to capacity (the growth of 20 million tons per year capacity between 1960 and 1980 took place in the non-integrated so-called mini-sector of the industry), neither did the investment render traceable improvements in productivity or quality. Even a quite drastic increase in investment would be most unlikely to notably improve productivity (for example to a level to ensure competitive productivity levels for the US steel industry).

5   A most remarkable feature is the growth of steelworkers' wages and benefits. Already in 1960, the average hourly wage cost for a steelworker was 1.46 above the average of the hourly wage cost in manufacturing industries. The difference rose to 1.75 in 1980, against 1.5 in Japan, 1.1 in Germany and 1.25 in the European Community, including Germany. As the steel industry is rather labour-intense[5] (on average, more than 20 per cent above the average labour's share of value-added for all US manufacturing industries), the combination of low or non-existent growth in markets and productivity with rather strong increases in labour cost (essentially during periods of trade protection) in the integrated sector of the industry implied a further decline of this sector's competitive power.

6   It is often claimed that environmental as well as labour safety and health legislation have levied heavy burdens upon the US steel industry since the 1950s. Comparisons with cost/investment incurred by major competitors show that the US steel industry has not had to carry higher burdens, despite the generally older vintage of US plants, which would have led one to expect higher cost/investment for the improvement/protection of the internal and external environment. The Office of Technology Assessment estimates the cost of full compliance to environmental regulation to be a maximum of 6 per cent[6] on cost and prices of raw steel.

7   The integrated US steel industry is a slow and, over time, ever slower adopter of modern technology[7] to improve yield, producti vity and quality. The industry was particularly slow in adopting the major new processes of basic oxygen furnaces during the 1950s and 1960s. It was still slower in adopting continuous casting during the late 1960s and 1970s. The approximate adoption rates for basic oxygen furnaces in 1960 were 4 per cent for the United States and 10 per cent for Japan; in 1970 50 per cent for the United States and 80 per cent for Japan. The adoption rate (per cent of total output) for continuous casting was as follows:

	Japan	USA	EC
1970	8	5	5
1979	52	17	31

8   The US stock market for steel stocks turned downwards at the beginning of the 1960s (and never recovered after that, except for the two boom years in the middle of the 1970s) in recognition of the fact that the US steel industry's cost situation was getting out of control. The ratio of market to book value of steel equities fell from 1.6 in 1960 to 0.5 in 1974 to 0.4 in 1979,

whilst the average for industrial issues was still improving and plateauing near 2.0 during the 1960s. It started a declining trend in 1973 to about 1.1 in 1979. Thus, the stock market discounted the general weakness of the US integrated steel industry quite early and correctly.

9 The high value of the US dollar against other currencies has contributed temporarily to the deterioration of the steel indus try's competitive power. However, during the crisis years of the second half of the 1970s the dollar reached unprecedented lows, which, nevertheless, did not improve the US steel industry's competitive power, neither at home nor on foreign markets. When the dollar regained strength in the early 1980s, this may have affected imports to the United States more than the subsidies given by some European countries (Great Britain, France and Italy) to their ailing steel industries. The market shares of the US imports from the subsidising countries are rather small. Lower prices because of lower labour cost have had a much greater impact on the US steel users' propensity to import. To a certain extent, the non-availability of high quality steel and steel products from US mills has accounted for increasing imports.

10 The integrated sector has not only been losing to imports but also had to give away a considerable share of the domestic market to mini and midi-mills because these operate at lower c.i.f. cost to the customer and are thus also highly competitive in relation to imports. As the size range of mini-mills is gradually increasing (from 100,000 tonnes per year to 1 million tonnes per year or even higher) such mills are gradually approaching the average mill size of the integrated industry (approximately 1.6 million tons) but at c.i.f. cost considerably below those quoted by integrated mills. The market share of mini-mills, which in 1960 was nil, in 1980 was equal to the share of imports (15 to 16 per cent, whereas the market share of the integrated steel industry was 63 per cent). The market share of the mini-mill sector is forecast to grow to approximately 25 per cent by 1990.

11 US governments, in their attempts to fight inflation, were no longer prepared to let prices grow, including those of steel.

This entirely new situation triggered two activities: a modernisation programme of considerable magnitude, and a political and lobbying campaign for which the industry gained the support of the United Steelworkers Union (USW).

The Washington administration became prepared to protect the industry, and has been doing so since the mid-1960s (albeit with short interludes, at the industry's initiative, as for example during the 'all-time-high' for steel in 1973–74).

The protective measures – to be penetrated in more detail later – were the so-called Voluntary Restraint Agreement (VRA) on carbon steel (1969–74) and, later (1976–80), on specialty steel; the trigger price mechanisms (TPM) of 1978 and 1982 (prescribing minimum prices, based on Japanese steel prices c.i.f. Californian destination, for imported steel); the promulgation of 'buy American laws' for steel and steel products; the 1982 reduction of the steel import quota for 11 product categories from the European Community countries;[8] plus further measures to come, for example a similar quota arrangement for specialty and alloy steels as well as quota deals with other countries of supply.

The efficiency of the measures taken to temporarily shelter the industry against foreign competition whilst being in a vulnerable stage during reconstruction for higher efficiency and thus increased competitive power will be scrutinised in some detail below. However, it may be said now that protective measures will have to be taken again and again for many years to come and that they will render continuously adverse effects on the industry and its customers. The industry has not improved its efficiency whilst enjoying protection at high costs to the users or the consumer in general.

In most years since 1959, US mill prices of comparable steel products have been above those of their European and Japanese competitors, even with due regard to fluctuations caused by shifts in exchange rates.[9] This applies to integrated mills, whereas there were small price differences between US mini-mill products and comparable foreign products.

In general, the integrated steelmakers in the United States have traditionally responded slowly to new challenges, both from the market and from new technology. The slow adoption of new technology during the last build-up phase in the immediate post-World War II period is often mentioned as one of the factors causing the decline of the industry during the 1960s and onwards. The tardy adoption of the BOP and, more lately, the obvious hesitance to employ the cost saving and quality improving continuous casting technique, are apparent examples of this general strategy. It now has nothing to do with cartelisation; it is simply part of a tradition which developed during this century and which favours defensive investment and market strategies over offensive ones.

During the 1960s, the big steel corporations also followed the general diversification movement of big industry. In hindsight two conclusions may be drawn from this phase:

171

1   Big mono-industry enterprises in general have not been overly successful in their diversification attempts. The more the business that they enter was unrelated to their industrial experience, the harder were usually the lessons to be learnt. The steel industry was no exception to this general experience, although this is not to say that diversification may not be both a necessary and successful strategy to insure a firm against industry-related risks.

2   The steel industry is highly capital-intense[10] and would thus need all the capital to which the steel firms may have access for investment within the industry.

Diversification has diluted big firms' capital available for steelmaking capacity. Although, as Mueller (monograph no. 25, 1980) emphasises, the US steel industry between 1950 and 1971 outspent its European and Japanese competitors per ton of capacity added or replaced. It thus was not insufficient capital spending but money spent on piecemeal stocked-up investment rather than well-planned organic restructuring. 'Technological progress is an organic rather than a piecemeal process' (Adams and Mueller, 1982).

Structural problems of the US industry

The total US raw steel capacity is currently approximately 160 million tons. Given the fact that presently almost 20 per cent of the market is in the hands of mini-mills[11] and assuming that the optimal plant size is around 5 to 6 million tons of raw steel per year for an integrated plant, there would be space for between 20 and 24 modern integrated mills. In 1980, 37 integrated plants were operating, of which only 9 had a raw steel capacity of 4 million tons and above. The average size of integrated plants operated by the largest American firms is less than 3 million tons of raw steel per year (Adams and Mueller, 1982). In comparison, the five largest Japanese producers cover 80 per cent of Japan's integrated capacity, operating 17 integrated plants, each of which is designed for a raw steel capacity of over 7 million tons a year. (The optimal scale size of a modern integrated plant today seems to be in the range of 8 million tons per year, rather than 5 to 6 million tons as assumed above. The modern Japanese mills seem to break even at approximately 70 per cent capacity utilisation.)

The majority of the US plants are rather old, most of them being of the pre-1945 vintage. Only two major integrated plants have been established since 1950. The modernisation strategy of the American steel industry has been the retrofitting of old plants to new technology.

This, however, is a difficult and obviously less successful strategy. It has to be carried out in plants running in continuous operations and at brownfield locations with given layout constraints because of the older technology and plant size. The results rendered by this investment strategy have been surprisingly low, which partly explains the industry's continous decline in competitive power.

A further problem for the US steel industry is that it has been almost continously suffering from overcapacity, with the exception of the boom years around 1974. The existing overcapacity has been an important brake on productivity growth, and the industry has been trying to 'muddle through at lowest expense' (Crandall, 1981).

Dirlam and Mueller (1981) explain the low capital productivity achieved by US producers by poor structural characteristics (small plant size and deficient lay-out) leading to costly duplication of investment efforts, installation of under-sized equipment as well as new equipment not being used optimally. Mueller[12] carries this analysis further by criticising management (of US as well as of European steel firms) for poor technical and market management behaviour. Probably, the long periods of absence from competition has made possible the nursing of old, inefficient and ill-balanced plants. Mueller also points to the importance of short-term profits to appear in the firms' balance sheets. Only seemingly (if not really) profitable firms will attract enough capital for patchwork renewal and additions; only they will be paying good salaries to their directors. Risk-taking by radical plant re-design and/or relocation is not encouraged. The strategy pursued has thus been and is[13] to invest in limited improvement as long as the average unit cost of an old plant stays below the total unit cost of a new plant. This strategy, however, also gradually lowers the firms' ability to compete in the higher price quality ranges, as recently (1981–82) was the case in steel supplies to the oil industry. Thus, in the longer term the mini-mills are at an advantage, and the integrated mills are losing market shares in at least those two ways.

So-called greenfield investment (investment in integrated steelmaking plants at new locations) obviously seems no longer economically feasible and justifiable in US locations. This, however, also seems to apply to major so-called 'brownfield' improvement investment projects, that is major replacements of, or additions to, processes in existing integrated plants.

One conclusion drawn by highly competent steel economists is that a revitalisation of the US steel industry with the aim of securing domestic supplies of carbon steel at competitive prices (although the justification of this 'need' is rigorously questioned by many of the independent steel economists) would – even in the USA – today require capital investments which most likely would not, or could not, be raised by private owners. This constitutes a new situation and a new – although nowadays well-known – insight. What was possible in Japan,[14] the Netherlands and Germany only ten years ago and resulted in some of the world's most competitive integrated steel plants was already then not attractive and has in the meantime become financially and economically impossible in the United States. And here, the US steel economists are divided into two groups:

The *independent economists* say that steel industry modernisation requires too high a contribution from the US taxpayers, who already have given away huge sums to both domestic and foreign producers because of protective measures taken in the USA. They claim that steel today is available from so many geographically dispersed and independent sources that there is no justification for sacrificing huge sums in order to maintain steel autarchy. They claim that domestic supplies will not decline under 'reasonable' levels and that protection is hurting the industry rather than helping it. Recent history has demonstrated that top carbon qualities – as for instance, those temporarily needed in the oil and energy industries – are no longer available from US producers.

Only *industry-related steel economists* today claim that the USA needs an industrial policy for steel. The claim is that the Federal Government should engage in programmes similar to those adopted by some European states, like France, aiming at a thorough reconstruction of the industry and to modernise it at maintained capacity. (The EEC or Davignon plan has a different aim: drastic reduction of overcapacity and maintenance of a highly modern, productive core industry.)

In the meantime, whilst the discussion goes on, Japanese prices, but not necessarily qualities, are adopted as reference prices for the US industry, to at least maintain some inducement towards mobility. But the industry claims better protection. The strategy is only a short-term one: let's try to survive until we see light at the end of the tunnel. But the tunnel only gets longer and longer: no wonder, since the industry itself has piled up some veritable rocks where the tunnel's end might have been. The steel-using industries are severely handicapped in their competition with a growing horde of competitors from abroad, because of the high prices they must pay for the steel they have to buy from American steel plants.

There lies the big danger not only for the US economy, but for the industrialised world altogether: many investment goods' prices have grown so high that the investment-led business cycle improvement simply does not take off. It would be unfair, however, to cast all the blame on the steel industry.

The Congress of the United States, Office of Technology Assessment's Study 'Technology and Steel Industry Competitiveness'[15]

In October 1977, the US House, Ways and Means Committee's Sub-Committee on Trade requested the Office of Technology Assessment (OTA) to examine how technology might be used to improve the steel industry's international competitiveness. OTA concludes that a number of technical opportunities exist for improving the competitiveness of the domestic steel industry. It also investigates non-technological factors shaping the environment in which new technology is created and adopted. A major conclusion is that sectoral policies are needed to revitalise and modernise the steel industry. A basic – unquestioned – assumption is that 'the domestic steel industry remains vital to the economic well-being and national security'. This assumption is questioned by several recent studies of the US industry. Crandall (1981) thoroughly analyses various claims for maintaining the US steel industry and the related cost to the taxpayer. He, like other economists (Gold, Mueller and others), sees no valid reason to artificially maintain the industry, although he also outlines some cost-optimal options to demonstrate what a political maintenance of the steel industry might imply in subsidies. Crandall's proposals stay considerably below the OTA proposals. One major source of disagreement is that Crandall foresees a rather limited (10 per cent) further reduction of the US steel industry's operative capacity, whereas the OTA study exposes deeper pessimism and, thus, recommends federal support of US $ 600 million (1978 dollars) per year over ten years, and to increase the industry's capital spending by US $ (1978) 3 to 4.9 billion per year during the same period. Crandall claims[16] that if, for whatever reasons, one wants to maintain self-sufficiency in carbon steel, this would require the subsidisation of only two large (6 to 7 million tons annual capacity each) integrated plants. The subsidisation would have to cover nearly 30 per cent of the new facility costs, or about US $ (1978) 2.4 billion. Trade protection would be far more expensive.

There are differences in the demand forecasts: OTA assumes (minimum) continuation of post-war growth in apparent steel consumption, averaging at an approximate 1.5 per cent or (maximum) 2 per cent per year. Crandall assumes the continued 1.5 per cent growth but with doubts. The 1.5 per cent average annual growth covered a period with exploding automobilism and construction activities – in particular of the US highway network. It is unlikely that future demand will be

175

carried along by similar strong factors. Rather, steel utilisation intensity will decline for several good reasons, for example the decline of the steel-using, mainly machinery, industries and the general tendency to replace steel by lighter and more resistant materials.

There is also considerable disagreement between the OTA study and contemporary reports by economists (including those represented in the present volume) about the causes of the US steel industry's recurrent predicament.

Crandall, for example, demonstrates[17] that the impact of governmental policy measures is of marginal importance, compared to labour cost increases, low productivity and patchwork modernisation, which all have a greater bearing on the gradual deterioration of the US steel industry's international competitiveness. The environmental legislation surcharges at actual compliance standards are estimated at less than US $ 10 per ton of output, whereas full compliance would be at approximately US $ 20 per ton of output. The OTA estimate is twice as high. Occupational health and safety programmes, according to the industry's own estimate, are only about one-sixth of the environmental costs. This is definitely less than the equivalent effective cost increases to Western European and Japanese firms. It may, however, be higher in the United States than in the newly industrialising countries.

In summary, a number of the assumptions as to market growth, productivity increases, energy-saving effects and price developments made in the OTA study[18] seem to be overly optimistic. It is also very hard to believe that the industry, or rather its shareholders, would be willing or even able to raise the sums proposed by the American Iron and Steel Institute or the OTA, given both the essentially negative cash flow during the last six years (with the exception of 1981) and, in particular, in view of the fact that the experience from the investments over the last 20 years or so is rather discouraging.

Steel users, who usually have little say in investment decisions, most likely would not be very happy if the proposed massive investments in US productive capacity were accompanied by further trade restrictions putting the burden on their shoulders and thus further lowering their competitive power both at home and abroad. It is of utmost importance to the manufacturing industry to have access to alert, competitive, service-minded and flexible suppliers. Thus, an open market is of vital importance to them as well as to the entire economy. Even if the OTA proposals were fully supported by the US government and by the shareholders as well as by financial institutions, the US steel industry would have little chance to get out of its disadvantaged situation because of diseconomies of scale, vintage, balance of subprocesses, location and labour cost. The capital expenditure required to implement the programme would raise the capital service charges quite drastically-

and thus the total unit cost of US steel production. The US steel industry would continue to run at high cost levels for repairs and maintenance.

Why do American steel users buy foreign steel?

Some of the reasons why US steel users rely on imports are:

1 The f.o.b. prices[19] are lower, 30 per cent for Japanese, 20 per cent for European qualities, than prices f.o.b. US integrated plants. Domestic transport prices in many high consumption regions of the United States sharpen the difference. (It should be remembered that mini-mills are selling at prices competitive with those of foreigners and at lower transport costs to regionally constrained markets.)

2 For qualitative reasons, either because of certain qualities not being available at all in the United States or by generally higher quality levels of steel being produced in modern foreign mills. (See the pipe and tube example below.)

3 Because of greater flexibility, higher preparedness to respond to customers' wishes and because of better customer services. The uncertainties surrounding import regulations and even the sometimes incredible arbitrariness of the regulation administration to some extent offset the service advantage offered by foreigners.

For the US market it is neither good, nor even feasible, to be cut off from imports since:

1 Complementary products are needed. The advantages of specialisation should neither be forgotten nor be cut off, as it is hardly feasible, neither economically nor technically, to produce all varieties needed within the boundaries even of a large country.

2 Certain regions would suffer if cut off from imports, as transport costs put great burdens upon f.o.b. prices. Freight rates from Chicago to Los Angeles range between US $ 86 and 106, from Japan to Los Angeles between US $ 30 to 45 for plates and structurals (see Dirlam and Mueller, 1982).

3 Open, largely unregulated markets will produce new solutions and innovations to problems, once they are recognised and once they have been assessed as being sources of (not too uncertain or short-lived) profitability. An example of this is the

existence of mini-mills serving certain regions as markets, or certain sectors of the market over considerable time, but not necessarily for ever.

4   Constraining imports quantity-wise (that is by imposing maximum percentage import shares in tons) will, by necessity, induce foreigners to shift their export strategy to the United States market to high price per ton qualities, that is, to market niches with low price elasticities. There is empirical evidence from the recent past after the TPM was introduced that Japanese imports fell to between 6 and 6.5 million tons per year. At the same time, the per-ton value has more than doubled. This, in the long run, will hurt the US industry, or make it increasingly difficult for it to compete in the higher quality ranges like, for example, the quite flexible border zone between commercial and quality/alloy steel. This border zone will be firmly occupied by highly competent, technologically advanced foreigners, and to a degree also by advanced domestic midi and mini-mills.

Pipe and tube as an example of imports

The temporary slight upturn in US steel industry profits during 1981 stemmed from a 'sectoral' boom in the demand from the oil-producing countries and industries. Jones and Loughlin and Republic made 50 to 60 per cent of their pre-tax 1981 profit from oil industry goods, for which the profit margins averaged at about 25 per cent. The US industry had been able to obtain a larger share of oil project orders than it could fill from its own production, quantity and quality-wise.[20] Thus a substantial share was ordered by US Steel Co. and Armco in Europe and elsewhere. Both steel producers were amongst the largest steel importers of the USA.

With slackening oil demand and consequently falling oil prices, however, the new investment market in the oil industry withered away. This not only severely hit many of the wholesalers–importers, of which many went bankrupt, but also the above-mentioned producers (as well as Wheeling-Pittsburgh Steel). This is one of the triggering factors behind the anti-dumping claims filed by US Steel Co. against European steel producers during spring 1982. For the second time, US Steel Co.'s action terminated the TPM system – this time the revised second version. It had also a bearing on the quantity restriction claims against German Mannesmann high quality pipes and tubes, which became part of the final settlement of the US–European steel quarrels which lasted for eight months during 1982. The pipe clause is symbolic rather than real, because the market for oil pipes in the free world has been dead since December 1981 and inventory available in the USA – the greater share of which is of European origin – would suffice to cover two years' demand, were there any.

The case of quality pipes is of particular interest insofar as it urges a thorough correction of the US industry's claim that the US steel market is a market place for foreign steelmakers' dumping (there was no dumping in pipes and tubes either). When pipe imports accounted for 30 to 40 per cent of total US imports in tons, it was because of a genuine market demand and the inability of US producers to meet the demand. Even the remaining 70 to 60 per cent of US imports, between approximately one-third and one-half are in qualities not available at all, or not in sufficient quantities or grades, from US producers. There will always be a need for imports of such quality steel, as it is neither possible nor profitable to adjust domestic production to all possible shifts in demand for all types and grades of steel.

Some effects of protection upon the United States' steel industry structure

Regulation

Steel imports to the United States have been regulated with short interruptions ever since 1969, when the first Voluntary Restraint Agreement (VRA) was concluded in order to give the US steel industry time to re-organise during a three-year period. The VRA was prolonged in revised form to 1974, and the restrictions were removed during the boom years of 1974–75 at the request of the industry.

The year 1978 brought a number of new regulations. The trigger price mechanism (TPM), a means of controlling minimum import prices for imported steel, started in February 1978 (becoming effective mid-1978) and was terminated at the beginning of 1980, at the request of the industry. In 1978, also, 'buy American laws' for steel were promulgated. The trigger price mechanism was reinstated in autumn 1980 and terminated in January 1982. During 1982, a new quota arrangement for carbon steel from European Community countries was worked out. It became valid in autumn 1982 and is supposed to last to the end of 1985, that is as long as the Davignon plan for the EC steel industry is planned to be valid.

During 1983 and 1984 restrictive arrangements for imports from Japan as well as from other countries plus quota arrangements for alloy and specialty steels were introduced.

The trigger price mechanism was the first comprehensive plan to involve government in the support of the steel industry by granting it minimum prices. During the 1960s and until the middle of the 1970s, the steel industry had often been blamed for pushing inflation on the US market.

Methodological observations on the assessment of regulation efficiency

How efficient, in retrospect, were the protective measures of the Voluntary Restraint Agreement 1969–74, and the first Trigger Price Mechanism system, 1978–80?[21]

The ultimate aim of the protective measures was to provide a temporary shelter to the industry under which it would have a unique opportunity to adjust and rejuvenate itself, in order to become competitive and, thus, fit again for life without artificial shelter.

When attempting to assess the efficiency of a policy instrument, care must be taken when a) estimating the lead times for major decisions to be taken and implemented, b) understanding the expectations held in the industry about the long-term outlook both in the market (or rather as to the long-term profit expectations), and c) concerning government interventions (at home but also to a certain extent abroad): how will the business climate be for the industry, not only market-wise but also regulation-wise. In other words, it is difficult to answer the question if the industry expects a fairly stable competitive climate to be established, or expects regulations to spoil the competitive climate. Will the industry's behaviour change to the positive or to the negative? Usually it changes in both directions. It is often claimed that negative consequences of regulation have a greater impact.

Certainly those two sets of expectations – market and regulation – will have a tangible impact upon the propensity to invest, as they are central decision parameters concerning long-term external conditions.

Another influence causing a lag between a political decision in favour of the industry, and consequently expected responses from its firms, is to be found not only in the size and complexity of major physical investment, but also in the pace of technological change or modernisation. If the throughput time (from its initiation to a plant, or part of it, becoming operative) for a major investment decision is five to seven years, a time span which seems to be quite relevant for the steel industry, then one cannot expect great changes to become visible in output/ price/quality and profitability figures in the shorter term. This is a well-known phenomenon not only from major process industries like the iron and steel industry, but generally from the application of medium-term economic policy instruments aimed at coping with business cycles. Firms do not, as a rule, keep prepared investment project plans in their drawers, to be implemented immediately a decision on economic policy is taken by government. One thus should keep in mind that there will be lags, for plausible reasons.

One difficulty when it comes to estimating the size of the transferred sums is that regulations – as regulations usually do – stimulate a range of counter-intuitive behaviours, aimed at bypassing them. Other difficulties stem from lack of precision of product or quality definitions.

With such reservations in mind, one may now ask how the industry reacted.

The Voluntary Restraint Agreement

The VRA lasted for five years, terminating during the 'all-time high' year for the steel industry of the industrialised world, in 1974.

The evidence is discouraging. The industry invested much less than during the five to six years before VRA became valid and virtually no steps were taken in restructuring. Here one must remember, and this is an important message to economists and political scientists, that there is no decision body acting at 'the top of an industry'. Economic and thus investment decisions are taken at firm level, or in some instances (as in the British Steel Corporation) at the level of a major holding corporation (which will not necessarily mean shorter decision-making time).

At the time of the termination of the VRA in 1974, the industry was in about the same shape it had been five years before when it was implemented, being about as split up, with about as many obsolete, but also undersized, plants in operation. A number of new hot strip mills were taken into operation, which led to duplication of capacity in a quite narrow range of products.

Dirlam and Mueller claim[22] that the industry lacked in innovativeness against its major competitors, but also that it lost in labour as well as capital productivity. According to Adams and Mueller (1982) the transfer effects of the VRA may be estimated to range between US $ 386 million and 1 billion.

The Trigger Price Mechanism[23]

Many of the effects of the (first) TPM came through the appreciation of the yen against the US dollar in 1978 and 1979. As long as the triggers moved upwards, the TPM was accepted by the industry. It collapsed in March 1980, when the US Steel Corporation, the leading producer, filed anti-dumping petitions against all the major steel producers within the European Community. The TPM was consequently revised and, towards the end of 1980, brought about a new regulation together with substantial price increases for certain steel qualities. It stipulated automatic investigations of imports if they exceeded a certain share of the US domestic market, provided the US steel industry was operating at less than 82 per cent of capacity utilisation. This arrangement lasted only for a few months, given the rapidly slackening demand during 1981. During late autumn, the US Government began negotiations with the European Community for new lower quotas, to be 'voluntarily' imposed upon Community exporters.

Crandall (1981) estimates the sums transferred by means of the first trigger price mechanism at US $ 1 billion for each of the years 1978 and 1979, but this obviously underestimates the effect of waiving discounts of unknown sizes on official list prices granted before the TPM became valid. Mueller and van der Ven (1982) claim that, while the TPM was enforced, the premium of hourly employment cost in the steel industry over the average for the country's entire manufacturing sector had risen from 28 per cent in 1967 to 64 per cent in 1977, climbing further to 75 (78) per cent in 1980 (1982). For the steelworkers within the European Community this premium amounted to 21 per cent, for German steelworkers to 12 per cent and for Japanese steelworkers to approximately 50 per cent in 1981.

General conclusions

The VRA and the TPM seriously distorted patterns of consumption[24] in that:

1   Certain regions would have consumed substantial amounts of cheaper imports.
2   Due to the (Japan to US destination-based) artificial transport cost levies, c.i.f. prices in the Great Lake region were kept relatively high, whereas domestic steel was further disadvantaged in the west coast region.
3   The TPM favoured purchase of low quality US products at the expense of higher quality products in the same product line, foreign rolled products offered on the US market generally being of higher quality, at the same or lower prices because US rolling capacity is more vintaged compared to foreign rolling equipment. Imported steel usually goes to those segments of the market in which domestic steel is inferior. Thus the TPM favoured lower grade domestic products at the expense of higher foreign grades in the same product ranges. This is damaging the industry's competence in the long run.
4   TPM is claimed to have punished customers by favouring US firms, which are generally less alert to customer service, and less flexible in relation to customer requirements than are un-

protected foreign suppliers.

5   As a result of the above-mentioned factors, US firms in the steel consuming industries were subjected to losses of income and market shares where competition with cheaper and even better steel, both at home and abroad, is hard. Particularly affected were industries producing offshore and other oil handling equipment, wire products and so on. An obvious example of the cost disadvantage from using US-produced steel are US automobile producers. Using US-produced steel makes the US-produced car cost well above US $ 100, the US-produced truck well above US $ 200, more expensive than comparable imports. To the US auto mobile industry this means increasing headaches and, to the US buyers, it means an extra cost in the neighbourhood of US $ 1 billion per year (see Dirlam and Mueller, 1981).

6   A more long-term consequence is a speeding-up of the trend away from steel as a material for construction and other industrial uses, in favour of other lighter materials. This long-term trend, being well under way, was accelerated by the protective measures. It is, however, difficult to measure its impact in more exact terms. Such measurements request longer time series to become available.

As far as investment and capital stock are concerned, the conclusions are similar. Aylen (1981) concludes 'the industry has broadly maintained its output while running down its capital stock', and 'the run down has been delayed by low energy prices and protection'.[25] It is not that no capital was available for modernisation, rather that many major firms attempted to diversify out of the steel industry, by means of the additional liquidity made available through protective measures. The investments made in the industry were of replacement character (compared to expansion investment in Europe and Japan). Dirlam and Mueller (1981) give the following estimates for the years 1960 to 1979, in million tons.

	USA	EEC	Japan
New capacity investment	11	65	107
Replacement investment	75	44	17

The consequences of protection became well visible in the short unprotected period between 1974 and 1977, when the steel market was quite open, and after the first TPM was relaxed between 1980 and 1982.

A considerable number of plants, including some major ones, were shut down for ever with the loss of many jobs. At the same time, however, a number of mini-mills were started, and these are the major competitive force on the US market when it is sheltered against foreign competition. Their cost efficiency has caused the closure of several larger company-owned mills in those market segments where the mini-mills are competing (carbon steel rods and bars as well as light shapes).

Dirlam and Mueller (1981) claim that US mini-mills are taking market shares of US integrated mills at a good pace. For 1990 they, as well as OTA, expect the mini-mill share to be 25 per cent of the domestic market. Dirlam and Mueller also claim that closures of integrated mills in segments in which non-integrated mini-mills compete have been caused by the competing domestic mini-mills rather than by foreign competitors.

A most severe impact of the protective measures granted is that they have largely sheltered the US steel industry from foreign competition, implying retarded speeds of adjustment to a then restricted competition, as well as to the depreciation of the US dollar. Instead of using the protection to improve the structure of the industry, to close obsolete plants, and to modernise plants worth modernising, the industry continued to distribute the slack created a) to the workers and b) to shareholders, as the industry must compete for capital with other investment opportunities. (Steel industry shareholders, however, have not been spoilt by their holdings; see Chapter 12.) c) A third way to use the slack created by the regulations was for big steel to invest outside the steel industry.[26]

In Chapter 25, Gold claims that estimates of the effects of the VRA and the TPM neglect the negative effects of maintaining less efficient management policies, which considerably postponed necessary efficiency measures and thus contribute to the quite remarkable losses suffered by US customers, due to an overaged, improperly managed industry. In addition to this, one may claim that the protective measures also made it possible for the unions not only to maintain conservative policies and behaviour but also to take a perhaps greater share of the surplus produced by the protective measures than any other steel industry-related group interest was able to harvest.

Whether the protective measures introduced by the US Government since 1968 have raised the prices and reduced the quantity of imported steel and by how much is difficult to say with precision.[27] There is no doubt that prices of US-produced steel going into domestically manufactured steel products have increased more than comparable import prices, and that purchasing power has therefore been transferred from the general public of the USA to the steel industry, its employees, and possibly to its shareholders.

In summary,[28] it seems plausible that the temporary shelters did have a number of counter-intuitive effects: the industry at large (even at the micro level exceptions from the average are only too rare) did not use the shelter, the additional cash flow, and the time for sheltered action it rendered, to reorganise its enterprises and works, to reshape its structure, or to improve or change its management. It was and is obviously not the shelter of protection but only competition which during its hundred-year-history has moved the industry towards improved competitiveness.

The crisis worsens

1983 was a critical year for the US steel industry with steel mills operating at less than 30 per cent of nominal capacity. Unemployment amongst steelworkers was creeping above 60 per cent. To avoid bankruptcy, two firms merged. A third intended merger had to be withdrawn for anti-trust reasons. The boom of 1984 brought some relief to the industry, which however urges further import restrictions to be introduced.

Obviously, the industry has lost approximately US $ 3.5 billion on its 1982 steel operations. (Bethlehem Steel alone lost 1.5 billion on its steel operations, United States Steel 850 million, Republic Steel 240 million.) American Steel Industry is selling steel at an average loss of approximately US $ 125 per ton. To end the hidden discounts of approximately US $ 125 per ton on valid list prices, a cartel agreement was reached to increase prices and to abandon the discounts.

The situation for the machine producers is even worse. Whereas the steel industry between 1978 and 1982 experienced a growth of only 8 per cent in sales, it was able to improve its profits by 122 per cent. During the same time, the machinery industry's sales grew by only 5 per cent and profits fell by 26.4 per cent.

The steel industry's paramount problem is that it not only has reached the bottom but that it may also remain there for some time to come. The much hoped-for upturn during 1983 will not affect major steel customers. Indeed, the steel-intense machinery industry has been pulled into a still deeper crisis, to a considerable extent because of the artificially high steel prices it has to pay to US suppliers. (This is reminiscent of similar problems amongst European machinery producers, which seem to be the major victims of the Davignon-EEC steel cartel.)

The financially weakest large steel producers in the USA are Republic Steel[29] and National Steel, who, given present conditions, will hardly be able to survive to the end of 1985. Wheeling-Pittsburgh Steel (merger of 1968), the eighth largest US steelmaker, has been living on the brink of bankruptcy for years already. It has been able to hang on mainly because of sacrifices made by all ranks of its employees. In

December 1982 wage cut agreements were again reached between the corporation and its employees, and this agreement became the model for the general agreement of March 1983 between the United Steelworkers and the 'Big Eight', covering 43 months to come and implying temporary cuts by approximately 9 and up to 11 per cent as well as withdrawals of a range of fringe benefits. Part of the cuts go towards unemployment benefits for jobless steelworkers. The cut temporarily brings the industry average down from 75 per cent to 68 per cent over the average of the hourly labour cost for the entire manufacturing sector. The total hourly employment cost for the steel industry would fall from US $ 23.45 in 1982 to about US $ 22 in 1983. (Non-unionised companies pay up to US $ 7 less.) The steel industry average wages would increase by about 14 per cent over the following three years again. The agreement also includes a profit-sharing agreement, which, however, will become effective in a few years to come.

Summary

The United States' steel industry has been and is adapting to its 25 years of continuous decline and occasionally acute crisis with a retrofitting and *ad hoc* strategy. Despite 15 years of protective measures, the industry has not been able to regain competitive power. Rather, the negative trends observed have been, and still are, at work. The industry is now greatly inferior to its major competitors in Japan, but also to modern European (in particular German) competitors. The most remarkable conclusion one can draw from comparing the three competitors is that the unsheltered, unprotected and unsupported competitors remain the fittest, even if they are temporarily suffering losses which, however, mainly seem to be due to market distortions caused by protection and subsidisation.

Against the optimistic forecasts of the industry (AISI) but also of the Office of Technology Assessment's progammatic plea, there seems to be little, if any, gain to be expected from so-called industrial policy programmes. As Crandall and others (Gold, Mueller etc.) have demonstrated, the return on proposed investment, for example on the OTA proposal, would be negative. And, as Gold claims in Chapter 25, still the industry would not be operating at performance standards nearly comparable to best or even average Japanese ones.

However, given the politically inconvenient consequences of declining national self-sufficiency in the steel sector and, in particular, drastically growing unemployment in crisis-stricken steel regions, the future outlook points to continued or even increasing protective measures. Their effects will most likely differ from the effects of the protective measures during the last 15 years: the industry will gradually lose still greater shares of the market segments, which should be of interest to traditional high labour cost industrialised countries. The steel using industries will continue to be losers, not only in similar ways to the steel industry, but, as they are not subject to protectionism, their losses will be much more tangible. The ultimate loser is the American taxpayer. But as there are many more taxpayers than steel industry employees and steel industry dependants, the burden on the taxpayer will be less obvious to the individual than the gain from the income transfer to the immediately concerned steel people.

Notes

1   The term 'the (steel) industry' refers regularly to the so-called integrated steel mills, but excludes the new type of low cost-of-entry technology competitors, i.e. the so-called mini-mills.

2   See Adams, W., Mueller, H., 'The Steel Industry' in Adams, W. (ed), *The Structure of American Industry* , MacMillan, New York, 1982 where a vivid description of Judge E.H. Gary's policy can be found. In 1911, the Federal Government filed a suit against the US Steel Corporation, demanding its dissolution because of its monopoly of the market. When the case was decided after World War I, in 1920, the Supreme Court concluded that 'mere size was not an offence'. This was quite a unique decision.

3   c.i.f. = 'cost, insurance and freight', that is cost at user's point of use.

4   1 US ton (short ton) = 0.907 metric tonnes

5   Between 8 and 9 manhours per shipped tonne at full capacity against slightly above 6 manhours in the Japanese steel industry, as an average. The most modern mills operate at considerably higher productivity.

6   For comparison, the share of labour cost is above 35 per cent; cf the GAO Steel Report.

7   As the contributions by A'cs and Rosegger (Chapters 16 and 17) to this volume as well as Adams and Mueller, op.cit. 1982 (and Mueller in earlier papers) demonstrate, the US producers lag behind their European and Japanese competitors' innovative activities. It must, however, be remembered how difficult it is to measure 'innovation'. Nevertheless, the indicators used by the above-mentioned authors are convincing.

8   During which year the US Steel Co. and the so-called steel caucus demonstrated their power in Congress.

9   Dirlam, J.B. and Mueller, H.G., 'Import Restraints and Re-industrialization: The Case of the U.S. Steel Industry', Middle Tennessee State University Conference Paper Series no. 67, Murfreesboro, December 1981.

10   Even if it is surprisingly labour-intense, one should rather say that the steel industry's investment cost per installed unit of capacity is very high in as far as integrated facilities are concerned. A frequently quoted low estimate of today is that a US $ 1,000 investment – compared to an approximately US $ 100 investment cost per new tonne installed in non-integrated mini-mills – is needed to establish the best available integrated technology capacity for one tonne per year raw steel capacity, and that the optimal economy of scale ranges around 6–7 million tonnes per year. Reasonable economy may, however, be achieved at 1–1.5 million tonnes lower annual capacity per plant – not per firm!

11   Crandall claims that the raw steelmaking capacity added between 1960 and 1980 in effect is that of the mini-mill sector. This means that the integrated firms had declined in reality, during a time in which the US gross national product has doubled.

12   In a private communication of a manuscript, spring 1983.

13   See also Crandall, R.B., 'The US Steel Industry in Recurrent Crisis', The Brookings Institution, Washington DC, 1981.

14   Between 1960 and 1979, Japanese firms added eight times more capacity than the integrated US steel industry for approximately the same amount spent (if corrected for non-steel investment and differences in labour costs, the Japanese spent 30 to 40 per cent more), according to Adams and Mueller op. cit.

15   June 1980, Washington, DC (OTA Steel Report).

16   Crandall, R.B., op. cit., Chapter 6.

17   Crandall, R.B., op. cit., pp.38–40.

18   As before in the American Iron and Steel Industry's Report *Steel at the Cross-Roads: The American Steel Industry in the 1980s* , Washington, DC, 1980.

19   f.o.b. = free on board = factory prices.

20   The share of quality pipes of the total steel import was in 1979 approximately 17 per cent. By 1981 it had grown to 30 per cent, during the first five months of 1982 to 39 per cent. More than 50 per cent of the imports in 1979 and 1980 came from Japan. When US imports grew by 74 per cent in 1981 – to approximately 6 million tons – the Japanese share fell to 43 per cent. During the first five months of 1982, the European share of US pipe imports rose to 30 per cent, of which seven-eighths came

from Mannesmann.

21 It is too early to estimate the effects of the revised TPM because the statistics are either not available yet or cover too short a period of its validity.

22 Dirlam, J.B. and Mueller, H.G., op. cit., Appendix, Tables 4 and 5, p.18.

23 For a description of the original and revised TPM cf. Dielmann, 1981. For an account of the GATT Anti-Dumping Regulation and the US Anti Dumping Statutes (as well as related problems of definition and measurement) see Kawahito, 1982.

24 As demonstrated by Dirlam and Mueller op. cit.

25 As will be shown in Chapter 14, the US industry lags far behind the European and in particular the Japanese steel industries in taking energy-saving measures.

26 See Adams, W. and Mueller, H.G., op. cit., 1982, on steel industry diversification.

27 Dirlam, J.B. and Mueller, H.G., op. cit. 1981.

28 For the sake of justice it should be emphasised that many of the measures taken by the European Community since 1978 will have similar effects on the steel industries of EC member countries (cf. Dicke, 1983) as the many protective measures have had on the US steel industry.

29 Republic and Jones & Laughlin LTV Co. decided to merge during autumn 1983.

12 Investment and productivity growth in the steel industry: some implications for industrial policy

Robert W. Crandall[1]

Industries rarely adjust smoothly and quietly to adversity and decline. The loss of comparative advantage by US shoe producers, clothing manufacturers, or television receiver manufacturers was not accepted quietly. Trade suits were filed, 'orderly marketing' agreements were signed and Anti-dumping duties were assessed. The results were predictable, however: the industries continued to decline even though American consumers were saddled (and, in some cases, continue to be saddled) with excess costs due to protection.

No one could have foreseen that the recession of 1958 signalled an irreversible turning point for the US steel industry, but from the vantage point of 1982, we now see that the 1958 recession and the 1959 strike mark the beginning of a secular decline for the industry. As the industry entered the 1960s, it had approximately 140 million tons of capacity. By 1980 its capacity had increased to only 154 million tons, 20 million tons of which was owned by small, non-integrated firms[2]. Since the non-integrated capacity in 1960 was less than 4 million tons, it is clear that in two decades the large, integrated firms have actually declined, despite a doubling of real gross national product in the United States. Since 1980, integrated capacity has continued to decline, and speculation concerning further plant closings has increased.

In the midst of this painful, contracted decline, US steelmakers have predictably sought government assistance – particularly in the form of trade protection. Their cries have not gone unheeded. From 1968 through 1974, a set of Voluntary Restraint Agreements (VRAs) were negotiated with Japan and Europe. From 1978 through 1981, a 'trigger price' (minimum price) system for imports was imposed upon all importers of carbon steel. Obviously, these two bouts of trade protection have not 'worked' – that is, they have not reversed the industry's decline. Since 1977, numerous plants have been closed and many more face the padlock. The industry now acknowledges that its capacity will have to shrink, but it still argues for a variety of government policies to assist and protect it from foreign competition. Citing its need to modernise, it argues that, with temporary increases in its cash flow, it can increase its investment in modern labour- and energy-saving equipment[3]. This accelerated investment programme will allow it to compete successfully with efficient producers, such as the Japanese, in the US market.

Is there any evidence to suggest that the US steel industry will be able to increase its investment rate profitably? Is it likely that such investments will have a pay-off sufficient to lower costs substantially and thereby restore the industry's ability to compete with imports? This chapter investigates the effects of past investments in the steel industry and their relationship to productivity growth and returns to stockholders. If the industry has been constrained by a myopic capital market, we should expect to see very handsome returns from surges in investment permitted by temporary increases in cash flow. But if investment opportunities are limited because of the overbuilt condition of the industry, temporary surges in investment may reflect undue optimism, not the exploitation of a backlog of profitable projects that had been delayed by the inadequacy of funds.

The steel industry in decline

Elsewhere, I have argued that the US steel industry lost comparative advantage to the more advanced, less developed countries beginning in the late 1950s and early 1960s[4]. Prior to this time, the advantages of low cost, proximate raw materials offset the disadvantages of labour intensity in the indusry. Contrary to popular notions, the steel industry is not capital-intensive. The labour required to produce steel has always been substantial, particularly in maintaining the primary facilities and in operating the steel-finishing processes. Even with the shift to basic oxygen furnaces and greater automation of rolling technology, the labour intensity of the steel industry exceeds the average for all US manufacturing. In 1947, the Census of Manufacturers reported that labour's share in steel value added (excluding fring benefits) was 22 per cent above the average for all manufacturing. (See Table 12.1). Twenty-five years later the steel industry was still paying a larger share of its value added to labour than the average US industry. In 1972, the share of all labour costs, including fringe benefits, in value added was 23 per cent higher for steel than for all manufacturing. Even in 1974 when steel demand was extremely strong, the steel industry paid a larger share of its value added to labour than the average manufacturing industry and nearly as large a share as the average clothing manufacturer[5].

Table 12.1

The share of labour expense to value added in US manufacturing and the steel industry, 1947-76

	All manufacturing			Steel industry (SIC331)		
Year	Value Added ($million)	Payroll ($million)	Labour Share	Value Added ($million)	Payroll ($million)	Labour Share
1947*	74,290	39,696	0.534	2,657	1,732	0.652
1954*	117,032	62,963	0.538	4,755	2,470	0.519
1958	141,541	73,875	0.522	6,863	3,571	0.520
1963	192,083	93,283	0.486	8,424	4,168	0.495
1967	261,984	123,481	0.471	10,170	5,022	0.494
1972	353,974	160,415	0.453	12,116	6,389	0.527
1976	511,471	212,998	0.416	17,274	9,166	0.531

Source: Bureau of the Census, *Annual Survey of Manufactures and Census of Manufactures.*

*Data for SIC 331 in 1947 and 1958 are not strictly comparable with data for later years, due to changes in SIC classifications. The changes in industry definition for SIC 331 are relatively minor.

With the world iron ore market becoming more competitive in the late 1950s and 1960s and real shipping costs falling, countries with no indigenous natural resources could become steel producers if they had good coastal locations, low wage rates, and the willingness to import foreign technology. The Japanese are an obvious example, for they expended their steel industry more than tenfold between the mid-1950s and the mid-1970s[6]. During this period, their basic materials costs for steel production fell by approximately 50 per cent while US material costs remained essentially constant[7].

In today's market, foreign steel producers with deep-water port locations can obtain their iron ore, coal, limestone, and scrap at costs that are comparable to the best located (inland) US plants. In many of these countries, plant construction costs are even lower than in the United Staes; hence, their labour cost advantages loom large. It is not surprising, therefore, that the United States began construction on its most recent integrated plant in 1962 and has no plans for any new integrated plants as far as one can see into the future.

In the face of this decline in competitiveness, US integrated producers have been required to attempt to manage their existing asscts by modernising where they can and phasing out plants where they cannot. Bethlehem has closed parts of its Johnstown and Lackawanna plants in the past five years. US Steel has closed its Youngstown works and faces similar decisions in the Monongahela Valley, Geneva (Utah), and Fairfield (Alabama). Youngstown Sheet and Tube (now merged into LTV Corporation) closed its Youngstown works. National Steel appears ready to sell one of its three basic steel plants and has closed half of another. Kaiser made the misguided decision to build a new basic oxygen shop in Fontana, California, in the late 1970s. It is now closing this facility. McLouth Steel is in bankruptcy, and Ford Motor Company is agonising over the future of its Rouge steelworks.

Some of the recent difficulties of the US steel industry derive from the combination of deep recession and a high value of the dollar. Steel consumption in the US and most of the developed world has never returned to its 1973-74 levels. Moreover, the high value of the dollar has exaggerated the differences in production costs between US producers and their foreign rivals. While the Japanese have produced and sold steel in their domestic market at prices averaging 20 per cent less than US prices in the past twelve years[8], the differentials have not been so great between the US and Europe until the 1981 appreciation of the dollar. At current exchange rates, however, the US integrated industry has average production costs that are substantially above those in Germany and France, and even the grossly inefficient British industry now has costs not much different from those in the US plants[9]. Therefore, the extreme pressure on US producers in 1981-82 is more than the continuing effect of excess capacity due to a loss of comparative advantage. Rationalisation takes place in recessions (when driven by market forces), and the US industry is currently being rationalised with a vengeance. Economic recovery accompanied by lower interest rates and a depreciating dollar will relieve much of the current pressure, but it is unlikely to reverse the continuing decline of the industry.

The erratic pattern of investment in the US steel industry

The stock market was remarkably prescient concerning the fate of the steel industry. After bidding up steel stocks in the late 1950s, the market turned bearish on steel stocks for the entire 1960s. (See Table 12.2. Real investment in steel followed, declining from $2.1 billion annually ($1972) in the late 1960s. This investment surge coincided with the Vietnam era of economic expansion in the United States and continued into 1969, the first year in which voluntary steel import restraints were in place. In the first half of the 1970s, despite continued trade protection and the devaluation of the dollar, steel investment sank to an annual level of $1.3 billion per year once again, reflecting the low rates of return in the industry.

With the surge in steel demand in 1974, steel industry profits soared, rising to a level at which the return on equity was greater than the average in manufacturing for the first time since the 1958 recession. Another mild surge in steel investment followed this year of 'steel shortage', as annual investment rates rose to $1.7 billion in 1975- 77, but the collapse in the world steel market in 1977 sharply reduced US steel investment for the rest of the 1970s.

The sharp swings in investment in the steel industry provide fertile ground for examining the pay-off to capital formation in this industry. If the industry was being squeezed by the capital market during periods of low cash flow, it might be forced to postpone a number of invest-

Table 12.2

Stock market performance and investment in
the steel industry 1956-77

Year	Average of high and low annual values for:		
	Standard & Poor's 400 Industrials	Standard & Poor's Steel Index	Investment in the steel industry (thousands of $1972)
1956	49.5	64.8	1692
1957	47.6	60.9	2068
1958	51.2	71.4	1451
1959	61.2	92.5	1002
1960	60.2	81.8	1733
1961	68.8	80.3	1202
1962	65.0	61.7	1074
1963	72.4	57.0	1333
1964	85.5	68.4	1847
1965	92.5	64.1	1944
1966	89.2	57.4	2201
1967	95.7	55.7	2404
1968	106.5	54.6	2363
1969	107.0	54.5	1965
1970	89.2	42.7	1605
1971	107.6	39.3	1155
1972	122.6	43.5	1059
1973	119.0	42.1	1218
1974	90.6	46.1	1609
1975	92.6	55.6	1730
1976	111.3	73.6	1712
1977	109.4	56.7	1671

Source: S Standard and Poor's, Analysts Handbook, 1979; Bureau of the Census.

ment projects that could potentially yield high returns. When cash flow increased during more prosperous times, these projects would be funded with salutary effects upon operating costs or product quality. (Expansion would seem not to have been a serious option for the past twenty years.) If the equity market responds to short- term changes in operating results, we would expect a return from holding steel equities to respond with a one or two-year lag to changes in industry investment. And if this investment adds to the capital stock effectively, it should generate increases in the rate of growth of labour productivity.

Alternatively, if the industry's investment surges in the late 1960s and mid-1970s were simply excessively optimistic reactions to short-term, but ephemeral, changes in its fortunes, these surges should have been followed by negative excess returns in the stock market as investors realise that the money had been ill-spent.

Finally, there exists the possibility that much of the investment since 1959 has been misapplied because the industry failed to realise that it had lost competitiveness. Some plants may have been kept alive in the hope that the adverse events of 1959 and beyond would turn out to be transitory. Instead, they were simply a set of continuing forces that we now realise were signalling the inevitable decline of the industry.

The evidence on productivity

In a publication released in 1980, the American Iron and Steel Institute (AISI) argued that the steel industry would need an average of $4.4 billion ($1978) in replacement and modernisation investment per year for the decade 1978-88 if it was to improve labour productivity and energy efficiency sufficiently to remain competitive with imports[11]. The AISI report forecasts a 2.1 per cent annual increase in labour productivity if this rate of investment materialised. Since the AISI 'capital requirements' scenario posited at least a doubling of the recent investment rate, we might ask if there is any evidence to support its projection of the potential for productivity gains.

The long-term trend in productivity growth in steel, all manufacturing, and the non-farm private business sector of the economy is shown in Table 12.3. The data are presented for a 23-year period beginning in 1956, two years before the 1958 recession and extending through 1978, a year of moderate US steel demand. These data demonstrate quite clearly that steel has exhibited a lower productivity growth rate than the general economy or the manufacturing sector of the economy. In fact, as the analysis below will show, there has been very little change in the trend rate of growth in productivity for the steel industry throughout this period. It has remained at about 60 per

196

Table 12.3

Productivity growth in steel, manufacturing and the private non-farm business sector (1967 = 100)

Year	Steel	All manufacturing	Non-farm business
1956	86.4	73.5	74.9
1957	84.3	75.0	76.2
1958	77.9	74.6	78.0
1959	87.5	78.1	79.3
1960	82.3	78.8	81.2
1961	84.9	80.7	83.6
1962	89.2	84.5	86.6
1963	93.2	90.4	89.4
1964	97.2	95.2	92.9
1965	101.1	98.2	95.8
1966	103.3	99.7	98.2
1967	100.0	100.0	100.0
1968	103.5	103.6	103.3
1969	104.0	104.9	103.0
1970	101.3	104.5	103.3
1971	106.2	110.4	106.8
1972	112.7	116.0	110.8
1973	123.5	119.4	113.6
1974	123.5	112.8	110.9
1975	107.6	116.3	113.2
1976	114.5	124.2	116.9
1977	115.6	126.9	119.1
1978	120.7	128.0	118.9
Average annual rate of growth 1956–1978	1.5%	2.5%	2.1%

Source: US Department of Labor, Bureau of Labor Statistics (BLS)

cent of the manufacturing average.

It is possible that the sluggish growth in productivity in the steel industry is a reflection of a low rate of investment in capital. With declining fortunes, steel companies have not been expanding and have been forced to close some plants and undermaintain others. As a result, the capital-labour ratio may not have been growing as rapidly in the US steel industry as in growing steel industries.

To isolate the contribution of changes in the capital-labour ratio to the growth in labour productivity, I have estimated an equation of the form:

(1) $\text{Log}(Q/L) = a_0 + a_1 \text{ TIME} + a_2 \text{ Log}(Q/\text{CAPACITY}) + a_3 \text{ Log}(K/L)$.

where Q/L $\qquad$ = output per manhour
$\quad$ TIME $\qquad$ = a time trend
$\quad$ Q/CAPACITY = capacity utilisation
$\quad$ K/L $\qquad\quad$ = the ratio of capital to labour

This equation was estimated using BLS indices of labour productivity, a measure of capacity utilisation based upon interpolations between peak years of output, and a measure of capital stock based upon the perpetual inventory method of accumulation.[12]

The results of the estimation demonstrate that the changes in the rate of capacity utilisation, not the capital-labour ratio, influence output per manhour in the steel industry. (See Table 12.4.) The results are shown for 1956-76 (because data for capital stock are not available past 1976) and for two sub-periods, but there is little shift in the contribution to the time trend. Therefore, the rate of technological progress does not seem to have changed much over the two decades. But the growth in the capital-labour ratio does not seem to influence productivity growth in any of the specifications. One can only conclude that obsolescence was a problem throughout the period, rendering the capital stock measures meaningless, or that investment outlays did not improve labour productivity, or both.

Table 12.4

The effect of capital intensity on productivity in the steel industry
(t-statistics in parenthesis).

Productivity measure	Time trend	Capacity-utilisation elasticity (Q/CAPACITY)	Productive capital-labour ratio elasticity (K/L)	Time period
BLS	0.021 (5.53)	0.242 (2.30)	− 0.001 (1.27)	1956-76
BLS*	0.021 (1.13)	0.241 (0.48)	0.00003 (0.01)	1957-65
BLS	0.021 (5.23)	0.480 (2.04)	− 0.0008 (0.54)	1966-76

*Adjusted for serial correlation.

198

One cannot quibble with the proposition that additions to the industry's capital stock could improve labour productivity. what we have shown is that such increases have not necessarily contributed to productivity growth in the past. Perhaps this is a reflection of our counting obsolete old capital with new capital accumulation, but it is also possible that much of the 1956-76 investment was misspent. Modernisation expenses at Kaiser's Fontana, California or Youngstown Sheet and Tube's Youngstown works obviously were ill-advised. Investments in BOF furnaces at Alan Wood Steel or McLouth, now bankrupt, may fall into a similar category. But we cannot know how much of the investment by the major companies at such far-flung plants as Geneva (Utah), Lackawanna (New York), or Fairfield (Alabama) was also misspent.

Some evidence that the industry has not moved aggressively to adjust to declining comparative advantage by scaling down coastal plants and targeting the more efficient plants for modernisation and expansion, may be found in the Census of Manufactures data for 'integrated' steel plants in the 1963-77 period[13]. The average size of an integrated plant has remained remarkably constant, as Table 12.5 demonstrates. The average integrated plant shipped $294 million in products in 1963 and $306 million in 1977 (in $1972). This translates into shipments of less than 1.6 million tons, given a price of $180 per ton for the average carbon steel product in 1972 and much higher prices for alloy products. Unfortunately, changes in blast furnace technology and in rolling mills has increased the minimum efficient scale (MES) of an integrated steel plant to at least 6 million tons of raw steel or perhaps 4.5 to 5 million tons of finished products. Thus, the average US plant in 1977 shipped only one-third as much as a plant of MES operating at full capacity despite the fact that the US industry was operating at 80 per cent of its

Table 12.5

'Integrated' steel plants in 1963-77* (millions of $ 1972)

Year	Number of plants	Shipments	Investment	Average shipment per plant	Average investment per plant
1963	46	13,536	900	294	19.6
1967	46	15,577	1,388	339	30.2
1972	52	15,583	741	300	14.2
1977	52	15,912	931	306	17.9

*Includes all steel mills with blast furnaces, steel furnaces, and rolling mills.
Source: Census of Manufactures.

capacity. Apparently, the industry has been slow to adjust to the need to close the smaller, poorly located plants and to expand the more efficient, better located plants.

Investment and the return to stockholders

If surges in investment spending in the late 1960s and the mid-1970s allowed the steel industry to initiate and complete highly remunerative modernisation projects that had been left on the drawing boards because myopic capital markets and insufficient cash flow, we should expect the returns to stockholders to respond with a lag to these surges. In fact, the stock market, which had turned bearish on steel stocks in 1960, does not suggest such a response. It is easy to demonstrate the absence of a stock market response. The capital asset pricing model predicts that the excess return on holding a given asset should be related to the excess return on the entire market:

$$(2) \quad r_{it} - r_{rft} = \alpha_i + \beta_i \left(r_{mt} - r_{rft} \right) + u_{it}$$

where r_{it} = the rate of return on holding the ith asset in period t (dividends plus capital gains divided by the value of the asset at the end of period t-1).

r_{rft} = the rate of return on a 'riskless' asset in period t (the 1-year treasury bill rate).

r_{mt} = the rate of return on the entire market of risky assets in period t.

u_{it} = a random error term.

If surges in investment increase the excess return $(r_{it} - r_{rft})$ on steel companies' stock, we should expect the residuals, u_{it}, to be positive following such a surge.

To estimate (2), I used annual returns from Standard and Poor's index for steel and the S & P Industrials' index as a proxy for the entire market. The results are produced in Table 12.6 with the residuals for each year. It is clear from these results that the returns did not turn up with the increase in investment in the late 1960s or in the mid-1970s. The excess returns were positive in the 1955-59 period and in the 1974-76 period. The latter rise in steel equities reflected the reaction to the unusually strong world demand for steel in 1973-74. Investment outlays responded in 1974-76, but the excess return on steel equities fell sharply in 1977. There was no perceptible reaction to the 1964-68 investment surge, however, as the residuals demonstrate. The residuals hovered around zero in 1969-70 and then fell sharply in 1971.

Table 12.6

Residuals from the capital-asset pricing model estimates
of annual steel industry returns (equation (2)) (per cent)

Year	Residual
1955	+ 23.5
1956	+ 12.2
1957	− 0.8
1958	+ 12.6
1959	+ 12.4
1960	− 9.3
1961	− 14.0
1962	− 17.1
1963	− 17.2
1964	+ 4.6
1965	− 13.0
1966	− 5.8
1967	− 7.6
1968	− 10.2
1969	+ 1.9
1970	− 3.7
1971	− 24.8
1972	+ 0.4
1973	+ 2.6
1974	+ 36.4
1975	+ 22.3
1976	+ 15.6
1977	− 21.2

Coefficient estimates: $\alpha_1 = -0.4268$ $\beta_1 = 0.9227$ $\bar{R}^2 = 0.408$
$(t = 0.12)$ $(t = 4.02)$

A more formal method for investigating the impact of investment on the stock market returns for steel equities is to include lagged values of investment in (2). When this is done, the results are as expected. There is no systematic relationship between the returns and investment levels for the industry lagged one or two years. Rather, the evidence suggests that investment responds to improvements in the stock market, but stock market returns do not respond to changes in investment spending.

201

Future prospects for integrated steel producers in the United States

The above results suggest that it is not the absence of capital that has impeded the growth of the integrated sector of the US steel industry. When investment rises sharply, there are no sudden surges in productivity or in the returns to stockholders. Rather, it appears that the investment funds are sprinkled over far too many plants of inefficient scale and poor location.

In 1980, the American Iron and Steel Institute (AISI) made its plea for government assistance sufficient to allow it to raise its investment outlays to $7.0 billion ($1978) per year, including $4.4 billion for modernisation and replacement[14]. In 1980 and 1981, the industry's investment spending was approximately 40 per cent of this desired rate despite the fact that it had rising net income during the period[15]. US Steel was able to raise more than $6 billion to purchase the Marathon Oil Company in late 1981, an amount nearly double the total investment outlays for all US steel companies. Obviously, steel companies are not investing at the rate AISI thought desirable in 1980 despite favourable changes in tax laws and environmental regulation.

A major reason for the continued slow rate of capital investment in the steel industry may be found in the depressed economic conditions of 1981-82 and the relatively high value of the dollar. But the low prospective returns from such investments, even under AISI's most rosy predictions, must be the major explanation for this sluggish performance. In its 1980 report, AISI predicted that its investment programme would reduce energy consumption by about 15 per cent per ton of production and labor utilisation by 20 per cent[16]. This would translate into a maximum net savings of $3.9 billion per year by 1988. Since these savings are achieved gradually over the 1978- 88 period, the savings in the intermediate years are obviously lower. Thus, the reader is asked to believe the industry's advocacy of an investment programme that would have the industry spending an average of $4.4 billion for modernisation and investment every year for the foreseeable future so that within ten years it might achieve savings of as much as $3.9 bilion per year. Of course, if the assets currently in place were producing steel profitably and much of this $4.4 billion in annual outlays were simply replacing these assets as they depreciated, the additional 3.9 billion in cost savings might loom large indeed. But if these assets have been yielding very low returns (as they have), the $3.9 billion in cost savings may be far too little to justify the more than doubling of the investment rate.

The last two years have brought even greater worries for the US industry. In my 1981 book, I argued that the large integrated plants in the Great Lakes area would survive, but that scattered plants in the

Monongahela Valley, the south-east, and the east would be much less viable[17]. I predicted that, over time, the integrated companies would modernise and perhaps even expand the Great Lakes plants, but that many of the other facilities would gradually close. Since then, the distress of the steel fabricating industry in the mid-west, caused by recession, high interest rates, and high US steel prices, makes this assessment of the future somewhat less secure. Even the Great Lakes plants may encounter difficulties in the next few years if the large fabricating industries – automobiles, farm machinery, home appliances, cans and industrial machinery – do not recover.

The futility of an industry policy that relies on raising product prices through trade protection is thus quite evident in the case of steel. The US fabricating industries cannot continue to compete successfully with imports if their steel costs $100 to $150 per ton more than the steel purchased by their foreign rivals. Import penetration of products made from steel will continue to rise, further reducing the demand for US-produced steel – particularly the flat-rolled products. This will lead the US steel producers to rely more heavily upon construction products, but many of these products can be produced more efficiently by the mini-mills. In short, protectionism threatens to place even greater pressure upon the integrated steel producers.

Conclusion

The US steel industry is being 'rationalised' by market forces. This rationalisation process may be accelerated by a policy designed to help the industry through trying times – the policy of import restraint. The hopes for a viable integrated steel industry in the United States must rest upon an acknowledgement by the steel companies that they have to close a number of poorly located plants and compete aggressively for the mid-western flat-rolled market with imports from Japan, the EEC, Canada, and various emerging LDCs. Protection for steel will only cause exporters to shift from steel mill products to steel fabrications. This, in turn, will further erode the market for flat-rolled steel in the United States.

It is quite clear that many past investments in the steel industry have not payed handsome dividends to share owners. Nor have these investments increased productivity growth in the industry measurably. Any thought of increasing the rate of investment in the US steel industry from its current rate of $3.5 billion per year to more than $8.5 billion ($1982), as suggessted by the American Iron and Steel Institute in 1980, should be abandoned. The value of the industry's entire outstanding common equity is less than $8 billion. There is no way that this declining industry can profitably invest at an $8.5 billion rate. It is more important that it concentrate on those plants that continue to enjoy a strong competitive position. But if it continues to invest money in plants that are doomed to be closed, shareholders can expect no better returns than they have enjoyed over the past twenty years.

Notes

1   The views expressed in this paper are the author's and do not necessarily reflect the views of the officers, trustees, or other staff of the Brookings Institution.
2   American Iron and Steel Institute, Annual Statistical Report, Washington, DC, 1980, and Crandall, R.W., The U.S. Steel Industry in Recurrent Crisis: Policy Options in a Competitive World, The Brookings Institution, Washington, DC, 1981, Table 2-4
3   American Iron and Steel Institute, Steel at the Crossroads: The American Steel Industry in the 1980's, Washington, DC, January 1980.
4   Crandall, R.W., op.cit., 1981.
5   Bureau of the Census, Annual Survey of Manufacturers, Industry Profiles, 1976.
6   Id., Chapter 2.
7   Federal Trade Commission, The United States Steel Industry and Its International Rivals: Trends and Factors Determining International Competitiveness, Washington, DC, 1977, Chapter 3.
8   Peter Marcus, World Steel Dynamics, Paine Weber Mitchel Hutchins, Inc., New York, quarterly issues.
9   Id. (1982).
10 See table 2.
11 American Iron and Steel Institute, op. cit., 1980, Chapter VI.
12 For a detailed discussion of the data and results, see Crandall, R.W., 'Steel Industry Productivity and Public Policy,' ms., November 1981.
13 Census of Manufactures, Summary Statistics, various editions.
14 American Iron and Steel Institute, op. cit. 1980
15 American Iron and Steel Institute, Steel and America: An Annual report, Washington, DC, 1982, Statistical highlights.
16 American Iron and Steel Institute, Steel at the Crossroads The American Steel Industry in the 1980's, Washington DC, 1980, Chapter V.
17 Crandall, R.W., op. cit., 1981, Chapter 8.

13 Restructuring of the United States steel industry requires new policies

Joel S. Hirschhorn

The problem

Since World War II the United States steel industry has declined in terms of economic and technical performance criteria, and domestic steelmaking capacity has recently been decreasing as imports increased and firms properly, but belatedly, closed down older inefficient plants. Creation and adoption of technological innovations have also lagged. But one bright spot in this otherwise dismal picture is that there has also been a significant restructuring of the industry which has instilled new vitality.

Steelmaking firms in the industry are not all alike. Dividing the industry up into three segments allows restructuring to be examined. While integrated steelmakers have received the most attention from policymakers and the media, and indeed have undergone serious declines in virtually all measures of industrial health, the non-integrated and alloy/speciality steelmakers have, for the most part, experienced much higher levels of profitability, growth and technological competitiveness. The decline in market share of the integrated companies is in contrast to the rapid growth of non-integrated firms, which will likely continue to provide a source of competition to domestic integrated producers perhaps more significant than foreign competition.

The present challenge to policymakers is to decide whether to let the US steel industry drift along as it has been under a set of Federal policies which have been of an *ad hoc*, uncoordinated nature and often inattentive to serious problems, or whether to make use of a number of recent major studies of the industry which offer many suggestions for a more comprehensive and future-oriented policy strategy for the industry. There appears to be a consensus that the US steel industry should, and can, exist in a more profitable, competitive and technologically innovative form than it presently does. Because steel is so critical for the functioning of society, the US steel industry, most studies conclude, should receive special policy attention and probably requires industry-specific policies for its survival and renewal.

The background

Revolutions are not always noticed while they are taking place. This is especially true when the shift in power is within an industry rather than an entire society, and when the profound change takes place over years rather than over days, weeks or months. Such is the case for the US steel industry. For the past ten years small steelmakers, often called mini-mills, midi-mills or market mills, have undergone phenomenal growth. Based on a strategy involving technology, marketing and management changes relative to past practices, these successful firms have captured a significant portion (about 15 per cent) of domestic production for steel products while making far higher profits than the well known large integrated companies which still account for the majority of steel production. That this restructuring of the American steel industry has been taking place during a period in which the poor performance of the large steelmakers has attracted considerable public and political attention, and the steel industry has been labelled sick, troubled and a loser, may explain, in part, why the revolution has gone so unnoticed.

The growth of small non-integrated steelmakers, also taking place in other nations, and other changes in the international steel industry, such as the growth of direct reduction, challenge government policymakers to make public policy that is keyed to the future rather than the past, and to fostering the growth of successful companies rather than preserving the existence of firms with poor performance through often costly distortions of the marketplace. This analysis is an attempt to develop a strategic plan for government policy formulation which, although done for the United States, may be useful for other nations also.

The structure of the discussion consists of four elements. First, important historical trends are summarised which describe the decay of the US steel industry viewed in aggregate as well as providing a brief description of the more recent restructuring of the industry explained in terms of three industry segments. Second, important and often controversial issues are discussed, which arise in any attempt to explain the historical decline of the US steel industry. Third, the present crisis facing the private and public sectors is scrutinised in terms of ideal solutions, emphasising long-term strategies and policies. Fourth, a number of major policy approaches now under consideration for dealing with the domestic steel industry in a specific manner are reviewed and compared, as well as the potential for utilising more broadly defined economic or industrial policies, perhaps in conjunction with sector-specific policies.

It would be easy to cite voluminous data about foreign and domestic steel industries, but since so many recent studies of the steel industry exist, there is no reason to do this here. Instead, a conceptual policy issue-oriented discussion will be emphasised. For those interested in more detailed and data-filled discussions two recent studies are recommended: one by the domestic steel industry's major trade association, the American Iron and Steel Institute, entitled *Steel at the Crossroads: The American Steel Industry in the 1980's*,[1] and the other by the Congressional Office of Technology Assessment entitled *Technology and Steel Industry Competitiveness*.[2] Unless noted otherwise, the OTA report is used as a primary source for factual material for this chapter, although the opinions expressed are totally the author's personal ones.

Historical trends

Production and trade

Up to and throughout World War II, the United States maintained an unapproachable lead in steel production, and its technology was considered first-rate. However, the post-war rebuilding stimulated the expansion of European and Japanese steel mills, and provided foreign producers with great competitive leverage. US firms did not build enough new plants or expand existing capacity sufficiently to capture a portion of the rapidly rising world demand for steel.

The dramatic decline in the growth rate of the US steel industry, compared to that of other countries, is revealed in world production figures. From 1956 to 1978, for example, the US share of total world output of steel dropped from 37 to 17.5 per cent and domestic production increased only by 10 per cent. During this period, Japan increased its production nearly tenfold. Japan and the European Economic Community (EEC) experienced a combined growth rate from 1950 to 1976 that was 10 times greater than that of the United States.

Steel exports from the United States have remained relatively constant during the past 30 years, even though worldwide exports increased more than tenfold during that time. In 1978, for example, the United States exported only 2.5 per cent of its total domestic raw steel production, while West Germany exported 53.7 per cent; Japan 36.8 per cent; Italy 37.6 per cent; and the United Kingdom, 21.5 per cent.

Many foreign industries built steelmaking capacity with the export market in mind, because their capacities far exceed the volumes needed to satisfy their domestic needs.

Steel imports into the United States since the late 1950s have grown at the rate of 10 per cent per year. The average for the past decade is approximately 15 per cent of domestic consumption. The increasing gap between domestic steel exports and imports has had a striking negative effect on the US trade balance. Steel imports exceeded exports in dollar value for the first time during the late 1940s, and in volume during the late 1950s. Since that time, imports have captured much of the growth in domestic steel consumption. Steel trade patterns have led to a very high annual trade deficit, second only to petroleum as a source of trade deficit, in terms of commodities. Although exports of ferrous scrap and coking coal have sometimes been high, they reduce this trade deficit by a relatively small amount, and large amounts of imported iron ore and, occasionally, high levels of imported coke have contributed to the trade deficit associated with the steel industry.

Profitability and investment

During the past several decades the US steel industry has had a far better financial performance than major foreign steel industries, according to US standards and measures. Only the smaller Canadian steel industry has consistently outperformed the US steel industry. However, international profitability comparisons should be made with caution since foreign government ownership and direct and indirect support by governments, all substantial for many foreign steel industries, make measures of profitability used for private domestic firms difficult to apply to all foreign firms. The interest paid to banks by the highly debt-leveraged Japanese steel firms makes comparisons to US firms difficult.

In relation to other US industries, however, the situation is quite different. In only four years (1955–57 and 1974) during the past 25 did profitabilily (after-tax profits as a percentage of stockholder equity) of the domestic steel industry exceed the average for all domestic manufacturing firms. Steel industry profitability has been lower than the prime interest rate for five years in the period 1967–78. The real rate of industry net income has declined to very low levels, finally becoming negative during the past few years as inflation rates exceeded steel industry profit margins.

With regard to capital use, dividend payments have been surprisingly stable, however, even in years of very low profitability. In addition, capital expenditures as a percentage of net internally generated cash funds have been relatively high. The industry's long-term debt increased tenfold between 1950 and 1978. In the same period, stockholder's equity increased only by a factor of three. As a result, the debt-to-equity ratio increased from 11.2 per cent in 1950 to 44.0 per cent in 1978.

The relative profitability of the US steel industry compared to foreign steel industries, the large size of the domestic market, the increasing costs of transporting foreign steel to the US, the existence of company stocks that are undervalued relative to book value, and, at times, exchange rates favourable to foreigners have made investments by foreign firms in the US steel industry attractive. However, the increased foreign investment in recent years has taken place mostly in the small steel firms and steel distributors rather than the larger and least profitable integrated companies. About ten small domestic steelmakers are owned partially or entirely by foreign interests.

Steelmaking costs

Through the 1950s the US steel industry could easily claim to be the world's low-cost steel producer. In the 1960s, however, several European countries and Japan became lower-cost producers of steel. In more recent times, production costs have become more volatile and difficult to assess, in part because of the role of foreign governments, in, for instance, eliminating debts. Since about 1973, the Japanese may have lost some of their cost advantage relative to the US, and European producers lost their advantage altogether. But recent strengthening of the dollar has reversed this effect. Compared to other major steelmaking nations, US raw material and employment costs per ton of steel are somewhat high and capital costs somewhat low. Several Third World, developing nations have relatively low production cost steelmakers, but their capital costs are often high.

Widely fluctuating exchange rates and substantially different utilisation rates among nations continue to make international cost comparisons difficult. Nevertheless, on the whole, the US steel industry is cost-competitive in most domestic markets, and for some types of steel even cost-competitive in international markets where, of course, other factors such as exchange rates may still preclude market competitiveness. It must be noted that the costs of exporting, including transportation costs, warehousing, sales and marketing, are also relevant when making international cost comparisons. This is why, today, lower-cost foreign steelmakers may not be competitive in the domestic market, especially for inland markets inaccessible to low-cost water transportation. Moreover, a relatively low rate of capacity utilisation substantially increases production costs, especially for large economy of scale integrated plants such as those in Japan, which have been operating at about 70 per cent of capacity for the past several years. But even at that level they can be profitable. US utilisation rates have been higher. It cannot be emphasised enough, however, that cost-competitiveness is usually calculated on an aggregate basis for the entire steel industry. Production costs vary substantially among steelmakers because of the wide differences in age and efficiency of facilities, plant management practices, local market conditions, product mixes and variations in regional costs of basic inputs.

Much attention has been given to wages and labour costs in the domestic steel industry and to problems with productivity improvements. Abroad, steel industry wages are often higher than the all-manufacturing average, but less so than in the US. In the United States, the gap between steel industry and manufacturing industry average wages narrowed during the late 1960s in response to increased import competition and reduced profitability in the steel industry; but in the 1970s, and particularly since 1974, the lead held by steel industry wages again increased significantly. The 1974 steel labour settlement which took place in a booming market has been criticised by many for being too generous, including its cost of living increase clause. Management was obviously eager to avoid any disruption of production. Since that time steel industry wages have been much higher than the average for domestic manufacturing, and hourly employment costs in the steel industry have increased by 10 to 15 per cent since 1960, on an annual basis, with higher than average increases during recent years.

The increases in wages have led to substantial increases in the contribution of labour toward the total costs of making steel because such increases in hourly costs have been offset by labour productivity gains to only a small degree. For the period 1960 to 1978, the average rate of change in hourly compensation was nearly 11 per cent annually in terms of actual dollars and 2.6 per cent in terms of real dollars, while output per employee–hour (productivity) increased at an average rate of 1.9 per cent annually. US productivity is comparable to that in Japan, and better than that in Europe. Productivity, however, remains at a high level for domestic steelworkers. Moreover, it must be noted that productivity is also a strong function of capacity utilisation rate, which is very cyclic, and of the character of technology chosen by management. Labour productivity is greatest in the non-integrated mills, where it is also increasing significantly.

While low hourly wages in some developing nations are an advantage which is often combined with new plants, steel industries in industrialised nations have, during the past decade, experienced increases in labour costs that were greater than those in the United States because of currency changes and because of wage increases that exceeded those in the United States. From 1969 to 1978 West German and Japanese employment costs increased 345 and 299 per cent, respectively, compared to 117 per cent in the United States; and these two nations are generally considered to be the most effective and technologically modern in the world. Nevertheless, in 1978, US hourly costs were still 30 per cent higher than West German costs and 40 per cent higher than Japanese costs, but annual employment cost increases in local currencies were much lower than in dollars.

Technology, innovation and research and development

The two most striking aspects of technology that have changed during the past several decades for the US steel industry are an increase in the age of facilities and a reduction in the amount of advanced forms of technology in use as compared to foreign steel industries. Moreover, there is ample evidence to support the claim that the US steel industry has not been an innovator in steelmaking technology in the sense that it has not played a key role in developing and introducing into commercial use important new technologies, although, in fairness, it has done better in the steel product development area than in process technology. The generally held view is that the Japanese steel industry is the world's premier steel industry in terms of technology, even though it took many research results and new technologies from other nations, including the United States. Given large amounts of new steelmaking capacity, the Japanese, like some other nations, have been able to build on a base of foreign technology by constantly improving and innovating. Japan has become a major source of new steel technology for the entire world and it has shifted its interest to exporting technology, rather than steel products, to nations building their own steelmaking capacity.

Because the US steel industry has not built major new integrated plants to the extent other nations have, the average ages of different types of equipment are relatively high. Estimates have indicated that 20 to 25 per cent of the integrated steelmaking facilities in this country are technologically outmoded or obsolete, a much higher figure than for most industries. In several important categories of equipment, such as plate mills, hot strip mills and cold strip mills, the average age is 20 years or more. Important recent advances in computer control and instrumentation which greatly influence quality control and productivity have not been widely adopted by the industry.

The record with regard to several important changes in technology varies. For example, the US steel industry is at the forefront of the world industries for adoption of electric furnace steelmaking, up from 10 per cent of steel production in 1964 to nearly 30 per cent today. However, unlike most major steel industries, the domestic industry still

211

has a significant amount of steel made in open hearth furnaces, while, with the exception of the USSR, most other countries have few or none of these furnaces in use any more. This is related to a low adoption rate for basic oxygen steelmaking.

More importantly, the United States has fallen behind in the adoption of continuous casting technology, which has spurred cost savings, energy savings, improvements in productivity and reduced environmental pollution by all the world's other steel industries. The use of continuous casting is also a major way to increase the efficiency or yield of steelmaking operations that is, to increase the amount of finished steel products made from a given amount of raw steel. Japan now makes about 60 per cent of its steel by continuous casting, the European Community about one-third, but the United States makes only about 20 per cent of its steel this way. And the Japanese are aggressively pursuing a goal of making 80 to 90 per cent of their steel in this manner.

While several nations, notably the Japanese, have very large, modern integrated steelworks employing all the newest technology, the United States has built only one major new steelworks in the past two decades, and even this has probably not matched the improvements made through incremental technological innovation in which the Japanese have become proficient.

The innovation process covers the spectrum from research and development through pilot testing and demonstration, and finally to introduction into the marketplace. The technological decline of the domestic steel industry includes a decline in research and development (R & D). R & D expenditures by the US steel industry, as a percentage of sales, have declined in recent years and are lower than in most other basic industries in the United States. For example, the level of R & D spending by steel is about half that of the non-ferrous industry and about one-seventh that of the chemicals industry. The industry's basic research effort is particularly small. Steel industry R & D has very little Federal support compared to other industries and is complemented by only a limited amount of steel R & D carried out by the government and the universities.

Foreign steel R & D, on the other hand, is generally in a more vigorous state because of large budgets, stronger government support, more positive attitudes towards future prospects for innovations in steel-making and more aggressive approaches towards the export of steel technology. A number of foreign steel industries benefit also from greater cooperative efforts among firms and between firms, universities and government facilities.

Industry restructuring

The US steel industry is no longer the homogeneous industry it once was. During the past decade the industry has been undergoing permanent and important changes in the character and competitive positions of its constituent firms. Important aspects of this restructuring include changes in technology, product mixes, geographical patterns of company locations, costs of entry into the industry, and raw material use. The competition among three distinct segments of the industry may be more important than the competition from foreign steelmakers. The three industry segments are integrated, non-integrated and alloy/speciality steelmakers.

Integrated steelmakers (such as United States Steel Corp. and Bethlehem Steel Corp.) start with iron ore and coal to make iron and coke, and go on to make a large variety of steel products. These plants, based on coke ovens and blast furnaces, are what most people think of when steel plants are mentioned. There are 20 such companies in the United States, with an average of about 2.5 plants per firm. Generally, the plants make from two to six million metric tons of steel annually. Most steel made in this country (and the world) is made in integrated companies. The decline of the integrated segment of the United States steel industry is, in fact, bleaker than the preceding discussions indicate, since the historical aspects of the two other industry segments are quite positive in many respects. The decline of integrated firms, particularly in terms of financial performance, has led some companies into considerable diversification out of steelmaking. Hall has presented an analysis of domestic integrated steel companies[3] and has concluded that the diversification strategy has generally been unsuccessful for most steel companies because their efforts have been too small and managed in too conservative a fashion. The exception is Armco; they continue to reduce the contribution of steelmaking (more so for carbon than alloy steels) to their corporate efforts and generally have the best financial performance of major integrated producers. Hall also notes that in steel and other basic manufacturing industries it has been possible for firms to be quite successful by staying in their original line of business and becoming a low-cost producer. This has been the case for Inland Steel, whose financial performance in recent years has been relatively good, even compared to other types of companies perceived to be profitable and successful. Thus, the loss of market share by the domestic steel industry in recent years is accounted for by the integrated segment and, more importantly, by the least successful integrated companies, who have reduced their steel making capacity and may continue to do so.

Non-integrated steelmakers, often called mini-mills, midi-mills or market mills, start with ferrous scrap and make simpler commodity items such as wire products and reinforcing bar. There are about 50 such companies in the United States; some of the more successful and better known firms are Nucor, Florida Steel and Korf Steel. Non-integrated steel-makers have increased their share of the domestic market from under 3 per cent in 1968 to about 15 per cent today. Much of the increase has come from penetration of markets formerly held by integrated producers and, in some cases, by being more competitive than steel imports. Many of the better performing non-integrated firms are the world's lowest-cost producers of steel. Non-integrated producers have capitalised on locally available domestic ferrous scrap and are almost dependent on it as their source of iron, whereas integrated producers use a small percentage of purchased scrap in their operations. The energy contained in such scrap enables scrap-based steelmakers to keep costs down. They use scrap from domestic and imported steel made in energy-intensive integrated steelmaking plants. Moreover, non-integrated producers rely mostly on local markets served with low-cost transportation (although some are exporting and some are becoming more regional in character), generally use non-union labour because of their concentration in non-unionised regions of the country in the 'sunbelt', and have capital costs per ton of annual capacity that are about 10 to 20 per cent of those of new integrated plants.

Technologically, the non-integrated steelmakers make use, for the most part, of a highly efficient combination of electric steelmaking furnaces, continuous casting and a relatively narrow product line for any particular plant. They have spearheaded the development of a number of technological advances even though they have modest formal R & D programmes. They are quick to adopt advances made available by equipment manufacturers and often do their own engineering and construction to keep capital costs down. Although many observers, particularly those from the integrated steel producers, point to the potential problems with the quality and availability of ferrous scrap, and possibly electricity, the non-integrated firms continue to expand their capacity. They do this more by building new plants than by expanding capacity at existing plants, although the latter is feasible at relatively low cost and is going on at some plants. The optimum size of these non-integrated plants appears to be about 500,000 tonnes of steelmaking capacity annually and is increasing. Several companies, however, now have total capacities in the million tonne or more range. Many of these firms are expanding their relatively simple and narrow product mixes to include more costly, sophisticated and higher quality steel products. Several major non-steel US corporations have invested in this industry segment.

Alloy/speciality steelmakers, such as Carpenter Technology, Lukens Steel, Cyclops and Washington Steel, start with scrap or iron ore and make higher-priced, more technology-intensive steel products, such as stainless steels, tool steels and high alloy steels for aerospace and other demanding applications. From an international perspective, alloy/speciality steelmakers possess the best technology, have pioneered many manufacturing and product innovations, and have low production costs. The use of their steel products has been growing at a much faster rate than for carbon steels. Some of these firms are beginning to capitalise on their ability to export. While these producers only account for a few per cent of the domestic production of steel, the dollar value of their products is several times greater. Like integrated steelmakers, the alloy/speciality producers have often faced stiff and probably unfairly traded (below cost) imports; nevertheless they have, for the most part, persevered with emphasis on new technology and capital investment to reduce costs, with the weakest product line being tool steels.

Some comparative data on the three industry segments of the US steel industry are given in Table 13.1 for 1978. While there is significant variation among firms in all three segments, there is little debate that the non-integrated and alloy/speciality segments exhibit far better financial and technical performances than the integrated segments. Similarly impressive financial comparative data for the past ten years are given in Table 13.2.

Table 13.1

Summary data on steel industry segments, 1978

Characteristic	Integrated	Non Integrated	Alloy/ Speciality
Steel shipments			
1,000 tonnes	75,522	11,291	2,014
per cent	85	13	2
Return on investment	6.9	12.3	11.1
Steel only- Pre-tax profit, $/tonne shipped	$9.60	$31.60	$81.33
Employment costs $/tonne shipped	$209	$138	$341
Per cent steel continuously cast	11.0	51.5	16.5

Source: Office of Technology Assessment, *Technology and Steel Industry Competitiveness* 1980.
(note: tonnages are metric)

Table 13.2

Comparative financial data for industry segments

Average	Non-integrated + alloy/speciality Composite (14 firms)	Integrated (six largest)	Dow Jones Industrials (30 stocks)	Shearson Manufacturing Composite (180 firms)
Pre-tax profit margin %				
1975-1979	9.80	3.18	12.26	9.64
1970-1979	9.96	4.90	13.25	9.75
Net return on assets (%)				
1975-1979	7.54	3.06	6.92	7.78
1970-1979	7.17	3.74	6.97	7.52
Net return on equity (%)				
1975-1979	12.86	5.78	14.02	15.32
1970-1979	12.19	6.87	13.47	14.39

Source: Joseph C. Wyman, *Steel Mini-mills – An Investment Opportunity,* Shearson Loeb Rhoades Inc., New York, 20 Nov., 1980.

Table 13.3

Summary generalisations on causes of historical trends

Historical Trend	Explanatory power of cause: + + + Very important + + important + minor and indirect		
	Foreign actions & policies	Steel industry management decisions	Federal policies
Production & trade	+ + +	+	+ +
Profitability & investment	+	+ + +	+ + +
Steelmaking costs (incl. regulatory costs)	+	+ + +	+
Technology, innovation, R & D	+	+ + +	+
Industry restructuring	+	+ + +	+

216

Understanding the past: causes and issues

To explain the previously described historical trends for the United States steel industry one must invoke many complex and often inter-related factors that frequently rest on subjective judgements rather than unequivocal facts. To simplify and organise this discussion, I will use three basic types of explanations for each of the five areas of historical trends discussed previously. These three types of explanations are foreign actions and policies, steel industry management decisions, and Federal policies. In this manner it is possible for government policymakers to understand those factors over which they have little control, of which they must be aware and may influence indirectly, and over which they have direct control, respectively.

As an introduction and overview of this discussion I have summarised the major impacts of these three types of explanations on the five areas of historical trends in Table 13.3. This is useful if one asks the question: how important, for example, have foreign actions and policies been to the decline in profitability and investment in the domestic steel industry? And how do foreign activity and policies compare to Federal policies or management decisions in explaining this decline? The answer, in this particular case, is that both management decisions and Federal policies have been very important, while foreign activity and policies explain this decline in only a minor and indirect way.

I have also listed the five historical trends in order of decreasing importance from the perspective of the domestic steel industry as manifested by their emphasis in recent years when they have brought their case to the public and political arenas. The industry (mostly through the American Iron and Steel Institute) generally emphasises trade and import issues first, then its problems with declining profitability and ability to re-invest in steelmaking and then costs, including the burden of regulatory costs. The areas of technology, innovation and R & D and industry restructuring, for the most part, receive little attention from the industry. In examining Table 13.3, therefore, it is important to note that as one attempts to understand the causal factors for the five historical trends, going from top to bottom (from most to least important from the perspective of the industry), there is a shift from foreign actions and policies and Federal policies to management decisions. In other words, the industry emphasises those areas of difficulty for which it can externalise the causes. This is perfectly reasonable from the industry's perspective, but, from a public policy viewpoint, the consequence has been, in part due to the successful lobbying efforts of the industry, that Federal policies have been much more responsive to problems facing the industry which have remedies

217

outside the industry than to those problems or trends (positive in the case of industry restructuring) that are best understood in terms of the decisions made by industry management. The lack of a trade association representing the interests of the non-integrated firms has been noteworthy.

Production and trade

Why has the US steel industry, since the 1950s, experienced a substantial decrease in steel exports, a loss of domestic market share due to rising imports, and a loss in position in terms of fraction of total worldwide steel production and capacity? The answer lies in a combination of newly built steel industries in European nations whose steel industries were devastated in World War II and in Japan and the Third World developing nations meeting both domestic needs and capturing export markets.

Foreign steelmaking industries are generally viewed by their governments as strategically crucial to satisfy the immediate and long-term growth needs of industrial infrastructures. Moreover, steel exports have been viewed as a means to obtain foreign currencies, gain positive trade balances, maintain high levels of employment in domestic steel industries, particularly in times of slow domestic demand, facilitate the construction of large economy scale steelworks, and in some cases to make greater use of domestic raw materials such as iron ore, coal or natural gas. Major steel industries have also been used to spur the development of other industries that use steel, such as shipbuilding and automobile manufacturing, to develop a base of skilled workers and technical professionals, to develop technology for export and, in a more vague way, to gain national prestige and influence. Rarely have foreign steel industries been required or expected by owners to make profits in the sense that US steelmakers are. And for the most part, foreign steel industries have not been profitable in terms of return on investment or equity, but they have served the goals noted above, even to the extent that they have often cost considerable money to their governments who, to a large extent, own or subsidise their steel industries.

Has the United States been singled out as an export market by foreign steel industries? Yes, to the extent that US markets represent the single largest and one of the most open and highest priced markets in the world for steel products. In some ways there has been considerable restraint, particularly by the Japanese, not to fully utilise available capacity at times and deluge the American market with competitively priced steel. In summary, it is difficult to 'blame' foreign steelmakers for their expansionist policies, for defining the value of steel industries in ways other than the United States has, and for taking advantage of legitimate export market opportunities. With the growth of steel industries in the Third World developing nations, formerly lucrative export markets, there is increasing interest in exporting to the United States. Balancing this, however, is the inability of many nations (such as the United Kingdom) to continue to sustain large economic costs for maintaining steel industries with far more capacity and often with technologically inferior facilities than can be justified on the basis of domestic needs or other social benefits, such as maintainig employment.

This brings us to the second most important cause of the problem of production and trade – Federal policies. The issue is: to what extent have Federal policies been responsible for increasing imports and, less importantly, for decreasing exports? With regard to imports, the industry's argument is that Federal policies have been ineffective in preventing unfairly traded steel imports. The Federal Government has been responsive to steel industry arguments that steel imports have been unfairly traded by, for example, instituting the trigger price mechanism to detect unfairly priced steel imports (on the assumption that any foreign steel priced below Japanese steel is suspect since Japan is the low-cost producer) and imposing quota restrictions in some cases. Reasonable evidence has been found to indicate that some products from some foreign industries during limited periods have been traded unfairly in the sense that they have been priced below costs or below market prices in the market of the foreign industry, but not for the majority of steel imports over past years. In particular, the bulk of the evidence suggests that Japanese imports have not been traded unfairly for the most part. However, the evidence (substantial firm losses) does indicate that some European steel industries have dumped steel in the United States at times. The Federal Government appears to have been relatively tolerant of European dumping of steel because of other international trade considerations, such as the positive net trade from the United States to Europe, and for political considerations. Considering the excess capacity of the Japanese steel industry that has often existed, however, it can be argued that, even if unfairly traded European steel from some European sources had been more effectively eliminated, the Japanese could have increased their penetration of the United States market.

With regard to exports of steel from the United States, there appears to be merit in the position that the government could have played a stronger role in promoting exports of those steel products for which the country is especially competitive. In particular, the inhibiting effect of anti-trust laws on the formation of joint trading ventures, used effectively by other nations, appears valid. Finally, for both exports and the inability of domestic steelmakers to compete effectively in the domestic market with some foreign steel, Wyman[4] has presented a convincing argument that the negative impact of Federal monetary policies, linked to overly rapid credit generation, growth of United States budget deficits and an overvalued dollar, has put domestic steelmakers at a disadvantage. With regard to imports, however, during 1974-78, when the value of the dollar declined substantially against currencies of major trading partners, there was a significant rise in imports. This can be seen as resulting from increased demand for imports leading to weakness of the dollar or, as the industry sees it, increasing unfair trade of steel.

To what extent have steel industry management decisions been the cause of the production, capacity and trade decline? Considering that domestic steelworks survived World War II without damage, that the United States generally wanted both its allies and former enemies to rebuild their industrial infrastructures and that it has been in the nation's best interests to maintain and promote free international trade, especially for United States exports, it is difficult to attribute much responsibility to steel industry management. At the most, it can be argued that management generally pursued relatively conservative strategies after World War II. They did not, but perhaps could not, respond to demographic shifts in the United States which moved steel demand away from the traditional north-east to the west and south and created opportunities for import penetration. The settlements made with the steelworkers union have been criticised, but previous strikes did create opportunities for import penetration.

Users of imported steel and foreign exporters of steel have often argued that domestic steelmakers have not reduced prices sufficiently during periods of worldwide sluggish demand to remain competitive. But since the domestic steel industry is far more profit-oriented and dependent on internally generated cash to sustain itself, this pricing strategy makes sense. After all, domestic steelmakers are not able to turn to the Federal Government for financial assistance in the same manner that many foreign steel industries can, particularly in some European countries.

Profitability and investment

Both steel industry management decisions and Federal policies are the chief causes of the decline in profitability and investment in the domestic steel industry. Management can be criticised for:

a)   poor long-range strategic planning which would have allowed faster and more appropriate responses to changing market conditions, available new technologies to reduce costs, major changes in the costs of inputs, the availability of low-cost ferrous scrap, and a host of social and political changes such as the environmental and worker health and safety movements;

b) failing to rationalise steel plants to obtain a better and more efficient match between technology and a limited product mix, rather than compete across a broad mix in the same market areas;

c) developing a strategy based on attracting investors through high dividends rather than through appreciation of stock value without making equity investments attractive;

d) failing to emphasise cost reductions through technological innovation;

e) failing to take measures to reduce the capital costs of steel making facilities, such as doing more in-house engineering and construction of equipment, and thus obtain more productivity from their investment capital;

f) poor labour relations;

g) poorly planned and executed diversification investments. Non unionised non-integrated firms have demonstrated how union work rules can be quite detrimental.

The industry correctly has placed considerable blame on Federal policies for its problems with declining profits and capital; however, it must also be noted that a few steelmakers operating within the same policy environment have performed markedly better than the industry average because of different management decisions, even without pursuing a diversification strategy. The Federal influence on profits has mostly been caused by limitations placed on steel prices, often through informal 'jawboning' at times when steelmaking costs were rising sharply. The steel industry has been selected as an industry for which price controls are especially important due to the basic nature of steel and the influence on other domestic industries. However, while this logic may have been true at one time, it is no longer compelling. The intensity of steel consumption in the United States has declined and the contribution of steel costs to the total costs of products has decreased for many large-scale applications such as automobiles and consumer appliances.

Domestic steelmakers also argue that Federal trade policies have contributed to the profit problem by allowing unfairly priced imports to force domestic prices down. And to a limited extent, this has some merit. But the recent trigger price mechanism has provided a means to increase domestic prices, although not sufficiently to bring about any significant increase in profitability, nor has the system stemmed imports and, therefore, it has not helped to increase capacity utilisation.

On the investment side, capital formation has been hurt by the overly long depreciation times. A number of nations have used short depreciation times to spur steel industry investment, particularly Canada.

The industry has maintained that Federal regulations in the environmental area have led to high capital costs which have limited investment in steelmaking. While these costs have certainly not been trivial, neither can they be considered to have been a prime determinant of the decline in the industry's steelmaking investments. The data in Figure 13.1 indicate that the changes in total and steelmaking capital investments and in long-term debt do not mirror the changes in capital investment for pollution control. In terms of a percentage of total capital spending, the pollution abatement costs for the steel industry, although high, have been only slightly greater than for some other domestic industries, such as chemicals and petroleum refining, and similar to those of the Japanese steel industry. The steel industry has often made considerably more note of its projections of pollution abatement capital costs than its actual costs, and retrospective analyses have revealed that the actual costs have been markedly less than previous projections. For example, an analysis performed by A.D. Little in 1975 for the industry for the period 1975–77 estimated capital costs of $(1972) 2.92 billion, which would equal 25 per cent of total industry capital spending, but the actual costs were $(1972) 0.9 billion and 13.5 per cent of the total.

The OTA study revealed that close to half of the industry's past pollution abatement capital costs have been met through industrial development bond financing (borrowing through governmental units at low interest rates). This mitigates the cash flow problem and provides a low cost for borrowing. Thus, Federal policy has also reduced the burden of regulatory costs.

However, there are a number of other types of regulatory costs which the steel industry has had to face, including the uncertainty of future Occupational Safety and Health Administration (OSHA) standards and regulations.

Another example of Federal policy which has contributed to the capital formation problem is the refusal to grant special energy-saving investment tax credits for continuous casting facilities. Narrow bureaucratic interpretations appear to be the cause.

To the extent that there has been below-cost pricing by some foreign steel industries, foreign actions and policies may be blamed for a relatively small role in the decline of profitability and investment in the domestic steel industry.

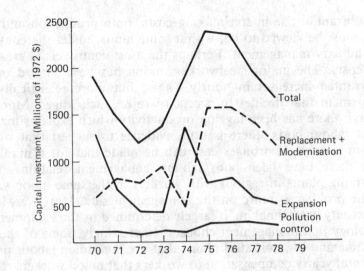

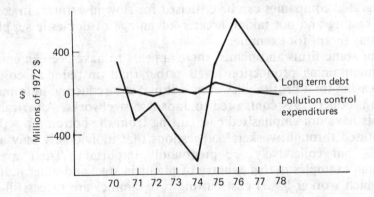

Figure 13.1
TOP: Capital investment in constant dollars from 1970 to 1978 for domestic steel industry in three categories plus total
BOTTOM: Long-term debt and pollution control capital costs from 1970-1978 in constant dollars for domestic steel industry.

Source: P. Meier and S. Brown, *The Impact of Energy and Environmental Legislation on Industry: Problems of Regulatory Cost Assessment,* 89th National Meeting of American Institute of Chemical Engineers Portland, Oregon, August 1980

The substantial rise in steelmaking costs, both production and other costs, must be viewed to be, within some limits, under the control of steel industry management. Perhaps the most controversial area is labour costs. The major steelworkers union has been singled out for unwarranted increases in hourly wages, but more realistically it is management that decides to accept or reject such wages. Moreover, although there has been lagging productivity which explains increases in unit labour costs, there is little evidence to suggest that workers perform poorly. A stronger case can be made that for a number of plants there have been poor labour–management relations. Better performing plants in regions with workers in the same union suggest that the problem is more on the management side. Moreover, labour productivity in steelmaking is largely determined by the equipment and technology that management chooses. Interestingly, some of the most profitable and low-cost steel plants that use non-union labour pay the same total yearly compensation to workers that union workers receive, but the management style, work rules and facilities are different.

Apart from labour costs, technology also determines the costs of other factors of production such as energy and raw materials. Management can be criticised for inappropriate selection of new technology. Integrated companies can be criticised for slow investment in continuous casting and not taking greater advantage of domestic supplies of ferrous scrap, for example.

For some firms and plants, there appears to have been an emphasis on maintaining production levels rather than on reducing costs and increasing total profitability even if it meant a reduction in tonnage of steel produced. As contrasted to Japanese steelworks, American steel plants have not emphasised cost cutting from the bottom up – that is, instituted through workers' suggestions that individually may appear small, but collectively are profoundly important. There are now enough examples in this country of plants in steel and other industries in which worker-based cost cutting programmes are successful to indicate that such an approach can and could have been used in the domestic steel industry pervasively.

Management can also be cited for placing more emphasis on technology and the use of technical personnel for product development rather than on steelmaking processes. Process technology appears to have been viewed in a static manner. The relationships between R & D personnel and plant management and personnel have not generally been structured to promote steady cost cutting through better use of technology and the use of the plant itself as, in limited ways, a laboratory for process improvement. One can see these preferred approaches in other domestic industries, such as chemicals, and in foreign steel industries.

To what extent can Federal policies be accused of increasing costs? The industry has placed considerable emphasis on the high costs of Federal regulations, particularly those in the environmental and worker health and safety areas. Pollution control costs have indeed been high for the domestic steel industry, but they have also been high for other domestic industries, such as non-ferrous and cement, for example, in terms of energy requirements to run pollution control equipment. Myers and Nakamura[5] find that the energy requirement as a percentage of gross energy consumption for pollution control in 1976 was 2.1 per cent at most for the steel industry, 2 per cent at most for the cement industry and 2.8 per cent for the aluminium industry. The Japanese and a good part of the European steel industries has similar pollution abatement costs. The contribution of pollution abatement costs to production costs in the US steel industry have been estimated by industry to be over 6 per cent, while EPA has estimated it to be about 5 per cent, including capital, operating and maintenance costs for air and water pollution abatement facilities. Considering the low profit margin for the industry, these costs are significant. To the extent that other domestic industries and some foreign steel industries do not incur such costs, the domestic steel industry is at a competitive disadvantage.

A controversial issue is the extent to which Federal policies have contributed to erratic and sometimes high prices for ferrous scrap, an important resource for the steel industry. The scrap industry maintains that a free market should exist in which foreign purchasers of scrap should have complete access to US scrap, the world's largest supply. The steel industry's viewpoint is that export of ferrous scrap, which has been substantial (about 10 million tons per year), should be limited in ways similar to those used by other industrialised nations. They argue that ferrous scrap is a valuable national resource which is highly energy-intensive, a means to ensure greater competitiveness for the domestic steel industry, and that it is used by government-subsidised foreign steel industries to make steel which is traded unfairly in our market. There is no doubt that exports of domestic scrap are substantial, but the link between export levels and domestic prices is difficult to quantify because, in part, prices vary considerably among different geographical regions. Nevertheless, there is some evidence to indicate that scrap exports do influence domestic prices (see Figure 13.2). The government has moved to a limited degree to at least monitor scrap exports in order to assess unusual trade that might be detrimental to the United States.

Offsetting the above considerations is the fact, often noted by foreign steel industries, that the domestic steel industry has benefited in the past by artificially low prices for energy because of government price controls.

To argue that foreign actions and policies have been responsible for increases in the domestic costs of making steel is not particularly persuasive. The US steel industry does import considerable iron ore, but there is no indication that unfair actions have increased its cost. For relatively brief periods the domestic steel industry has had to import coke to sustain integrated steelmaking, but here too there is little indication that anything other than market conditions influenced costs. Finally, the domestic steel industry is dependent on a number of foreign sources for alloying elements, such as chromium. In this case there is evidence to suggest that at certain times foreign sources of domestically unavailable metals and minerals have taken advantage of their market position. However, the contribution to the bulk of the domestic steel industry has not been large.

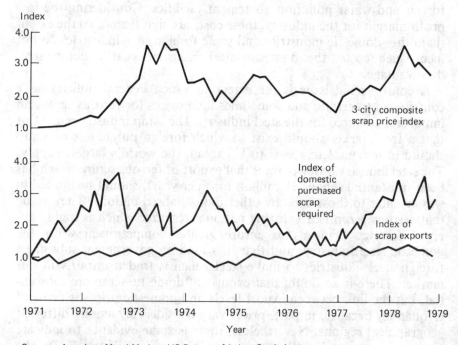

Source. American Metal Market, US Bureau of Labor Statistics

Figure 13.2: Scrap price and export levels; domestic purchased scrap requirements, 1971-79

The general decline in the scientific and technological base of the industry must be attributed primarily to management decisions and attitudes. The industry's position, however, is that insufficient financial resources prevented those existing technological opportunities from being fully exploited. To some extent, as already discussed, factors outside the industry have contributed to declining financial resources. But other facts shift the chief responsibility to industry management. If the industry had greater financial resources the evidence suggests that their emphasis would not have been on technology.

Here are two examples of such evidence. First, with regard to development and adoption of new technology as well as investment in R & D, it is possible to find companies within the domestic steel industry who performed in a manner counter to the general trend of the industry. What differed among companies were attitudes toward risk taking and the use of technology as a problem solving tool. Management often used available capital for technology rather than for modernisation or expansion based on older, well proven technology, or for diversification, or for dividends. Recent history for the nation's two largest steelmakers, Bethlehem Steel in 1977 and US Steel in 1979, has shown that closing of inefficient plants can have a positive influence on company profitability. While such plant closings have gone on, other companies have opened new plants using more efficient technology.

Second, to cite the report by the American Iron and Steel Institute in 1980, a major study and policy statement by the industry (albeit mainly representing integrated steelmakers), the executive summary contains no direct reference to technology and innovation, nor R & D. Some attention is given to these topics in two appendices.

The state of management has been well described by one of the industry's own R & D executives in an unusually candid statement:

> There is a trend toward more defense type research ... more time being spent on shorter range projects and projects designed to meet government mandates and regulations, and less time being spent on the kinds of long term, high risk, innovative projects which will lead to the new ways of making steel in the future. Part of the problem is that what we are doing with this money is not what everybody would call research and development ... but is pointed more toward short term objectives for a variety of reasons and not so much on the real innovative work and the fundamental re-

search work that you might define as research and development.

In addition to the nature of R & D changing in the industry, the amount of resources devoted to R & D declined. For example, in the 1960s 0.7 per cent of sales was spent on R & D, but by the late 1970s the level was only 0.5 per cent, one of the lowest levels for any domestic industry. Historical data suggest that the industry viewed R & D as the lowest priority discretionary spending of funds. In recent years the sum of money spent on diversification out of steelmaking and for dividends was about four times greater than the sum spent on R & D. The OTA study found a statistically significant relationship, with higher levels of pollution abatement capital spending leading to lower levels of R & D spending.

Steel industry management seems to have convinced itself that economic arguments favour trading leadership in innovation for less risky follow-the-leader adoption of well proven technology. They do not appear to follow a strategy of gaining technological advantage. Instead they seek, at best, parity. But while they argue that there is no knowledge gap between the domestic and foreign steel industries, because they have at least some of the latest technology in place, they miss the intrinsic advantages possessed by true innovators as well as the relationship between the source of innovation and the nature of the technology created.

US steelmakers have shown an increasing proclivity towards buying new technology from abroad, but whatever new technology is purchased from foreign sources still leaves the purchaser one step behind the originator. By the time all is learned about the innovation, the foreign source is well on its way to exploiting the next one by moving up on the learning curve for the generic technology. Knowledge about innovations is not equivalent to innovating. It takes years for new steelmaking facilities to be built, and those who innovate tend to stay ahead of their competitors, even those who purchase 'new' technology. Moreover, domestic steelmakers undervalue the importance of developing unique process technologies which are responsive to domestic resource opportunities, regulatory requirements, existing plant facilities and layouts, and market opportunities. Domestic steelmakers often spend considerable sums to purchase foreign technology, and then still more money to adapt the technology to domestic needs and constraints. Thus management's strategy of buying rather than developing technology is not necessarily optimal from either an economic or a technological viewpoint. It should be emphasised that development of process innovations requires considerable risk taking by management, since very expensive pilot and demonstration testing of new process technologies is required to evaluate economic and technological costs and benefits. These risks and costs have not won out, in most cases, against the apparent safety and costs of purchasing innovative technology from foreign sources, and neither has any competitive advantage been gained.

To what extent have Federal policies contributed to the scientific and technological decline of the domestic steel industry? The most important area is direct Federal funding of steel-related R & D. Historical data clearly show that for the past several decades Federal support for the ferrous area has been extremely small compared to other industrial areas. However, there is some correlation between the level of Federal support and the level of R & D spending by industry itself. Thus, the argument can be made that the Federal role has been one of partnership rather than unusual subsidies for selected industries.

The steel industry has also suggested that anti-trust policies have reduced the opportunities for joint programmes in high-cost process technology work. However, this has not appeared to have affected other basic industries, such as the chemicals industry.

To attribute some causality to foreign actions and policies for the decline in the scientific and technological base of the domestic steel industry is not easy. Neither the use of available information from the United States, nor the sale of technology to the United States can be viewed as extraordinary or unreasonable. In a number of nations the role of government in fostering both basic science and technology of use to the steel industry is very strong. Comparatively, therefore, the US steel industry has been at a disadvantage, but balancing this is the greater profitability it has exhibited.

Industry restructuring

The chief explanation for the decline of the integrated segment of the domestic steel industry as compared to the growth of the non-integrated segment and the healthy posture of the alloy/speciality steelmakers is the area of management decisions. Operating within the same domestic market, which accounts for most of all three segments' business, the same set of government policies, the same labour pool, and the same sphere of technological opportunities, management made widely differing decisions and chose different marketing and technological strategies.

The impressive growth and profitability of the non-integrated companies can be attributed to their fundamental decision to make greater use of domestically available scrap, to quickly adopt continuous casting and to emphasise narrow product mixes for individual plants which take advantage of nearby market opportunities. These companies emphasise reduction of production costs to a far greater extent than integrated companies. This has led to their being very competitive against both integrated producers and imports. Labour costs per ton of steel shipped are lower than for integrated producers because of greater productivity and not because of lower wages resulting from mostly non-unionised labour being used. Even in a sluggish domestic market, non-integrated firms have found particular niches on which to capitalise. Low capital costs have resulted not only from the elimination of ironmaking, but also from astute management that has made use of internal design and manufacture of equipment and sometimes lower-cost equipment from foreign sources. A number of recent studies of how American manufacturing firms must adapt their management and strategies to changing economic, political and social conditions can be applied to the non-integrated companies to explain their success.

The one characteristic of the alloy/speciality producers that stands out is their emphasis on technological innovations. In contrast to the integrated firms, who also have received various forms of trade protection in recent years, the alloy/speciality firms took advantage of the protection to greatly improve their cost and technological competitiveness. While the alloy/speciality producers have not been exposed to the growth of the non-integrated firms, they have been able to both maintain good profitabilities for the most part and compete effectively with very aggressive imports attempting to move from carbon commodity steels to the more profitable and technology-intensive alloy/speciality steels.

Industry restructuring must also be considered a consequence of the faults of integrated company management already discussed. It is reasonable to argue that whatever non-integrated and alloy/speciality firms have done to achieve success could also have been done by integrated companies. Indeed, the question has often been asked: why didn't the integrated companies, for example, build non-integrated plants pursuing the same strategies that non-integrated companies have? For the most part, the answer is that integrated company managers did not correctly assess the opportunities nor the changing economic, political and social climate in the nation. There was and, to a large degree, still exists a lethargy often associated with large corporations that makes management defensive and oriented to the past rather than forward-looking and positively aggressive in their attitudes and styles.

Have Federal policies been responsible for the restructuring? I believe not. While it is true that a number of Federal policies, *ad hoc* and poorly coordinated with each other, have presented serious problems to integrated companies, the success of the non-integrated and alloy/speciality companies demonstrates that there were ways to cope with such policies. Technology, for example, could have been used more aggressively and enthusiastically to deal with the burden of environmental regulations, capital formation problems and rising costs. The actions of Bethlehem Steel and US Steel have demonstrated that the closing of inefficient plants should have been done earlier. This process may continue during the next decade with further loss in market share. It should be emphasised that no Federal policies have contributed to the success of the non-integrated firms. If anything, policies tend to be biased in favour of large integrated steelmakers, including the most recent changes in tax policies.

It is equally difficult to seek foreign actions and policies which can truly be found to be the causes of the restructuring of the US steel industry. The pressures from imports could have been used as a stimulus by integrated steelmakers for reducing costs, improving steel product quality and improving marketing and customer services. Mature domestic industries are indeed vulnerable to foreign competition, but the restructuring of the domestic steel industry has already demonstrated that there are ways to cope with such competition as well as slow growth in demand, other than seeking Federal help or pursuing diversification.

Ideal solutions to present problems

Studies of the US steel industry, particularly those made by the industry itself, tend to speak of today's problems in terms of a crisis facing the industry. In reality, it should be noted that the trends which have been discussed do not reflect abrupt changes in the circumstances of the industry. Both the negative and positive developments have been taking place, for the most part, in a steady manner over many years. Crisis, however, is one effective way to get political attention. But there is not so much a crisis today for the US steel industry as there is a set of possibilities for the future which may lead to crisis conditions. There are, however, so many uncertainties about so many important variables, particularly future demand, that predictions of crises are highly speculative.

Ideally, what we need now, I believe, is a well co-ordinated *sector* approach for the steel industry that addresses both public and private policies which are responsive to past trends but are aimed at long-range strategic plans for the future. Changes in Federal policies which are macro in nature and oriented toward improving economic conditions in general and affecting all American industry are not likely to reverse negative trends in the steel industry. There must be industry-specific policies, but these do not necessarily imply, as some would think, that large-scale Federal funding or extensive trade protection are required. It is more a question of requiring responsiveness to particular problems, as well as systematic approaches and extensive coordination of policies. At the same time it is foolish to think of improved and changed Federal policies without also requiring changes in the industry itself, particularly in the policies, attitudes and decisions of management and, to a lesser extent, in labour.

Long-range strategic planning is just as much needed in industry as in government. Ideally, government and industry must recognise common goals and needs, as well as those that are different. The Federal government could provide:

a) the means to reach agreement on data and analyses upon which both public and private policy changes must be based;
b) a forum in which the various components of the steel industry, the government and suppliers and consumers associated with the industry can come together to reach mutually acceptable goals for the industry and government; and
c) a context to relate steel industry specific policies to those of other industrial and to economic and industrial policies of a more general nature.

Long-range strategic planning does not imply government intervention in the marketplace. While there has not been a *free* market system in the United States in recent history, Federal policies can be designed to facilitate *market forces* being the prime determinants of success and failure in the steel industry. Nor does long-range strategic planning imply that detailed and specific choices of a more operational or tactical nature are being made by government or by any consensus among corporations by themselves or in conjunction with other interests, such as labour. Planning on a macro (national or industry) level which reconciles the needs of both the private and public sectors can be compatible with freedom on micro (firm) levels.

The ideal solutions do not involve *saving* the steel industry. From whom is it to be saved – from itself, from the government, from foreign competition? Simplistic labels for industries have already done more damage than good. It is not a question of the US steel industry being sick, a loser or a winner. What is needed for the steel industry is a future-oriented, comprehensive approach that recognises changing conditions in the industry itself, in the nation and in the world, and is aimed at taking advantage of opportunities rather than preserving what objective analysis tells us is no longer effective, efficient or useful. This of course means that there will likely be dislocations affecting

232

companies, workers and communities. But the goal should be to develop cost-effective ways to deal with the dislocations, rather than to fight the changes which lead to the dislocations. The US steel industry of the future *will* be different from the one we have today, just as today's industry is incredibly different from the one which existed here twenty years ago.

Setting some long-range plans for the steel industry by setting goals, agendas and means reached through rational and equitable discussion among the interested and affected parties requires no abandonment of US economic, political or social institutions and principles. the United States has in fact done it before with certain industries, usually in time of crisis. The challenge for the nation and the steel industry today is not to wait for such a crisis, such as a major steel shortage. The best reason for not waiting is that while crises have tended to come upon us faster than ever before, the time to actually build a steel plant or develop a commercially feasible innovative technology has not been dramatically reduced, and although we have learned how to use other materials for many applications, steel remains a most vital material to keep our society functioning in times of peace and war. Although some may speculate about how much steel will be needed and from whom it may be available in the future, the US has learned about the risks involved in being overly dependent on foreign sources for commodities which are essential to the running of society.

Having long-range strategic plans (such as for technology development and adoption and restructuring) for the steel industry upon which public and private policies are based will not provide certainty for the future, but they would reduce major uncertainties which are partly debilitating for investment decisions, risk taking, innovating, and personal commitments. There are more than enough uncertainties thrust upon our nation from outside our borders. What is done for the steel industry could serve as an excellent case study of what is needed for other major domestic industries. It is a climate, a framework in which all the factors which by themselves are necessary but not sufficient to ensure health for the industry — such as technology, capital, skilled labour, technical personnel and good management can be focused on in a systematic way to make both a more efficient and more profitable industry. However, it is also necessary to recognise that no matter how good are strategic plans and public and private policies for a specific industry, success still depends on the general economic climate in the nation and, to an increasingly greater extent, in the world. They too are necessary but not sufficient for a healthy domestic steel industry.

Criteria for evaluation of policies

The preceding discussion provides a basis for developing several criteria which can be used to evaluate proposed policies or strategic plans for the domestic steel industry. The five criteria are as follows.

Comprehensiveness Does the policy plan address all major historical trends and chronic problems of the industry? It has become increasingly recognised that to act in one functional area, such as trade for example, without solving problems in other critical areas is not likely to be effective. I suggest that, at the minimum, there are six critical areas: capacity, trade, profitability, capital formation, cost (including regulatory), and technology. General national security and economic interests demand consideration for developing an understanding of a minimum acceptable level of domestic steelmaking capacity. The issue of unfairly traded imports and the potential for promoting more exports of technology-intensive steels remain important policy areas. Though capital formation is a major problem, declining profitability for many firms is just as important if capital is to be invested in steelmaking facilities. Reducing the costs of regulations must be balanced by goals of reducing production costs. The role of technology is crucial in improving the health of the industry in the long term, if the need for government assistance is to be minimised.

Feasibility Is the policy approach designed in a politically and economically feasible way for obtaining both short- and long-term solutions to fundamental causes, rather than short-term fixes for symptoms? There must be a pragmatic approach that recognises current limitations for Federal assistance and intervention. Wherever possible, policy instruments should not involve Federal spending. Where Federal spending is required there should be some attempt at assessing costs and benefits. The relationship between steel industry-specific policies and broader policies should be analysed.

Restructuring Does the policy strategy acknowledge and facilitate future restructuring of the industry because of the economic benefits offered? Recognising the problems of dislocations in society, there must be effective policies to cope with the dislocations rather than impede worthwhile restructuring.

Innovation Does the policy approach place emphasis on future technological innovation (often neglected for basic, mature industries) for the reduction of production and capital costs, and for domestic circumstances in order to improve economic and financial performance? There must be attempts to facilitate as well as to provide assistance for major innovative changes in technology which are responsive to the particular needs of the industry and domestic resources. Is innovation losing ground to diversification?

Equity Does the policy package distribute costs and benefits equitably across all interested and affected parties in society? An effective approach requires broad public support. Analyses must reveal how particular policy measures are, in their totality, fair in all ways.

Comparative analysis of available policy approaches

It is difficult to find an American industry which has been more studied than steel. Even though there is a vast literature on the steel industry, at present there are four policy studies which have recently been publicly disseminated. One is regarded as the plan proposed by the industry itself, the study by the American Iron and Steel Institute mentioned previously; the second is the OTA study already mentioned; third is the set of findings of the Steel Tripartite Advisory Committee formed by President Carter[7] in which big steel companies had a major influence. These findings were submitted to President Carter and, to some extent, were adopted by him and used in his announced programme for the steel industry. While the findings of the Tripartite Commission are likely to remain of interest, I do not believe that the Presidential programme remains viable and therefore it is not included in the following discussion. Lastly, there is a recent report by the General Accounting Office.[8] Where appropriate, I will note recently enacted changes by the Reagan administration.

There is of course a fifth option – that no new policy approach be pursued. This possibility is considered in both the OTA and AISI studies in the form of a scenario which paints a very bleak picture for the industry, and for the nation in terms of lost employment and risky dependence on imports. Since the OTA study does not make recommendations, the *status quo* is treated as one of three scenarios. However, the other three studies take the position that a new policy approach should be taken by the Federal Government.

The four studies are difficult to describe and compare for a number of reasons. First, each is, strictly speaking, not exclusively or even primarily a policy study. Each work attempts to analyse the problems and needs of the industry in varying degrees and from a particular perspective. In all cases there is very little detailed analysis of specific policy instruments, options or measures in a rigorous and quantitative manner. Rather, they attempt, for the most part, to speak in broad terms about critical policy choices or goals. Moreover, there is considerable detail which cannot be described here. Therefore, even though the four studies will be reviewed by scrutinising them in terms of the five criteria discussed previously, it should be emphasised that their shortcomings often stem from the lack of emphasis on policy which is their characteristic.

235

The AISI study

On the basis of comprehensiveness, this study and its policy recommendations are deficient primarily in the area of technology. Its methodology is primarily assertion, rather than analysis. The study emphasises the need for effective policy measures in three critical areas:

> Policies that would encourage and permit steel companies to achieve competitive rates of return, accompanied by provisions for accelerated capital recovery.
> Modifications of government-mandated regulatory programs (notably environmental) that would reduce non-income-producing capital demands.
> Firm assurances that imported steels, either by excessive volumes or unfair pricing, will not disrupt the domestic steel market, particularly during the industry's revitalisation effort.

The study also asks for removal of government interference in steel company pricing decisions. In an appendix there are recommendations for Federal support of research in the areas of energy and environment, new steel products and for development of formcoke processes.

In general, the political and economic feasibility of the proposed policy measures is good. There is attention to both the short and long terms, and a strong case is given for the need for some government costs, but the emphasis is on measures which do not involve outright government spending. The orientation is more on removal of disincentives for investment, on removal of government interference in the marketplace, and on the needs of integrated firms.

Both restructuring of the industry and technological innovation are absent in the analyses and in the formulation of policy approaches. The issue of equity is not addressed directly, and the absence of a number of topics, such as what to do about high labour costs and adjustment to plant closings and declining employment, indicate that, not surprisingly, this industry-generated programme did not delve into questions of equity.

The general economic, fiscal and regulatory policies of the Reagan administration are congruent with the AISI's goals and desires.

The OTA study

The central theme of this study was how technology might be used to improve the competitiveness of the industry. With regard to comprehensiveness, therefore, there is limited coverage on trade and a

236

limiting perspective on most issues. The primary emphasis is clearly on technology and innovation and on those policy options (without giving recommendations) that can use these as a means to achieve other goals, or on policy options in other areas which could have an impact on technology and innovation. The policy options which are discussed, and which could be instrumental in improving the industry's competitiveness' are aimed at: increasing R & D and innovation, encouraging pilot- and demonstration- plant testing of new technologies, facilitating capital formation, reducing the adverse economic costs of regulatory compliance, improving the availability of scrap, and constraining steel imports and facilitating certain exports.

There is much attention given to the potential benefits and risks of a steel industry sector policy by the Federal Government, as follows:

> The most critical policy option may be that of a governmental steel industry sector policy, that is, for a coherent set of specific policies designed to achieve prescribed goals. The present state of the industry and the need for a critical examination of policy options are, in large measure, a consequence of a long series of uncoordinated policies. These policies have not been properly related to each other or to a well-considered set of goals for the industry, goals that satisfy the needs of both the Nation and the industry. The lack of a sector policy and the designation of a lead agency to implement such a policy has led to policies that often conflict with one another, create an adversarial relationship between Government and industry, and fail to address critical issues. At present, a large number of people and agencies in the Government deal with steel, but they do not reinforce each other's work nor do they provide an accessible source of expertise and guidance for the industry or facilitate its efficient interaction with the Government.

The report also emphasises the need for comprehensiveness in policy by noting that 'Neither technology nor capital, alone, will solve the steel industry's problems. Substantial trade and tax issues exist with regard to the steel industry, and Federal policies on these issues need examination.'

Unlike the other studies, the OTA study emphasises analysis of the advantages and disadvantages of various policy instruments to achieve certain goals, such as greater capital formation. Table 13.4, taken from this study, illustrates this approach for the case of several options for increasing capital formation; and Table 13.5 presents its review of a number of regulatory changes.

Table 13.4

Features of four federal options for increasing capital formation in the domestic steel industry

Federal option	Government cost	Administrative burden	Bias against small firms	Promotion of new technology	Applies to steelmaking only
Accelerated depreciation					
Jones-Conable	High	Low	Yes	No	No
Certification of necessity	Moderate	Low	Yes	No	Yes
Investment tax credit					
Increase capacity	Moderate	Low	No	No	Yes
Modernization	Moderate	Low	Yes	No	Yes
Innovation	Moderate	High	No	Yes	Yes
Loan guarantee					
Increase capacity	Slight	Moderate	Yes	Yes	Yes
Modernization	Slight	Moderate	Yes	Yes	Yes
Innovation	Moderate	High	Yes	Yes	Yes
Subsidized interest loan					
Increase capacity	Slight	Moderate	No	No	Yes
Modernization	Slight	Moderate	Yes	Yes	Yes
Innovation	Slight	High	No	Yes	Yes

Source: Office of Technology Assessment

Table 13.5

Regulatory change: policy options and consequences

Regulatory change	Social impact[a]	Promotion of new technology	Regulatory cost impact	Capacity
Bubble Concept	Modest	Yes	Reduction	Facilitates replacement
Distributing cost of tradeoff requirements (offset policy)	None; increased equity among expanding firms in nonattainment areas	No	Reduction	Facilitates expansion
Extension of limited-life facilities policy while replacing steel facilities or otherwise providing for regional economic growth	Modes; at least partially offset by strengthening regional economy	Yes	Reduction	Replacement/ expansion
Fugitive emissions	High	No	Slow down growth rate	NA
Use of administrative penalty payment for environmental technology R&D fund	None, but goal change in favor of R&D	Yes	Transfer of costs	NA
Improved coordination of OSHA compliance deadlines	Modest	Yes, if given as condition for extended deadlines	None	NA
Improved coordination of EPA innovation waivers	Modest	Yes	None	NA
Cost/benefit analysis	Varies with cost-benefit trade off	No	Potential reduction	NA

NA – not applicable

a Social impact is defined as increased environmental degradation or occupational risk resulting from regulatory relaxation

Source: Office of Technology Assessment.

As some of the above comments suggest, the OTA study gives considerable attention to political and economic feasibility, to short- and long-term solutions, and to assessments of economic costs and benefits.

Perhaps more than in any other study, there is substantial emphasis on industry restructuring. There is a detailed analysis of the economic benefits of this restructuring, for example in reducing future capital needs of the industry. There is also a forecast for future shifts in market share among the three industry segments, and on what factors this restructuring depends. Policy options are generally evaluated in terms of their impact on restructuring, and policies concerning ferrous scrap are dealt with in some detail because of the dependence of non-integrated firms on scrap as a feedstock.

The need for technological innovation to improve competitiveness, and prospects for specific innovations are focused on. The problem of meshing both government and private sector policies in order to achieve innovation is discussed. A strong case is made for greater support by government for pilot and demonstration plant testing of new steelmaking processes which can reduce capital and production costs, reduce pollution and make greater use of domestic resources such as coal.

Equity issues are treated in a limited way. There is emphasis on dealing with the needs of industry as well as national needs. There is, however, no detailed analysis of dislocations and adjustment policies related to plant closures and the restructuring process.

The Tripartite Committee findings

An added difficulty when discussing the policy approach contained in this report is that the consensus findings of various sub-committees did not always become the consensus, conclusions and recommendations of the entire committee, but the latter are the ones described here. With regard to comprehensiveness, most major areas (with the notable exception of pricing) were considered, even though there were not original analyses performed by this group in all cases. The following were the major policy objectives recommended.

1   *Modernising the US economic base* 'Modernization of the American steel industry must be regarded as a key part of a larger effort to revitalize our overall industrial base and to increase productivity with our economy. Government assistance for steel will be most effective if it is part of a program aimed at stimulating business investment in general. The Committee ...

endorsed tax changes that would encourage more rapid capital recovery.'

2   *Reducing the burdens of adjustment* 'A central element in a steel program must ... be aggressive programs to assist workers to retrain for new jobs or to relocate to areas where jobs are more plentiful. The programs must also attract new employers to those communities where steelmaking jobs have declined.'

3   *Responding to unfair import competition* 'A steel industry program should, consistent with our overall trade policy objectives, work toward a situation in which trade in steel products is free of barriers, determined by economic costs — not government inducements — and conducted without injury attributable to dumping or subsidization. Our import competition laws are fully consistent with the obligations we have undertaken under the Multilateral Trade Agreements. The government must enforce these laws rigorously. The government should be prepared to administer them expeditiously.'

4   *Environmental and safety and health improvement* '. . . to achieve compliance with environmental goals as well as to encourage modernization, government should find ways to allow the industry to make investments in steel production which will assure that both objectives can be attained. Government reexamination of the reasonableness of its regulatory procedures may be required, too. We should expect that industy, for its part, will use any funds that are saved through changes in regulatory requirements for investment in modernization and commit firmly to full and timely compliance with all environmental and occupational safety and health requirements.'

5   *Adopting the best technology* 'it is not sufficient . . . for a steel program to aim at helping domestic producers match the extent to which foreign producers use the latest of available technology. Rather, the aim must be to assist our industry in the development and testing of the possibilities so that they are positioned to incorporate new technologies into their operations as soon as they are proven. The great share of research and development should properly be carried out and financed by the steel companies themselves. A steel industry program should set into motion multi-firm collaboration and industry-government cooperation in R & D to find solutions in these common problem areas.'

6   *Competition should set capacity* 'The size of the industry should be determined by market forces. Government's role is to do all it can to assure open and fair competition at home and in the international marketplace.'

7   *An integrated long-term approach* 'A piecemeal approach to the steel industry will accomplish little. The problems of the industry cover a number of areas including capital formation, trade, environmental regulation, technology, and the adjustment of workers and communities to changing industry conditions. With or without Government assistance, measures directed only at one of these areas cannot set the industry on a new path. A coordinated and integrated set of initiatives, maintained for a 3 to 5 year period, or longer, is required to remedy the industry's situation.'

8   *A partnership commitment* 'A commitment by management, labor and Government to support and contribute to a steel industry program will be an absolute prerequisite for its success. Changes in Government policies alone cannot make the industry modern and competitive.'

Generally, the Tripartite policy recommendations rate high in terms of feasibility, innovation and equity. However, there was little effort to focus attention on the restructuring of industry and the impact of various policies on that process and on steel consumers. This is due to the absence of representation from non-integrated firms on the Committee. For example, delayed compliance with the Clean Air Act and refundable tax credits promote the existence of integrated firms with the poorest performance. Faster capital recovery and delayed compliance with regulations have recently been enacted by Congress.

The GAO report

While the GAO study included examination of most critical areas, with the exception of technology, for which it referenced, for the most part, the OTA study, its policy recommendations are skewed toward one particular conclusion. GAO recommended 'that Congress define a policy guidance performance objective for the domestic steel industry'. Such an objective would be 'in terms of efficient capacity'. The purpose would be to 'help determine appropriate Federal means toward achieving overall objective of a competitive industry'. The rather unusual recommendation of asking the Federal Government to set a capacity level for a major industry was justified on the basis that 'in the absence of a generally agreed upon performance objective for the industry it is just not possible to judge the adequacy of the set of proposals for industry revitalization, nor be able to suggest how they might be usefully amended'. There is, however, no detailed analysis to demonstrate that consideration was given to other measures of performance, in terms of efficiency or means, rather than mere level of output.

The report also deals with supportive policies which 'ought to be formulated for a range of important peripheral activities' which include:

- 'wage and compensation restraint and labour–management commitment to a sound revitalization strategy,
- measures to induce the entry and growth of new competitors,
- accelerating depreciation rates,
- improving administration of environmental regulation,
- eliminating discriminatory price restraints, and
- creating a trade policy yielding predictable and acceptable effects on imports with a minimum of inflation.'

The report also presents a case for 'the likely need for new means of policy administration to insure that a necessarily interdepartmental, multi-faceted steel policy succeeds in promoting industry revitalization'.

Although there are some very strong findings concerning the benefits of industry restructuring, there is little attempt to relate policy measures to this process. Since the report did not address technology, innovation is not covered. With regard to both feasibility and equity, there is substantial coverage, with the one important exception being the difficulty of the government setting a level of capacity for the industry, the study's main recommendation.

Agreement among the four policy approaches

It is important to recognise the impressive amount of agreement among the four studies described above, even though different methodologies were used and each study had different mandates or objectives. To recognise this agreement is crucial if there ever is to be a serious attempt at formulating and implementing a specific and detailed policy for the US steel industry.

First, in all four cases, a strong case is made that the steel industry either can be or should be restored to a healthier state. Moreover, in each case there are specific measures which are believed to be capable and cost-effective to do this job. There is little despair or hopelessness exhibited. Rather, there is a sense of urgency in getting all interested parties to come to an agreed strategy.

Second, there is agreement that the best policy approach is to deal in a comprehensive manner with a number of critical areas at the same time, and that to do otherwise would be ineffective and inefficient. Moreover, for the most part, there is agreement that changes in government policies must be accompanied by changes in the policies and attitudes among management and labour.

Third, there is considerable agreement on the need to deal with the problem of capital formation in order to modernise the industry, and to develop and implement innovative new technology in order to improve profitability and international competitiveness. Although accelerated depreciation has become a reality, it is also generally recognised that by itself this policy will not likely provide sufficient capital for a major turnabout in the industry even though there is some disagreement on the actual amount of capital needed in future years.

Fourth, with the exception of more Federal spending on R & D and innovation activities, there is little support for direct Federal spending on the industry, although significant spending is generally recommended for adjustment programmes for workers and communities. Most policy measures which are supported involve removing existing regulations or disincentives and doing other things to achieve greater efficiencies in both the public and private sectors, and to let normal market forces operate with less interference by the government. There is, however, strong recognition of the need to have Federal policies which in their own right are competitive with foreign policies toward steel industries.

Fifth, in the two studies (OTA and GAO) in which industry restructuring and the benefits and growth of non-integrated companies were examined, there is strong agreement on the need to avoid obstacles to this restructuring if the overall competitiveness and health of the domestic steel industry are to be improved. OTA, for example, forecasts that by 1990 non-integrated firms could account for 25 per cent of domestic production.

Finally, although there are elements of each of the four studies that can be used in future policy work for the steel industry, there is also considerable need for more independent analysis and data gathering for precise policy formulations. There are important criticisms of past government efforts and a general consensus that a new administrative mechanism is needed both for analysis of the industry and for discussing, formulating and implementing steel industry specific policies.

The connection between steel industry policy and industrial policy

There remains the issue of how steel industry-specific policy should be related to a possible Federal industrial policy, which many nations have. Such a policy affecting all American industry has been advocated by many under the rubrics of revitalisation, reindustrialisation, renewal, and so on. Here no review of all the formulations of industrial policy will be presented. Some take the position that the policy levers for change should not be targeted to specific sectors or industries, while others focus on industry or sector-specific means. What emerges from the four studies of the steel industry considered here is a strong case for the need of a steel industry-specific policy approach. But this does not imply that, by itself, such an approach would actually work. Neither a steel industry-specific policy nor some broad-based economic or industrial policy without targets would by themselves be sufficient. Each is necessary but not sufficient. Most of the objectives and quantitative analyses lead to the conclusion that broad-based, untargeted approaches would either not provide an adequate solution to a particular problem, such as capital formation, or that specific problems would simply not be addressed.

A subset of those favouring the targeted approach of industrial policy prefer to label industries as winners or losers so that Federal policy can facilitate the decline or promote the growth, respectively. An important consequence of the past studies of the steel industry should be the recognition that applying such labels is fraught with risk. In all known cases where such labelling has been done, even in those using rigorous, quantitative means, the steel industry is labelled a loser. This is inaccurate because there is no recognition of restructuring, of the potential for technological rejuvenation, and of non-quantifiable reasons for wanting and needing the industry, such as national security considerations. Any targeted approach should simply be attuned to the opportunities and problems of an industry, as well as to the need to reconcile public and private sector objectives.

This suggested that the lesson to be learned from the studies of the steel industry which could be of great value in the debate over industrial policy is that there are three fundamental rationales for government actions that affect industry:

1   *Remedial*: Past government policies have either caused damage directly or caused distortions or imperfections in the marketplace or sent it perverse signals. These must be corrected.
2   *Competitiveness*: The actions and policies of foreign governments have caused distortions or imperfections in the international marketplace or sent it perverse signals, and these have damaged the US domestic industry and must be corrected by implementing Federal policies which offset the damaging effects.
3   *Efficiency*: Federal policies should act as a catalyst in making an industry reach desirable goals of improved performance for its own sake and the sake of the nation in the most efficient

manner. A catalyst does not itself take part in a chemical reaction, but it causes the reaction to take place with less energy, with less cost and often with greater speed. Appropriate, catalytic Federal policies, therefore, do *not* imply intervention in, or subversion of, the marketplace or large amounts of spending. Rather, they offer the promise of smoother running institutions in the public and private sectors, and more efficient and useful exchanges of information and resources across the many interfaces within these institutions. To paraphrase Abraham Lincoln: Government cannot help industry permanently by doing for them what they could and should do for themselves. But that leaves a rich field of opportunities for government to act appropriately.

Notes

1   American Iron and Steel Institute, *Steel at the Crossroads: The American Steel Industry in the 1980's* , Washington DC, January 1980.

2   Office of Technology Assessment, *Technology and Steel Industry Competitiveness* , Washington DC, June 1980.

3   Hall, W.K., 'Survival Strategies in a Hostile Environment', *Harvard Business Review* , September–October 1980, pp.75–85.

4   Wyman, J.C., and Gold, B., *Technology and Steel* , Shearson Hayden Stone, Inc., New York, 21 February 1979.

5   Myers, J. and Nakamura, L., *Saving Energy in Manufacturing: The Post Embargo Record* , Ballinger Publishing, 1978.

6   Robbins, N.A. in *The American Steel Industry in the 1980's – The Crucial Decade* , American Iron and Steel Institute, Washington DC, 1979.

7   Steel Tripartite Advisory Committee on the United States Steel Industry, *Report to the President* , mimeo, Washington DC, 25 September 1980.

8   General Accounting Office, *New Strategy Required for Aiding Distressed Steel Industry* , Washington DC, 8 January 1981.

14 Japan's steel industry

Introduction

Japanese industry has surprised the world by moving out of mediocrity to high or even superior levels of quantitative, but most of all qualitative, performance. In 1952, Japan had recovered to pre-war levels of production, its gross national product then being about one-third of that of France or Great Britain. Twenty-five years later it was as large as that of France and Great Britain together, or more than half of that of the United States of America.[1] In 1978 none of the world's 22 largest modern blast furnaces was operating in the USA, but 14 were operating in Japan.

A country which, 25 years ago when the first signs of new competitiveness became visible, only too few took seriously, succeeded in reaching the top of the industrial world in an incredibly short time and in a surprisingly wide range of industries. The spectrum of industries was in general chosen by the principle of high value added and high intelligence (optical, electronics, micro-electronics industries) or by 'national needs' aimed at lowering the transportation cost dependence of an island empire. By the middle of the 1970s, Japanese-built ships were not only 25–30 per cent cheaper than those built in Europe, but also superior in operating economy and control equipment.

Using its comparative advantage first in labour cost rather than in scale economies and modern technology, Japan built up industry after industry. In 1962, the average Japanese performance was about 100 tonnes of steel per worker per year, compared to 400 ·tonnes per worker per year in Great Britain. In the mid-1970s Japan outperformed Great Britain four times over. By 1981 the best available Japanese technology rendered 1,800 tonnes per worker per year.

247

Steel and MITI (Ministry of International Trade and Industry)[2]

The development of Japan's steel industry dates back to the days of the Ministry of Agriculture and Commerce (MAC), MITI's predecessor. Under its auspices, and at an expense of approximately 4 million yen, the Yawata Iron and Steel Works[3] were established between 1896 and 1901, located in the northern Kyushu coalfields, with access to a deep-sea harbour for imports of then easily available Chinese iron ore resulting from the Japanese victory in the first Sino-Japanese war, 1894–1905. Yawata's primary mission was to produce iron and steel for armaments. Government-owned, MAC–managed Yawata Iron and Steel encountered private domestic rivals only ten years later, when Kobe Steel and Nippon Kokan (Japan Steel Pipe and Tube) were founded. The MAC-Yawata liaison was extended to MITI, founded in 1925; indeed, the Yawata Tokyo sales office was housed in the MITI building until 1934. Many MAC and MITI vice ministers and top officials have held top positions in major Japanese steel corporations (not only in Yawata or Nippon Steel).

Leaving aside the war and post-war history of steel, one of MITI's great post-war industrial restructuring achievements was the creation of some large coastal[4] industrial parks, one of which was the Keiyo Industrial Park (Chiba Prefecture) reclaimed from the Tokyo Bay. There the world's first and most modern post-war super-steelwork was built on free land, with cheap capital, by Kawasaki – against the objections of Yawata, Fuji and NKK, which claimed that it meant building up a tremendous overcapacity. Kawasaki's Chiba plant became operative in 1953.

Immediately upon the termination of the allied occupation, important steps were taken by MITI to make possible high speed industrial growth by changes in laws restricting industry and trade. Cartels were now permitted for various reasons (for instance rationalisation, joint export efforts and so on), but also for prices, such as the 1958 price maintenance cartel for steel. Steel had a most influential caucus within MITI at that time, and even later.

As MITI took over the ideology of French *économie concertée*, known as administrative guidance MITI-style, in 1961, special steel was chosen as one of the 'designated industries' along with automobiles and petrochemicals.

During the mid-1960s a kind of rivalry developed between the private four (NKK, Kawasaki, Sumitomo Metals and Kobe) and government owned Yawata and Fuji with the introduction of administrative guidance, believed by the private four to favour the state-owned two, and ·with the 1965 recession, when the steel industry's self-regulation broke down. In order to prevent cut-throat price policies and bankruptcies, MITI ordered a 10 per cent production cut during part of 1965. Sumitomo refused to obey, pointing to its still superior export performance. Subsequently, the so-called Sumitomo incident – an unusual, and thus famous, public quarrel between a firm and MITI – was settled. However, there were repercussions during the lengthy process of the Yawata–Fuji merger, which materialised in 1970 as the New Nippon Steel Corporation (now not only the industry leader but the world's largest steel company). Nippon Steel was nevertheless as much a child of a new policy merger-wave to produce a concentration of economically powerful conglomerates to match similar US and German giants as an internal steel industry settlement. Yawata and Fuji were easier to merge than other large Japanese corporations, as they had been one firm between 1934 and 1945. Nevertheless, it was a complicated matter for several reasons. MITI's central role became more publicly visible than MITI had wished and had reason to be content about since for the first time MITI was under public attack for other reasons as well, for example, several pollution and environmental scandals. From MITI's viewpoint, the public and highly critical discussion of its role in the steel merger came at a most unsuitable time. The next time MITI would become active in the steel industry was towards the end of the 1970s, in attempts to assist the industry in adapting to a severe international crisis.

An assessment of MITI's role

This short historical sketch perhaps overemphasises the role MITI has, or may have, played in the development of the Japanese steel industry. One should remember that, since MITI traditionally has been dealing with big industry, steel is thus a natural subject for its concern. MITI has played an important role in the post-war restructuring of the industry and in strengthening the industry's international position. Nor should MITI's indirect role as a general promoter of industrial growth and efficiency be underestimated; in a few cases – such as Kawasaki's Chiba work – MITI has channelled free resources to the steel industry. It would, however, be incorrect to overstress MITI's role: the Japanese steel industry – irrespective of its being in public or private hands – is a most progressive, 'record-breaking motivated' industry, highly capable of managing its own road to success or coping with crises. The most important and lasting MITI influence may have been the breeding of a competitive spirit in industry, even in the steel industry.

Japan vs. the United States steel industry

Many competent analyses have been devoted to the competitive situation between the Japanese and the US steel industries during ·the late 1960s and, most of all, during the 1970s and the beginning of the 1980s.[5] When the long-lasting structural crisis of the US steel industry became visible in 1958 and 1959 it was not, as is usually the case, recognised as a structural crisis. Imports increased slowly and came from Europe, which was not entirely unusual. But suddenly, towards the end of the 1960s, a new and much more dangerous competitor appeared in force on the American market. The US steel industry experienced its first real crisis and much attention started to be given to the comparative competitiveness between the US and the Japanese steel industries.

The analyses of Mueller and Kawahito concentrate on three longterm changes in the cost-competitiveness between the two industries:

1   In contrast to the American industry, the Japanese industry had quickly adopted, and even improved, the world's best technologies in steelmaking process facilities as well as in steelmaking operations. Both the location (coastal, cf. Yawata of 1896) and the plant layout of the Japanese steelworks were far superior. The plants were brand-new at well-balanced economies of scale and employed automation and computerisation as far as feasible. Their labour and capital productivity, yield and energy efficiency were superior.

This superiority did not come about by mere chance. Japanese industry had been experiencing a drastic increase of demand for high quality steel firstly from domestic users like the shipbuilding and construction industry and soon also from abroad. Thus the size and the consequent economies of scale of plants came to lie in the upper ranges, slowly creeping up to 5–8 million tonnes per year per plant.

This in turn meant that always and only the latest technology was being employed. In many cases Japanese engineers had to solve problems in steelmaking which nobody else had yet encountered. Thus the level of originality and innovativeness grew rapidly, making necessary the establishment of an intelligence monitoring system covering the most advanced industrial countries of the world. Here both MITI and the industry exposed hitherto unparalleled creativity. MITI and the entire Japanese infrastructure in various aspects created a supportive environment for rapid development of a highly modern industry.

2   Whereas Japan had rather low cost of labour, it was severely disadvantaged in the availability of fuel, particularly coke, and also of ore (access to the Chinese high quality ore had been lost).

However, over time, the Japanese industry was able to overcome those obstacles, for instance by actively developing coke production as well as ore-mining in Australia and elsewhere. Finally, the Japanese policy of making bulk transportation by sea cheaper gradually improved its competitive situation both in transports of ore and fuel to Japan and of finished steel to new markets. In America, particularly in growth regions like California, the domestic steelworks soon were no longer able to compete.

It is worth noting that the changes mentioned in this paragraph were in many cases deliberately and cleverly monitored by MITI and Japanese industries and institutions outside and inside the steel industry.

3   Contrary to what one normally would expect, hourly labour costs in the United States grew faster than in Japan, at least as far as the steel industry is concerned. Thus, between 1956 and 1976[6] the difference had grown from US $ 3.35 to 6.89.

Perhaps the most dramatic change in the long run was the shift of industrial leadership in steel away from the United States to Japan. Over the years the gap in industrial competence has grown to such an extent that it is very hard to imagine today how it ever could be closed or even reversed again.

What applies to the United States also applies to Europe. However, one should bear in mind that Europe is much more differentiated than it seems when viewing it from across the Atlantic distance. The non-subsidised steelworks of Holland and Germany have until quite recently, that is towards the end of the 1970s, been able to keep pace, perhaps not with the absolute Japanese top, but at least with the upper range of their Japanese competitors.

As the US steel industry was taken by surprise by the lower and lower prices quoted by Japanese steel firms supplying the US market, it was only natural to suspect dumping and subsidisation from the Japanese side. There have, however, been a number of very competent investigations into this matter revealing that the yield[7] of Japanese plants is much higher, as are both labour and capital productivity. For natural reasons it was a great shock to the American industry as it was to the Europeans – to learn that the Japanese were far ahead of their overseas competitors in plant location and design as well as in operating efficiency.

There is obviously only one blame which is valid when it comes to Japanese behaviour: there is very little, if any, import of foreign steel into Japan. The major reason for this is, of course, the still greater competitive advantage of Japanese steel prices in the home market compared to foreign imports. There may, however, also be other reasons. In 1982, imports to Japan from Taiwan and South Korea were still under 5 per cent of Japanese domestic apparent consumption. Nevertheless, exports from both countries have been able to sell at lower prices than their Japanese competitors' in the Japanese market. There may be a more or less silent gentlemen's agreement between decision makers in the three countries not to swamp the Japanese market with cheap steel goods, because it was Japanese assistance (technological, financial, construction and operational) which made possible the establishment of highly competitive steel industries in South Korea and Taiwan. Thus there may be a protective arrangement against 're-imports'. It is, nevertheless, almost as likely that Korean and Chinese exports will go to other markets, which give higher profit margins than would be available on the Japanese market.

There is still one important obstacle facing imports to the Japanese market: the channels for imports to Japan are mostly in the hands of financial and trade groups which also own steel companies in Japan. If there are independent importers, they may be hesitant to sell foreign steel in the Japanese market, fearing retaliations either from Japanese steel firms or from other members of the groups they belong to.

Statistics

The development of the Japanese steel industry can be subdivided into a number of phases. The first modernisation programme after the war lasted until 1956, the second until 1961, the third until 1966. The period 1967 through 1974 is characterised as enlargement of facilities whereas the post-1974 boom period is being devoted to cost reduction, that is, various types of improvement efforts.

The following major technological improvements were introduced:

1951	Hot rolling of silicon steel sheets
1952	Hot-strip rolling
1954	Continuous galvanising
1955	Electric tinning
1956	Basic oxygen furnaces
1957	Computer control
1958	Cold rolling of silicon steel sheets, vacuum degassing

1959	Sendzimir rolling
1961	Continuous casting
1962	Direct reduction
1963	High top pressure blast furnace operation
1967	Blast furnace stave cooling
1970	Pelletising
1972	Coke dry quenching
1976	Bottom-blown basic oxygen furnaces

Table 14.1

Development of the Japanese steel industry 1950–80

	1950	1960	1970	1980
Crude steel output (1,000 tonnes)	4,839	22,138	93,322	111,395
Iron-steel exports (1,000 tonnes)	586	2,507	17,980	30,327
Output of ordinary rolled steel products (1,000 tonnes)	3,486	15,675	66,691	87,227
Gross national product (in real terms) (1950 = 100)	100	281	666	950
Index of mining and manufacturing production (1975 = 100)	5.7	25.9	92.5	142.4
Steel stock (million tonnes)	71	114	340	559
Per capita crude steel consumption (kg)	49	207	670	540

Sources: Japan Iron and Steel Federation, Ministry of International Trade and Industry, Economic Planning Agency

Crude steel production in Japan grew practically uninterrupted until 1973, to 120 million tonnes per year and since then has been stabilising around 100 to 110 million tonnes. The oil price increase triggered break of the trend was caused by sharp rises in the prices of raw materials and energy, which, together with other factors, caused world steel demand to fall. The Japanese steel industry adapted quite early to the changes by improving its products to higher levels of sophistication, to save or even replace energy and by cooperating with customers in attempts to cut, improve and refine the utilisation of steel as a production material. Quality improvements, conservation of resources and energy and product yield improvements were the first major steps taken (for example, coal-brique blending, coke dry quenching, top pressure recovering turbines for blast furnaces, waste gas recovery in BOF; extended use of continuous casting, elimination of process steps, avoidance of cooling between processes – hot direct rolling, hot charging –).

253

The continuous casting ratio has grown from 6 per cent in 1970 to 60 per cent in 1980 and 72 per cent in 1981.

Examples of product development in response to customer demand are dual-phase low-alloy steel, for automobile weight reduction; one-side galvanised sheet and heat resistant steel, for improved corrosion resistance and durability and better shaping properties under high heat conditions; thin deep-drawing quality tinplate or two-piece cans for the packaging industry.

Japan's steel industry's share of oil-based energy has dropped from 21.3 per cent in 1973 to approximately 10 per cent in 1980. During the same period, coal-based energy has increased from 61 to 71 per cent. By 1982 oil had been practically eliminated in furnace operation – being replaced mainly by coke. During 1980, 30 out of the 44 blast furnaces in operation were all-coke operated. Oil consumption per tonne of crude steel produced fell from 128 litres in 1973 to 55 litres in 1980.

Raw materials beneficiation technology has been developed to the extent that the proportion of refined ore and pellets in the blast furnace burden ·is averaging at 90 per cent. In response to the worldwide heavy coking coal shortage, the consumption of that fuel is being reduced; non-coking coal fuels such as coal briquettes are used to a greater extent. In 1981 65 blast furnaces were in operation in Japan, 39 of which were units with an inner volume of more than 2,000 m³ each, 15 of them having an inner volume of more than 4,000 m³, the largest being 5,070 m³.

Amongst the process improvements are to be mentioned dynamic control of BOF, ultra-high power operation of electric furnaces, oxygen injection equipment, various kinds of automated systems, metallurgy refining facilities, vacuum degassing equipment, automatic gauge control, scrap preheating with waste gas, and so on.

During 1983, Japan's steel industry renegotiated its long-term contracts with suppliers of ore and coke/energy.

Still, the improvement drive is unbroken. To give a few examples: whereas in 1977 the best Japanese firm yielded 1,000 tonnes of finished output per man-year at 90 per cent capacity utilisation, the latest plant taken into operation has improved this figure to 1,800 tonnes at the same capacity utilisation. In both cases the improvement target is 3 to 5 per cent per year. Japan today uses approximately 30 per cent less energy per tonne finished steel compared to 1972. The most surprising difference, however, is in the capital investment cost per tonne finished steel output per year. Given the fact that about 20 per cent of the investment cost is for internal and external environment preservation purposes, thus not rendering higher output, the cost per tonne output per year is ranging around US $ 600 to 700.

The crisis reaches Japan

Japan's steel industry is not without problems of the same type as have been striking the steel industries of other highly industrialised countries. Steel stock has lost heavily recently. The year 1982 brought a rather drastic and unexpected decline in capacity utilisation. In 1980 the Japanese steel industry operated at an average of 76 per cent capacity utilisation; the autumn of 1981 brought only 65 per cent. The autumn and winter of 1982 brought the industry for the first time under the critical 60 per cent threshold. The industry is sliding into the red.

Several of the large steel corporations in Japan will not pay dividends. Nippon steel and several other big steel producers of Japan have been running into red figures in 1982. The raw steel production for the fiscal year 1982 is expected to be under 100 million tonnes, the lowest figure for the last ten years. Steel demand is still slackening,

Table 14.2

Steelmakers' efforts to adjust to low-growth economy in the 1970s

	1970	1973	1976	1977	1980
Sales (billion yen)	4,239	6,192	8,438	8,119	—
Net profit (billion yen)	208	392	147	36	—
Productivity index (1970 = 100)	100	145.9	151.0	154.8	197.9
Yield from crude steel of steel products (%)	81	85	87	87	91
Coke consumption (kg/tonne of pig iron produced)	478	438	432	434	450*
Heavy oil consumption (kg/tonne of pig iron produced)	42	60	50	40	17

Source: Japan Iron and Steel Federation

*About 50 per cent of the iron ore consumed stems from Australia, 25 per cent from Brazil and India; about 40 per cent of the coke used comes from Australia, about 30 per cent from the USA.

The prospects for 1983 have urged the Japanese government to label the Japanese steel industry a depressed industry. This makes it possible for the industry to apply shorter working hours. It also gives some access to government subsidies, provided no improvement is in sight at all.

Table 14.3

Capital investment by investment motives
(approximate percentages)

Motives	1973	1976	1979	1981
Capacity expansion	59	38	10	11
Rationalisation	25	32	40	32
Maintenance and repairs	–	13	24	32
Research	3	2	2	6
Energy saving	–	–	6	15
Safety and environment	–	–	12	6
Others	13	15	6	8

Source: Japan Iron and Steel Federation.

The causes of the rather drastic and sudden decline of the steel business are to be found firstly in domestic demand. Even Japan is today a slow-growth country. This hits the investment-oriented industries first.

The steel industry has not been able to increase its domestic prices, which have been partly stable, partly sliding downwards. The export business has been still worse. The steel crisis in the developed world has increased worldwide protectionism, making it very difficult now for the Japanese to defend their market shares. At the same time, Japanese steel is losing in neighbouring Asia, because of the aggressiveness of South Korea and Taiwan. And it looks as if recently the South Koreans have forgotten their code of good conduct vis-á-vis their master teachers.

The prospects for the industry are rather dull. No increase worth mentioning in steel demand is visible. Nevertheless, the big steel firms keep their personnel busy with rationalisation investment and productivity programmes. In addition, a range of new or improved products are in the process of being prepared for a new fight for old or new markets. In the long run, the prospects are not too bad, if only the recession does not last too long: the Japanese industry may, after all, be the only one which comes out of the crisis in a stronger shape than when it went in.

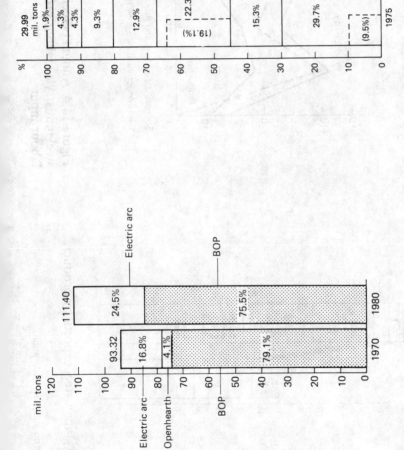

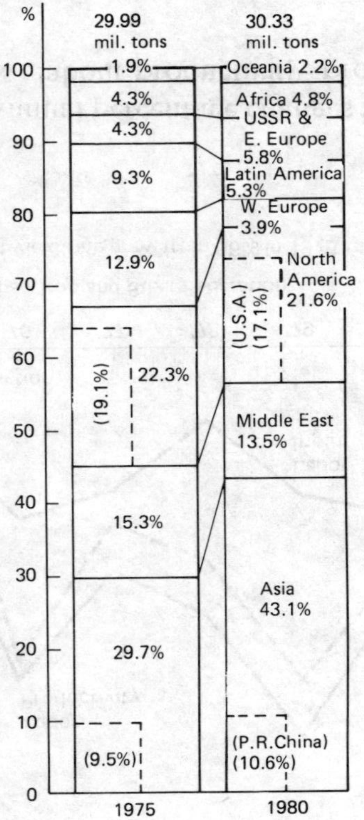

Figure 14.2 Japan steel exports by destination, 1975, 1980. In 1977 the export volume was 35 million tons, 1981 29 million tons. The export value was 11bn US$

Source: Japan Iron and Steel Federation

Figure 14.1 Japan's crude steel production percentages by type of furnace, 1970, 1980

Source: Japan Iron and Steel Federation

257

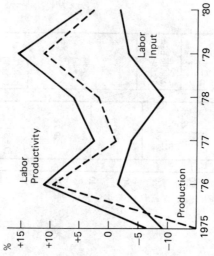

Figure 14.4 Annual percentage changes in the Japanese steel industry's labour productivity, production and labour input*

Source: Japan Iron and Steel Federation

*The product yield rate grew from 86% in 1976 to 91% in 1980

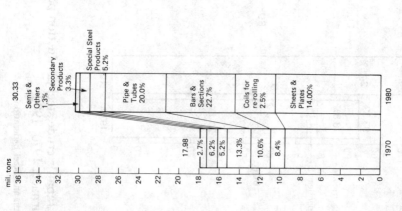

Figure 14.3 Japan steel exports by product, 1970, 1980

Source: Japan Iron and Steel Federation

Notes

1 Vogel, E.F. *Japan as No.1* , Harvard University Press, Cambridge, MA, 1979.

2 Johnson, C., *MITI and the Japanese Miracle: The Growth of Industrial Policy, 1925–1975* , Stanford University Press, Stanford CA 1982, provides the most competent and probing account of MITI's role in Japanese industry in general.

3 In 1934 Yawata Iron and Steel works merged to become Nippon Steel, which was dissolved in 1945 by the allied powers. On MITI's initiative Yawata Iron and Steel and Fuji Steel were again merged in 1970 to become the New Nippon Steel, generally known as Nippon Steel.

4 Six tonnes of raw material are needed to produce one tonne of finished steel. Thus access to cheap and direct – without reloading – bulk transportation is vital to modern large-scale steel industry.

5 See Crandall, R.B., 'The US Steel Industry in Recurrent Crisis', The Brookings Institution, Washington DC, 1981; Gold, B., 'Steel Technologies and Costs in the U.S. and Japan' in *Iron and Steel Engineer* , April 1978 (1978a) (see also Chapter 25 of this volume); Kawahito, K., *Sources of the Differences in Steel-making Yield between Japan and the United States* , monograph series no. 20, Business and Economic Research Center, Middle Tennessee State University, Murfreesboro, July 1978; and 'Japanese Steel in the American Market: Conflict and Causes' in *The World Economy* , vol.5, 1982; Mueller, H.G. and Kawahito, K., 'A Comparative Analysis of Structure, Conduct and Performance', Japan's Steel Information Center, New York, 1978.

6 Kawahito, K., 1981, op. cit.

7 Kawahito, 1979, performed a thorough analysis of the 'Sources of the differences in *steel-making yield* between Japan and the United States'. The analysis starts from the observation that the average yield in American steelmaking is approximately 15 per cent lower than the yield in Japanese plants. The question of yield is central to the TPM (trigger price mechanism) used for controlling mini-prices on the American market for foreign steel products (TPM was in operation between the beginning of 1978 and the beginning of 1980 and then again from the autumn of 1980 to January 1982). TPM prices, however, were not real Japanese c.i.f. prices but rather artificially calculated prices telling what the foreign c.i.f. price should be on the American market. In order to maintain an interest in the

American steel-producing industry to rationalise/modernise, trigger prices should be 'somewhat' (5 to 10 per cent) below American prices. In calculating those artificial prices assumed yield norms are used in order to convert raw steel prices to trigger prices for different types of finished products.

Kawahito identifies a number of major factors accounting for differences in US vs. Japanese steel yield: greater use of continuous casting; modernity of plant and major equipment; excellence in operational details, made possible by the unique labour–management relationship; comprehensive computer control; market structure; product composition; product quality; sophisticated customers and suppliers relations; statistical measurement of final products; yield of plates.

Each of the above-mentioned factors is further dissected in Kawahito's analysis.

The Japanese steel industry is far more research-oriented than its American or other foreign counterparts. To give an example, Nippon Steel on the average invests US $ 200 million a year in research and development. It employs approximately 2,300 engineers and scientists in R & D. Much of the research and development is done in cooperation with both suppliers (of production equipment but also of raw materials) and with customers. As for example deliveries to the automobile industry comprise about one third of Nippon Steel's output, there is a considerable degree of cooperation with the car manufacturing industry. Out of the 2,300 R & D personnel, approximately 600 are employed in product development, which is the level most closely connected to the users (their engineering or product development as well as construction departments). The cooperation with the automobile industry, for example, is emphasising efforts to reduce the weight of steel used in cars and trucks, and to improve the corrosion resistance properties of steel and plates used in cars and trucks.

The Japanese steel industry's R & D-related expenditure in 1980 accounted for approximately 1.5 per cent of sales and is steadily increasing. The comparative figure for the US steel industry is 0.5 to 0.6 per cent of sales.

PART IV
STEEL INDUSTRY
INNOVATION

15 Introduction

General considerations

When taking stock of the strategic options available for an industry in acute and deep crisis, *innovation* must be critically scrutinised. As innovation is normally a long-term strategy – it takes considerable time to mature, is surrounded by many uncertainties, and usually requires substantial sums of capital – it can hardly produce an upturn in the industry in the short run. Innovation is thus no crash strategy.

When strategies for restructuring an industry are assessed as to their potential efficiency, a central conclusion may be that, if one wants to maintain the industry, it is only worthwhile to invest in or allow to survive those firms or plants which are operating at the forefront of technology, or which are capable of serving customers (or rather a fairly secure segment of the market) with excellence, that is, with first-rate products at competitive prices, and which will be able to do so over a foreseeable future.

In the above-mentioned restructuring case it does not matter if it is the industry itself which will have to raise the money for the efforts, or if government is supposed to do it – as has been proposed for the US iron and steel industry, and as is under way in the European Community. In both cases government(s) are assisting the industries by different policies and means.

It is essential for corporate management to think in terms of market/product or technology/process innovation, particularly if the industry is acting in an environment of worldwide competition. Innovation is vital to the steel industry (as to any other industry) for some central reasons.

Technology as well as demand for steel products will shift with respect to qualities and quantities. This will trigger new steelmaking and refining processes by technology pull as well as by technology push.

Innovation aimed at defending and enlarging steel markets and improving steelmaking processes is necessary:

a) to produce new and/or better products;
b) to reduce cost;
c) to improve the internal (working) and the external (environmental) conditions.

It would, however, be unrealistic to assume that the steel industry's problem can be solved by innovation alone. It must therefore be regarded as one major subset of strategies needed to cope with the steel industry's structural crisis. Innovation is a rather complex, uncertain and also costly strategy, as will be demonstrated below. Nevertheless, firms neglecting this factor will face great difficulties in the medium and long term. The Office of Technology Assessment's study *Technology and Steel Industry Competitiveness* (1980) in an analysis of the United States' iron and steel industry's problems claims:

A number of industry difficulties might have been less severe, had there been a well prepared strategic plan for technological innovation. These problems include:

- increased production costs and declining eminence after World War II;
- lack of exports to meet rapidly rising steel demand in the Third World and industrialised nations;
- lack of emphasis on exporting proven technology, which could have justified new investments in R & D and innovation activities;
- the large integrated steel producers' lack of response to demographic changes and to opportunities for using scrap in local markets by means of mini-mills;
- and, lengthy and costly resistance to compliance with environmental regulations.

Innovation issues will be covered in this chapter sub-divided into two principal categories, namely: process innovations and product innovations, although it is obvious that this dichotomy is not clear-cut and well delineated. New processes make possible the production of new, improved or cheaper products; to meet the market requirements for new products, steel firms will have to innovate process-wise. The sub-division is thus artificial. It is motivated by the fact that the literature on innovation in the iron and steel industry essentially and almost

264

exclusively deals with 'major' process innovation. For the industry's viability, however, product innovations are equally important.

Process innovations

The OTA study (1980) defines a range of objectives for technology change, all of them concerning process innovations. This set of objectives may be used as the starting point for the discussion of this type of innovation. It is assumed that there are indeed differences in the emphasis on singular objectives in different countries, regions or firms. Nevertheless the OTA overview provides a good point of departure for a discussion of pertinent problems.[1]

> The U.S. steel industry needs new technology to cope with the changing nature of the economic, social, and political world in which it operates. New technology can improve the competitiveness of domestic steels with respect to quality and cost; it can also reduce industry vulnerability to inflation and other external factors. New and innovative technologies, some already commercially available and others with a significant likelihood of successful development and demonstration, offer potential for:
>
> - reducing energy consumption, including the use of coke;
> - making greater use of domestic low-grade coals;
> - reducing production costs as a result of improvements in process yield (although yield improvements will also put upward pressure on the price of scrap);
> - using more domestic ferrous scrap and other waste materials containing iron;
> - improving labour productivity;
> - reducing capital costs per tonne of annual capacity;
> - shortening construction time of new plants, and
> - allowing greater flexibility in using imports of certain raw materials and in importing semifinished rather than finished steel products.
>
> Although new and improved steel technology, alone, is not sufficient to reverse unprofitability and inefficiency, it is an essential ingredient for the future economic health and independence of the steel companies.

265

The OTA then lists three major strategic options for technological innovations:[2]

Possible technological solutions that might be considered for steel industry modernization are:

- to modernize existing operations by adding existing technology;
- to build new plants using the best available technology; or
- to develop and put in place at new plants radically innovative new technology.

These solutions differ in two major respects: in their capital costs, and in the amounts by which they can be expected to reduce production costs. The third, for example, is a high-cost, high-payoff solution; the second is somewhat less costly and somewhat less productive; the first is an incremental solution with incremental rewards. The choice among these solutions rests on how the costs and payoffs balance out.

The first solution, the extension of existing operations with available, improved technology (such as continuous casting), is generally considered to have the best balance between capital costs and reduction of production costs. The second option, involving construction of completely new plants using existing technology, would involve high capital costs that cannot be expected to be sufficiently offset by the limited production cost reductions it would bring. the third option, construction of new plants based on radically innova tive technology, will not be technically feasible for at least a decade; once feasible, however, there is a possibility that high capital costs could be sufficiently offset by significant production cost savings. Thus, the first option, complemented with a vigorous research program in radical steelmaking innovations, could prepare the industry now for short-term revitalization with the potential for long-term, fundamental modernization.

The most important technological innovations during the era of industrialised steelmaking are:

a)   Bessemer's invention of pneumatic steelmaking and its variations over time, for example the Thomas process, the Siemens-Martin oven, the open hearth and, after World War II, the basic oxygen process;

b) the so called hot-cast technology, permitting continuous production of liquid iron in blast furnaces;
c) the electric arc furnace;
d) the continuous casting process (originally patented by Bessemer in 1865); and
e) continuous rolling facilities for steel refining.

There has been a wide range of other developments in the production of coke, the benificiation of ore and also a wide range of analytical and control technologies. There have also been very important metallurgical developments under way during the last century.

OTA[3] mentions that during the period 1963-76 there have been more than 100 different developments adopted by the steel industry and draws the conclusion (with care):

- information on technological development in the steel industry is transmitted very rapidly;
- each new technology has characteristics that provide opportunities in markedly varying degrees to individual companies.

OTA uses a taxonomy for characterising new technologies which is very useful.[4]

Developments in the steel industry can be divided into four broad groups:

- radical and major technologies;
- incremental technology developments;
- regulatory technology developments; and
- developments from other industries.

The term 'radical' is used to describe a process modification that eliminates or replaces one or more of the current steelmaking processes or creates an entirely novel option for ironmaking and steelmaking. DR is thus a radical ironmaking change because it is an alternative to the traditional coke oven-blast furnace sequence. Direct steelmaking processes are radical changes beause they combine several processes into a single reactor. Continuous casting is radical because it replaces ingot casting and shipping, reheating, and blooming mill operations. Rolling of powders to strip is also radical.

'Incremental' technology developments include process modifications that improve efficiency, increase production, improve product quality, or lower operating costs. Energy conservation measures and the recycling of waste materials fall into this category. The economic impact and technological significance of incremental developments may be great.

Developments in environmental technology include add-on systems that do not alter the steelmaking process. Examples are biological treatment of waste waters, pipeline charging of coke ovens, and fugitive particulate collectors.

Developments from other industries can be used in many ways. Analytical and control techniques are transferred to the steel industry on a broad basis.

To provide an overview over the already mentioned wide range of technological options available and innovative technologies under way, but also to emphasise the uncertainties surrounding, as well as the technological potential connected with, different processes, a table[5] from the OTA study is reproduced here (Table 15.1). The OTA emphasises that the adoption projections in the table should be interpreted with care: a question mark indicates that a significant adoption is possible within that time period, if current pilot efforts show promise; an X-entry, that the technology should be significantly adopted within the time period; a dash (–), that adoption within the time period is improbable. A number of the technologies are currently adopted or near adoption.

Chapter 16 of this book demonstrates the often very long throughput time from invention to innovation in the case of steel industry technology. The OTA list contains examples of known technology which obviously still are rather far from being implemented, because some of their processes are not yet economically competitive.

One of the possible future technologies which is looked upon with high expectations is the hydrogen systems technology. Its feasibility was demonstrated in the laboratory about 40 years ago, but its industrial implementation still seems to lie beyond the year 2000, unless some major process problems are solved earlier. It may be interesting to remember here that the all-time giant innovator in steelmaking technology, Sir Henry Bessemer (1813–1898), not only designed the principles of the basic oxygen process (cf. chapter 16) but, some 120 years ago, in 1865, he also acquired the first patent for the continuous casting process. It then took almost exactly 100 years to commercialise the process (which first was done in Austria). The OTA study has been chosen as the primary source for this overview of technological development because it contains a rather thorough treatment of the subject and is unique in its way of presenting highly technical topics in a mode comprehensible to laymen – the study is written for a political audience. It also quotes many sources to which the technically interested and trained reader may want to turn for further detail.

The OTA study reports in deeper detail about the following new technologies:

1   *Radical major technologies*:[6] direct reduction; direct steel making; plasma steelmaking; and (very briefly) direct casting.

   Form coking and continuous casting are presented as cases of major technological innovations. Shorter cases are also provided for the Argon oxygen decarburisation and some other processes.[7]

2   *Incremental technology*:[8] From literally thousands of advanced incremental technological opportunities that are, or will be, available to the steel industry, OTA selects a few for illustrative purposes. The selection criterion is their expected wide applicability during the next ten years. They are: external desulphurisation; self-reducing pellets; energy recovery; continuous rolling; high temperature sensors; computer controls; processing of iron powders; and plasma arc melting.

3   *Long range opportunities* [9]

OTA identifies the two perhaps most significant opportunities for the steel industry as being

a)   complete elimination of the need for fossil fuels through the production of hydrogen by water hydrolysis, and
b)   adaptation of steelmaking to the potentials afforded by advanced nuclear reactors, high-temperature gas-cooled reactors and, further off, fusion reactors.

It singles out a number of innovations or technical developments as being particularly promising for the industry when attempting to solve the severe problems it is facing[10]:

● Alternatives to metallurgical coke as a blast furnace feed material;
● continuing improvement in coal-based direct reduction (DR) systems, including those that utilise coal gasification and those that use coal directly;
● continuing development of DR systems that allow for the use of alternative fuels and energy sources, including biomass, hydrogen and nuclear sources;

Table 15.1

Potential technological changes in the steel industry

Technological process[a]	Category	Significant adoption possible within:			Principal features
		5yr	10yr	20yr	
Plasma arc steelmaking[a]	1	—	—	?	Fast reactions, small units.
Direct steelmaking[a]	1	?	?	?	Eliminates cokemaking.
Liquid steel filtration	2	?	?	?	Improves product quality.
Continuous steelmaking[a]	1	X	X	?	Conserves energy and reduces number of reactor units.
Secondary refining systems[a]	2	—	X	X	Improves product quality.
Hydrometallurgy production of iron	1	—	?	X	Low-temperature processes.
Nuclear steelmaking[a]	1	—	—	?	Alternate energy source for steelmaking.
Hydrogen systems[a]	1	—	?	X	Alternate fuel/energy source.
Direct reduction processes[a]	1	X	X	X	Low-temperature solid-state reduction of iron ore to iron.
Coal gasification[a]	4	?	X	X	Alternate fuel/energy source.
Preheating of coking coal/pipeline charging	2,3	X	X	X	Reduces pollution and conserves energy in cokemaking operation.
Dry quenching of coke	2,3	X	X	X	Reduces pollution and conserves energy in cokemaking operation.
BOF/Q-BOP offgas utilization	2,3	X	X	X	Energy conservation measure.
High top pressure BF electricity generation	2	?	?	X	Energy conservation measure.

Process	Category			Description
Evaporation cooling	2,3	X	X	Improved cooling system, saves water usage.
External desulfurization[a]	2	X	X	Allows improved product quality and increased blast furnace productivity.
Induction heating of slabs/coils	2	X	X	Reduces scale formation, increases yield, and conserves energy.
Catalytic reduction process	2	—	?	Used with coal-based reduction processes to increase reaction rate.
Blast furnace fuel injection	2	X	X	Use of alternative fuel to replace coke, (possible energy conservation)
Direct casting of steel	1	—	?	Eliminates mechanical forming and heating.
Continuous casting	1	X	X	Direct conversion of liquid steel to solid slabs and squares. Major energy conservation measures and increased yield.
Formed coke	1	—	?	Replaces metallurgical coal/coke.
Biomass energy systems	2	—	?	Alternate fuel source.
Self-reducing pellets and briquettes	2	—	?	Iron ore/carbon flux is intimately mixed to allow reduction in the pellet.
Powder metallurgy steel sheet	1	?	?	No melting or reheating required, minimill concept.
Direct/inline rolling	2	?	X	Eliminates holding and reheating steps.
Computer modeling/control	2,4	X	X	Applies to any unit/process operation.
High-temperature sensors	2,4	X	X	Units to measure and control high temperature ironmaking and steelmaking process variables.

Categories: 1-radical, 2-incremental, 3-environmental, 4-transfer, ?-significant adoption possible if pilot efforts show promise, X-significant adoption possible.

a : includes a variety of processes.

Source : OTA 1980 (Table 75 page 193)

- improved methods of increasing scrap and sponge iron use;
- increased availability of direct reduced iron (DRI) as a substitute for scrap;
- increased availability of suitable iron oxide/carbon composites (pellets, briquettes) as 'self-reducing' materials;
- continuing development of systems for solid-state processing (the direct conversion of metallic powders into structural forms);
- alternative electric furnaces such as plasma systems;
- continuing development of high-speed melting, refining, processing, and transfer equipment;
- secondary or ladle steelmaking processes that allow separation of melting and/or primary refining from final composition control;
- improved instrumentation and control procedures;
- continuous casting and direct rolling; and
- improved methods for using waste materials and heat.

Technology suppliers

An aspect of innovation which has been paid remarkably little attention is the role played by technology suppliers. The dominance in steel of the traditional highly industrialised countries gradually disappears and there seems to be a parallel shift of technology suppliers from the mature industrial countries to younger industrial nations (paradoxically including India).

The low intensity of innovative investment in the USA has – despite the size of the market – implied that the US steel production equipment industry has been losing market shares rather quickly during the last 20 years. Most alarming, however, is the gradual loss in their competence relative to their overseas competitors. As the home market is the dominant one for them (and has become more so since the early 1960s), many firms have gone out of the market. Their outlook for the future is quite gloomy since, *inter alia*, physical investment plans in the modernisation of the US steel industry, in total comprising US $ 7 billion, originally to be implemented during the 1980s, had been shelved by mid-1982.

Equipment markets traditionally held by US technology suppliers have initially gone to Germany, Austria, Switzerland and Italy and later, from the beginning of the 1970s, growth has been most vigorous for the Japanese. Together with the Italians they use favourable financial conditions as a means of competition. In the meantime, however, it has almost become a must to invest in technology 'made in Japan'.

Perhaps the most remarkable shift since the late 1960s is towards the buying of more advanced technology from Japan, particularly in the large integrated equipment ranges.

The steel industry is one in which process development is often undertaken in close cooperation between the buyer–steel producer and the supplier of the machinery. It is not unusual for producers to develop and construct their own equipment. Once they have developed the skills, they will be inclined to sell their know-how to competitors, both at home and abroad. For the major Japanese steel producers the development and export of steelmaking technology is an integral part of corporate policy, aimed at compensating the sales of steel which are expected to level off in the future.

Most of the large and more advanced process machinery is produced upon customers' specification (or rather a jointly, but specifically, developed, individual specification). Thus there is constant incremental progress with experience from previous and recent installations being integrated into forthcoming projects. As is the case in other process industries, the machine producer will give away to competitors the know-how contributed by all previous buyers, unless it is patented (which, however, is seldom the case, as the supplier and the buyer often make joint and inseparable contributions to the development of the improvements).

As the Japanese have been the most vigorous investors on the one side, and as they, during the last eight to ten years, have been particularly interested – as well as clever – in developing processes and improving the quality of product by improving their equipment in order to meet their customers' needs and demands, it has since become more and more established amongst buyers of steel-producing equipment that if one wants to remain, or become, competitive, one has to buy the best available technology – which in most cases equals Japanese technology.

As demonstrated in the analysis of the effects of protective measures in favour of the US steel industry, the quota and mini-price control regulations contributed to the increasing penetration of the higher quality ranges by foreign suppliers, in particular the Japanese ones. This has therefore indirectly helped the Japanese to become the masters of technology supply to the modern steel industry. (See Figure 15.1.)

It seems apt to give credit to the Japanese steel producers for the interest shown in the steel consumers' functional needs and demands, when it comes to qualitative properties of steel (or alternative materials) as a material used in the manufacture of consumer or producer goods. Irrespective of whether this growing market orientation came as a reaction to the decline in the carbon steel market at large, or as a consequence of import restrictions, the Japanese producers have tried more deliberately than their foreign competitors not only to inform themselves about, as well as adapt to, steel users' needs, but they have also, by means of higher research intensity (measured for example by the share of steel prices spent on R & D, or by the number of researchers employed by the steel-producing firms),[11] been pushing the qualitative ranges of their product portfolio upwards – thus intruding into the markets of not only special or alloy steels, but also other materials. They are making steel more attractive again. They also act as problem solvers for steel-using industries: A remarkable example is in the 20–25 per cent improvement of pay load per square unit of steel – mm^2 or square inch – offered to the shipbuilding industry.

Thus when assessing the innovation potential of an industry it is not sufficient to merely analyse process innovations, both within the industry and with their technology suppliers, for its innovation capacity. One must also pay some attention to product innovation which may affect the future of the industry, and which will necessarily also imply consequent process innovation.

Product innovation: the future of steel as a material

It is neither possible nor adequate even to attempt sketching a future study of steel properties and uses in the context of a volume devoted to the present-day problems of the steel industry. In Part II it was emphasised that the intensity of steel consumption levels off and later declines with increasing GNP per capita. Several reasons for this were given, including the fact that new, lighter, cheaper or even better steel substitutes gradually become available or are even becoming more attractive on account of their better corrosion resistance, lighter weight, better elasticity, formability, malleability, workability or similar properties in production and/or use.

One example of competing new materials of high potential are carbon fibre armed[12] plastics (PAN), which are coming into use in air and surface transportation equipment production (aircraft, cars). This new material is lighter as well as stiffer and more durable in much wider ranges of temperature than other materials: it is lighter than aluminium, and at the same time is harder than steel. As PAN is a new material, it is still relatively expensive – about US $ 50 per kilogram. With increasing utilisation it will become cheaper – the average forecast is US $ 20 per kilogram by 1990. This is to give just one example of a material that is not only usable in products where steel may have no chance, but which may also intrude into traditional and future steel domains.

On the other hand steel may also be developed for novel uses or to correspond to new functional requirements, possibly also replacing other materials. One example is the 'super plastic' steel.[13] In temperatures between 650 and 800°C it is stretchable up to 1,200 per cent; in normal living temperatures it is tenacious and hard. It permits even rolling of cast iron. The discovery of the properties of this super eutectic steel permits new or drastically simplified modes of working and forming steel parts, in particular of complex and heavy workload parts.

The most important development as far as the steel-producing industry's *products* are concerned – in response to actual and expected requirements from steel manufacturing industries and their customers – may be summarised in the following way:

1   Increased durability, also implying lower total life cost or lower cost per unit of lifetime, by improved mechanical and/or chemical properties: higher resistance to abrasion and to corrosion. To the same category also belongs lower cost of maintenance.

2   The above implies lower cost of steel in application or use. There are other requirements to be met by lower cost of steel, in order to reduce the cost of input to the steel manufacturing industries. Examples are replacing alloy steel by improved qualities of carbon steel etc. There is a distinct tendency for lower priced qualities to intrude into the market shares of more expensive alloy steel.

3   Reduced weight whilst maintaining other properties, or combinations thereof, for example higher strength, tenacity, elasticity. The effects of lighter weight are for instance: lower fuel consumption for ships or cars, higher payload per unit of transportation space etc.

4   Extension of performance criteria – implying the extension of steel utilisation possibilities. The effects may mean reducing competitive pressure from other materials, but also development of entirely new or extended ranges of steel applicability, for example, under extreme temperatures and other requirements of so called 'frontier technology'. Close to this is the development of entirely new functional properties, hitherto not obtainable from steel or even other materials. The requirements may be achieved by new alloys or by novel production technologies.

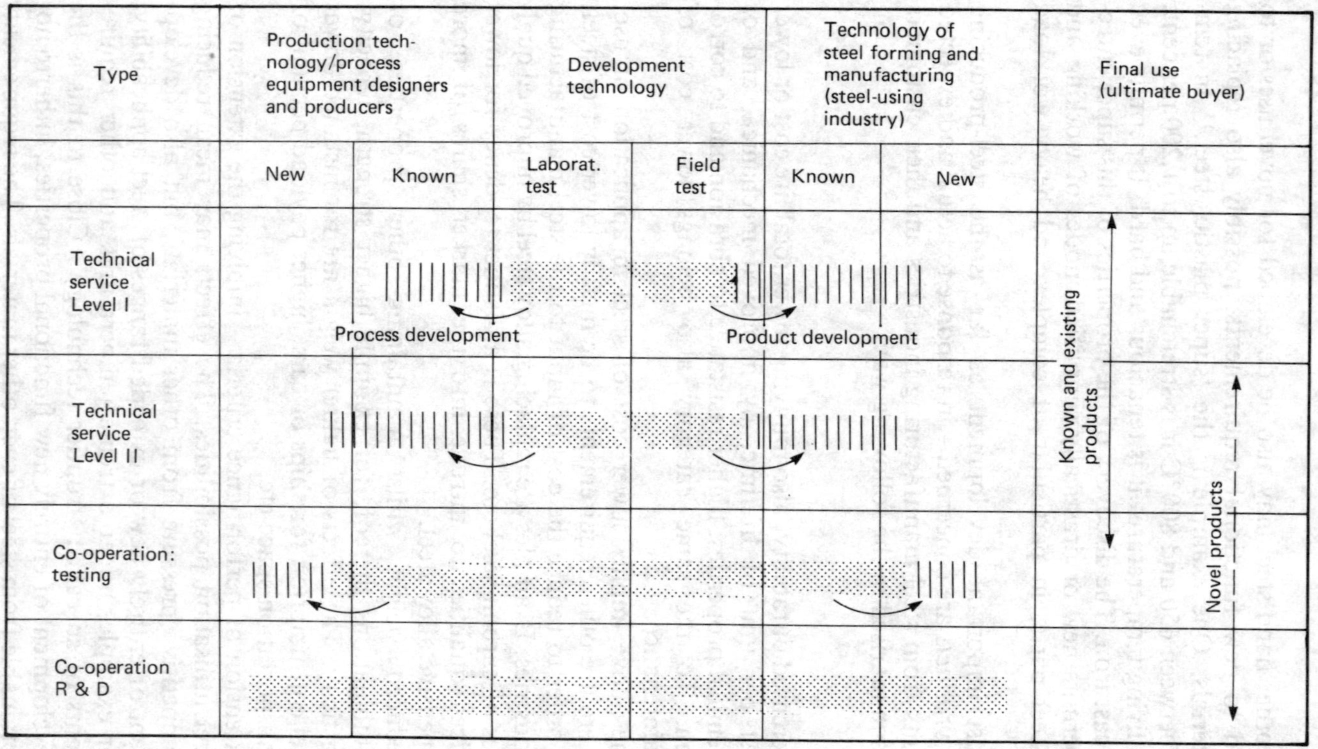

Figure 15.1

Type	Production technology/process equipment designers and producers		Development technology		Technology of steel forming and manufacturing (steel-using industry)		Final use (ultimate buyer)
	New	Known	Laborat. test	Field test	Known	New	
Technical service Level I							
Technical service Level II							
Co-operation: testing							
Co-operation: R & D							

Process development

Product development

Known and existing products

Novel products

Figure 15.1
Nippon Steels' technical service and R & D: categories

The above summary is schematic rather than a list of potential new products under way.

In principle the *market* may most likely offer many more opportunities and challenges to innovation than processes. Many products will cause necessary innovation or development in processing equipment as well, since new processes will open new opportunities to cover or develop the market's actual or potential requirements. It is likely that the emergence of new products will have an impact upon the industry's structure, particularly in the range of smaller and/or more specialised mills.

Chapters 16 and 17 by Ács and Rosegger

The dominant theme in the literature on iron and steel innovation during the last 15–20 years has been the US steel industry's rather mediocre performance when adopting new superior process technology (one generalisation says that when adoption time for the steel industry in general is 6–7 years, it will be 16 years or more in the US steel industry). The famous article by Adams and Dirlam (1966)[14] sounded as an alarm clock. By that time it was, however, fairly widely known, and had been for 5–7 years in the industry, in financial quarters, and in the business press, that the US iron and steel industry not only was generally running plants using old, if not out-dated, technological equipment, but that even recent additions to capacity had been done to pre-war technological standards.

As the theme of the 'transoceanic quarrel' has one of its focuses on the structure and performance of the US iron and steel industry, two chapters by invited contributors analyse this industry's innovative performance, from different points of view.

Chapter 16, by János Ács, summarises the development of the steel industry's technology from the viewpoint of innovation policy and also from the emerging innovation theory. The chapter intends to be a contribution to the development of some basic principles of innovation policy in general and for the innovation policy for the steel industry in particular.

Ács gives an overview of major innovations in the steel industry during the era of industrialised steelmaking. He pinpoints differences in innovative behaviour and tries to explain them, concentrating on the situation of the US steel industry. The author draws attention not only to the needs for hardware innovations and solutions, as well as potential solutions to production problems. He points also to the increasing importance of software and the consequent need for software innovation, for example, for control processes, simulation of technological requirements, and alternative strategies to meet new requirements. Further, the need for an availability of management aids to improve managerial controls, marketing, physical investment decisions, handling of uncertainty and so on is elaborated.

Ács outlines decision making problems concerning innovation and investment and discusses possible long-range strategies. He then draws a number of theoretical and pragmatic conclusions from observations made on innovative/conservative behaviour in the steel industry.

Rosegger, a highly experienced and competent student of technological innovation, particularly within the US iron and steel industry, presents a rather thorough analysis of the seemingly sluggish way the industry responded to the two major post-World War II technological innovations, the basic oxygen and continuous casting processes.

Against the criticism to which the industry has been subjected, Rosegger investigates the piecemeal approach to technological adaptation, by means of empirical studies of managerial decision making behaviour. Rather than *ex post facto* blaming the industry for not having invested in 'best available' or 'best practice' technique, Rosegger investigates the decision options and parameters at the time adaptive decisions were being made. A basic assumption is that one often takes industrial averages – and thus also the costs and benefits of technological innovations 'on the whole and on average' – whereas such decisions are always based on individual firm conditions and options available to the firm at a certain point in time.

There is no doubt that steel industry firms have been late in recognising the deep changes taking place in both markets – as far as, for example, the importance and long-term consequences of imports, but also of regulative policies, are concerned – and in technology. In their strategic moves, when adapting, they have often chosen options which in the medium and long run have resulted in smaller lot sizes or runs and consequently increasing unit costs. Spreading out the sales and service efforts has led to similar consequences.

When individual decisions about investment in new technology – for example, process innovations – are concerned, the *ex ante* determinants of the costs and benefits vary greatly from innovation to innovation, because of respective functional characteristics. Further innovations are modified and refined over time, implying that the uncertainty faced, when estimating the degree of maturity of an innovation, is rather high. A specific set of constraints are the vintage, the scale, and the technical properties as well as layout of existing facilities to which a novel technology is to be added. Remember that the issue is piecemeal innovation.

Rosegger investigates, for a set of plants at a sequence of points in time, the contextual contingencies of existing plans when innovative measures are at stake. He defines the functional characteristics of the novel technology in question, and scrutinises the decision making behaviour of management when quantitively and qualitatively assessing additions to available capacity, displacement of functional facilities and replacement of capacity. He further investigates the importance of the scale of existing plants and their technological characteristics upon decisions to be taken – or not to be taken.

For both oxygen steelmaking and continuous casting, Rosegger finds at the micro-economic level sets of plausible reasons for taking certain decisions as to whether or not to invest. He shows for instance that promised improvements by new over old technology could only be realised to rather limited degrees, or that, when the expectation and real effects actually were close to each other, the cost factors concerned were of rather marginal importance in the steelmaking process. Management did consider the sometimes rather rigid requirements on inputs, servicing, etc., upon which the novelties were contingent.

Rosegger finds support for the findings of earlier research in so far as:

a)   even major innovations, when placed into the context of an operating system, do not render sufficiently revolutionary cost savings to be worth major investment efforts; further

b)   management's bias towards incremental (rather than thorough) technical change seems to be based on largely correct assessments of the technical and economic characteristics of large material-dominated production systems;

c)   textbook-type capital budgeting procedure does not fit very well and thus is not used very much in systems having the said properties;

d)   data and information produced by intrafirm costing and accounting systems and practices often influence decision making processes in a conservative way (but also make interfirm and even intrafirm process cost comparisons rather meaningless).

Rosegger concludes that piecemeal innovation strategy implied a reasonable and sometimes necessary strategy, but not a sufficient condition for the traditional steel industry of the USA to remain internationally competitive. He elsewhere[15] investigates the 'strategy' of piecemeal innovation. Some of his conclusions are:

1   There is no indication that there is an 'optimal' plant size which increases over time, with cumulative experience. Instead, plants over a wide range of capacities remain economically attractive.

2   Scale increases are achieved by adoption of increasingly larger processing equipments; however, each such increment is decided upon under considerable doubts about the technical feasibility of scale-up.

3   Rosegger's empirical data point in the direction of scale-up decisions being 'accidents of necessity', that is, of *ad hoc* type rather than following a grand design strategy concerning incremental growth over time.

4   Major new technical innovations are usually met with a 'wait-and see' strategy by management of existing large integrated plants, that is letting the new technology mature, letting other firms carry the burden of experience as well as of improvement of the novel processes. However, waiting alone obviously does not necessarily mean learning and gaining experience from others.

It may be rather appropriate to recall one of the central conclusions drawn in the classic Adams and Dirlam article:[16] 'big steel is neither big because it is progressive, nor progressive because it is big'. They conclude their investigation of the slow adoption of the BOP by large US steelmakers with an open question which is well worth repeating now that the pace of adoption of the continuous casting technology is known. Departing from the plausible, and often stated, assumption that the leading members of homogeneous oligopolies may well compete in innovation rather than in pricing, and that the welfare gains from such competition may be as high, if not higher, than from price competition, they conclude that the history of basic oxygen process introduction (and we may now add, of continuous casting technology adoption) by large US steelmakers 'reveals the oligopoly as failing to compete in strategic innovations. What benefits, then, remain for large size in steel?'

Conclusions

Some of the central conclusions on the issue of innovation in the iron and steel industry may be summarised in the following way.

The presence or absence of innovation alone, is not the only, but is a rather important, factor behind medium and long term success or failure and survival capacity of the industry, its firms or its plants.

Innovative measures alone thus cannot turn the fate of the industry to the positive, but each firm should, for each of its products, markets, processes and plants, try to assess what the options available may render in terms of market shares, profitability and survival capacity in the medium and long term.

The emphasis must not only be on process innovation, however important it may be. The markets' actual and potential demand is an important source of ideas for product innovation. Still more important is the fact that – as in most industries mainly serving industrial customers – a close and active cooperation with the iron and steel manufacturing industry, that is the iron and steel industry's immediate customers, is a highly important activity strategy for survival and competition.

There are many options available for both product and process innovation. Some of them are contradictory (like the increasing level of economies of scale of integrated steel plants on the one hand, and the economies of small and medium scale for non-integrated producers on the other). This makes necessary careful scanning of opportunities and threats, as well as the design of well laid-out development plans to follow (where the dividing line between success and failure can often be very narrow, as for example shown by the harm the energy price increases have done to the producers who chose to put all their eggs into the direct reduction technology basket).

An innovative technology must always be assessed to its economic potential. It must also be evaluated as to its fit to the general market and physical conditions of the firm or plant. This may imply the danger of rejecting an innovative technology which should have been entered upon even at the expense of existing equipment becoming obsolete.

Even if piecemeal improvements may be a good strategy to postpone major commitments to one mainstream technology, it cannot carry up long-term competitiveness and survival for the industry or for its firms.

Notes

1   Office of Technology Assessment, US Congress, *Technology and Steel Industry Competitiveness* , US Government Printing Office, Washington DC, 1980, p.268
2   OTA, op. cit., p.270
3   OTA, op. cit., p.188
4   OTA, op. cit., p.192
5   OTA, op. cit., Table 75, p.193
6   OTA, op. cit., pp.194–210
7   OTA, op. cit., pp.282–298
8   OTA, op. cit., pp.211–214
9   OTA, op. cit., p.214.
10   OTA, op. cit., p.185.

11 Nippon Steel (turnover 1981–1982 almost US $ 12 billion; one third of its turnover goes into car making), employs 2,300 persons in R and D. Its annual R and D budget is approximately US $ 200 million.

12 Polyacrylnitrile, PAN.

13 'Super eutectic' (1–2 per cent carbon content instead of 0.7 per cent held by normal eutectic steel).

14 Adams, W. and Dirlam, J.B., 'Big Steel Inventions and Innovation', *The Quarterly Journal of Economics* , vol. LXXX, no. 2, May 1966.

15 Rosegger, G., 'Diffusion and Scale Dynamics; a case study', *Technovation* , vol.1, 1982, pp.201–304.

16 Adams, W. and Dirlam, J.B., op. cit.

16 The development of the steel industry from the viewpoint of innovation theory

János Ács[1]

Steel and industrial development

Although the basis of iron metallurgy has been well known for over 3,000 years, steel metallurgy only gained importance after the first Industrial Revolution. The total world production of iron and steel for 1830 has been estimated at half a million tonnes. By 1929 the production figures for iron and steel had reached 120 million tonnes. In the past 100 years the production of raw steel increased from 10 million to over 700 million tonnes per annum. Thus the predictions made in the 1950s by the United Nations Economic Commission in Europe proved to be quite accurate, a contrast to the present situation, in which predictions are no longer so reliable.

Iron metallurgy is an important factor in economic growth and in a nation's infrastructure. Our present technology is mainly based on steel metallurgy; steel has even become the symbol of industrialisation. Among the industrial countries, there are only a few that resort to importing the necessary steel products. Industrialisation starts in most cases with the building of steelworks and, in the course of the economic development, metals and metal products show a rising tendency in the relative importance in the manufacturing industries.

Chenery (1960)[2] and Maizel (1963)[3] conducted various econometric studies covering more than 30 countries. The results of Maizel's research, shown in Figure 16.1, indicate a steep decline in the relative importance of textiles, food processing and, after a growth phase, also of the miscellaneous group. Until World War I, steel production in the USA was growing at a faster rate than other industrial production. In the 1940s the growth was more or less equal. Thereafter the growth rate of steel production has remained below that of total industrial production. The other industrially developed countries experienced similar growth patterns for steel.

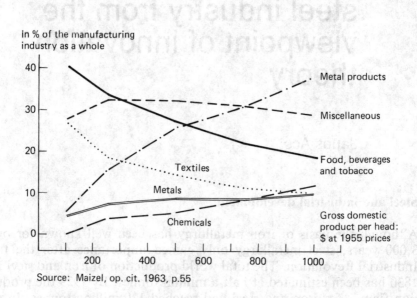

in % of the manufacturing industry as a whole

- Metal products
- Miscellaneous
- Food, beverages and tobacco
- Textiles
- Metals
- Chemicals

Gross domestic product per head: $ at 1955 prices

Source: Maizel, op. cit. 1963, p.55

Figure 16.1
Typical patterns of growth in manufacturing industries

The econometric studies applied here also serve to predict the estimated steel consumption for the next years. The per capita steel consumption has been calculated by means of double logarithmic equations deduced from the per capita income. The results show that steel consumption in developing countries is rising faster than in highly developed countries.

Another method shows the so-called 'steel intensity' of the national income. This coefficient shows how many kilograms of steel are necessary for the production of one unit of the national income. Table 16.1 shows the development of the steel intensity of the national income in the selected countries. Table 16.2 shows the probable trend of steel consumption per capita. As one can see from these tables the steel intensity of the national income is receding in the highly developed countries.

Table 16.1

Steel intensity by area/apparent consumption/gross national product (GNP) kg/1963 US $

	1955-9	1960-4	1965-9	1980	1985	1990
USA	189	170	179	161 (138)	156 (127)	151
Canada	190	174	201	173 (174)	164 (164)	157
Latin America	135	131	147	200 (190)	223 (200)	244
Oceania	191	216	228	216 (230)	207 (232)	198
ECSC	245	251	238	227 (212)	217 (195)	210
UK	256	243	234	213 (232)	206 (211)	198
Other Western Europe	163	196	217	245 (244)	245 (248)	240
USSR	251	282	291	281 (297)	270 (273)	260
Eastern Europe	240	286	293	302 (314)	289 (303)	276
Japan	310	428	474	444 (405)	432 (360)	420
India	111	156	144	132 (182)	138 (205)	144
China & North Korea	152	212	243	314 (392)	339 (443)	366
Other Asia	59	84	106	161 (162)	178 (168)	196
South Africa	321	304	342	377 (383)	390 (395)	404
Other Africa	90	81	90	128 (103)	139 (105)	154
Middle East	122	127	155	235 (185)	264 (208)	285
Total World	203	215	226	232 (225)	231 (217)	231

Note: The first three columns are taken from Projection '85. The figures for 1965-9 will be slightly revised. Figures in brackets are the forecasts shown in Projection '85. The quantity of steel consumed in a country is expressed in terms of crude steel.

Apparent steel consumption is the result of the following computation: production of crude steel + imports multiplied by yields factor – exports multiplied by yields factor. With the progress of the share of continuous casting the yields factor has been changing substantially, improving from 1,31 to 1,21, and will diminish further.

Source: International Iron and Steel Institute (IISI), *Projection 85* and *Projection 90*, p.26

Table 16.2

Steel consumption per capita 1955–90 before adjustment for continuous casting in kg

	1955–9	1960–4	1965–9	1980	1985	1990
USA	541	527	663	780 (723)	864 (751)	956
Canada	370	360	494	686 (652)	791 (733)	877
Latin America	41	44	55	114 (93)	152 (110)	202
Oceania	236	296	358	453 (508)	500 (571)	502
ECSC*	272	350	408	622 (640)	712 (737)	803
UK	355	382	413	492 (553)	514 (581)	598
Other Western Europe	98	140	189	335 (351)	406 (434)	481
USSR	245	326	407	631 (680)	737 (749)	853
Eastern Europe	188	276	342	621 (643)	758 (764)	909
Japan	121	260	453	935 (967)	1.145 (1.103)	1.380
India	9	14	13	15 (20)	19 (25)	26
China and North Korea	14	19	23	53 (58)	74 (80)	102
Other Asia	6	9	14	32 (32)	42 (39)	54
South Africa	144	155	211	289 (315)	338 (364)	420
Other Africa	9	9	10	19 (14)	23 (15)	30
Middle East	24	31	46	123 (79)	173 (100)	226
Total World	101	121	148	187 (210)	242 (231)	278

Note: The first three columns are from *Projection '85* and subject to revision.
Forecasts in brackets are from *Projection '85*.
See also the comments to Table 16.1.

*ECSC = European Coal and Steel Community

Source: International Iron and Steel Institute (IISI), *Projection 90*, p.28.

According to a recommendation from UNIDO, in the middle of the 1970s the steel-producing capacity in the developing countries in the year 2000 should be brought up to 420 million tonnes per year; this would amount to a quarter of world steel production. This means that within approximately 20 years about 100 steelworks the size of the works in Linz should be built, and would indicate a growth rate of per capita steel consumption of about 3.5–4 per cent. As a result, the developing countries together with the COMECON countries have become the main growth areas for world steel consumption. In the past decade the highly developed countries achieved a steel consumption growth rate of 2.3 per cent instead of the 4.7 per cent forecast in *Projection 1985* . The estimated growth rate of steel consumption for the 1980s is about 0.6 per cent.

The regressive tendency of steel intensity in highly developed countries is caused by a) rising new industries like electronics, chemical and pharmaceutical works; b) increasing use of light construction materials as substitutes for steel; and c) improved durability of higher quality steel products.[4] Based on these facts, the International Iron and Steel Institute predicted an average increase rate of 4.5 per cent until 1985; this means that for 1985 a world production of 1,100 million tonnes of raw steel is predicted (*Projection 1985*). As the present production figures show, this prediction has to be corrected and decreased due to the fact that the billion-tonne-level of world raw steel production cannot possibly be reached in the 1980s. (This is in contrast to the accuracy of the predictions from the 1950s.)

The most important basic innovations in steel-producing methods since World War II

In order to demonstrate the role of innovations in the development of steel industry until now, we will refrain from a discussion of the present state of innovation theory, but instead refer to the literature on this subject. Hinterhuber has given a useful definition of innovation.

Innovation is any important change

1   in the interactions of reciprocal actions of the enterprise with the natural or social environment on:

    a)   the introduction of new and improved products and services onto the market,

    b)   the application of new and improved methods of production and

c) the elimination of negative effects of the methods, products and services on the natural and social environment;

2 in the internal structure of the enterprise by changes in the organizational structure, introduction of an electronic information system, increase of productivity of the factors of production, etc.;

3 in the subsystems of the enterprise (realization of a new product/market combination, creation of a new subsystem, etc.)

An innovation problem is a 'non-reducible' problem, which is a problem that cannot be solved in a satisfactory way by conventional methods (by existing production processes, by existing products, etc.) but requires a new impulse of ideas, a relevant change.[5] (Translated from the German by J. Ács.)

Mensch proposes in his work a dichotomy of basic versus piecemeal innovations. A radical variation of the usual ways of solution constitutes a basic innovation, whereas a further development in this field represents a piecemeal innovation.[6] He emphasises that basic innovations are determined by a) the level of knowledge, b) the know-how in the different lines of industry, and c) the nature of the suppliers and buyers of the new products.

In addition, it is useful in many instances to differentiate between innovations in products, processes, and organisational structures.

Table 16.3 and its supplement show the most important basic innovations of steel production methods since the introduction of the coke blast furnace in 1796. As Adams and Dirlam noted in a classic paper,[7] which is still relevant, the three most important innovations in steelmaking the Bessemer, open-hearth, and basic oxygen processes (BOP) – did not develop out of the big corporations' research laboratories. In general, innovation literature does not consider the steel industry as research-intensive. The development of BOP is somewhat of an exception because it was developed in various university institutes and perfected by the Austrian steel firm VÖEST,[8] which had at that time a total steel ingot capacity of about 1 million tonnes; VÖEST, in other words, was less than one-third the size of a single plant of the United States Steel Corporation. (See Table 16.4 and history of BOP.) As early as 1856 Bessemer pointed to the possibility of the application of oxygen, but the level of technology at that time prevented him from carrying out the necessary experiments. In connection with the diffusion of process innovations in steelmaking it is noteworthy that, in the American steel industry, the Bessemer air injection process had thor-

Table 16.3

Some important technological basic innovations in the steel industry

	Invention AD	Innovation AD	Time Years
1 Blast furnace	1713	1796	83
2 Cast steel	1740	1811	71
3 Thomas steel	1855	1878	23
4 Open-hearth	–	1880	–
5 Continuous steel casting	1927	1948	21
6 Continuous hot rolling	1892	1923	31
7 Stainless steel	1904	1912	8
8 LD-process	1943	1952	9
9 Direct reduction	1928	1969	41
10 Coal reduction	1982	?	

Blast furnace
1713 Darby: Application of coke in reduction
1796 Reden: Builds the first blast furnace in Gleiwitz

Puddle furnace
1783 Cort: Invention of the puddle method
1824 Rasselstein: First German puddle furnace

Cast steel
1740 Huntsmann: Steel foundry in crucibles
1804 Fischer: First cast steel in Germany
1811 Krupp: Cast steel

Thomas steel
1855 Bessemer: Invention of the Bessemer converter
1866 Bessemer: General application of the Bessemer converter
1878 Thomas: Basic additive for processing phosphate ores leads to industrial breakthrough

Special steel
1771 Rose: Research on metal alloys
1856 Musket: Improvements of steel tools by adding tungsten, titanium and molybdenum

Rolled wires
1773 Rasselstein: First German sheet metal rolling mill
1820 Rasselstein: Wire rolling mill

Table 16.4

Development of Linz-Donawitz process
(basic oxygen process)

Year	Phase	Event
1856	1	Bessemer mentions the possibility to apply oxygen in steel production
1929	*	Linde-Fränkl-Method makes economic production of oxygen possible.
1943	2	Patent no. 735196 awarded to Prof. *v.* Schwarz, Berlin-Charlottenburg
1946	2	Belgian patent no. 468316 by John Miles
1948	3	Durrer-Hellbrügge, experiments in Gerlafingen prove the feasibility of the patents
1949	4	Start of the development by VÖEST-Linz-Donawitz
1950	5	Development and pilot plant to test the process
1952	6	Start of production using the new process

*important constraint removed

Notes: Phases in innovation processes (Technology push type)

1 Discovery or development of a new theory (new knowledge)
2 Realisation of a practical possibility of using it (discovery, basic invention)
3 Proof of feasibility
4 Start of market-oriented experiments
5 Decision to implement the technological invention (decision)
6 Start of production by the new method or introduction of the new kind of product in the market (basic innovation)

Source: Mensch, op. cit., 1972

oughly replaced crucible steel within a period of 10–15 years. The maximum share of the Bessemer process of the total US steel output was approximately 90 per cent in the 1880s. Since the Bessemer process presents quality problems (nitrogen in the blast air makes the steel brittle and less malleable) the open-hearth furnace was invented in 1880, but it took about 75 years before this process (also known as the Siemens-Martin process) had reached a share of approximately 90 per cent of the American steel output.

The open-hearth process had indeed offered an excellent solution to the quality problems of the Bessemer process, but it is relatively slow; one charge takes about eight hours, in contrast to half-an-hour with the Bessemer process. So it is not surprising that continuous efforts were made to find a method to apply oxygen to steelmaking. In 1929, only after the Linde Corporation was able to produce oxygen (Linde-Fränkl-process) at a favourable cost, could the technological experiments using oxygen in steel production begin. The Linde-Fränkl method of oxygen production provides one of the essential basic side prerequisites for the maturation of the innovation process. After a number of unsuccessful attempts, a patent was granted to Professor Schwarz for the method in 1943. In 1948 Robert Durrer was able to test the practical application in a 2.5 tonne pilot converter at Gerlafingen (Switzerland). In 1949 VÖEST carried out various experiments with a modified converter using this technology, and since 1952 raw steel has been produced in the steel plant in Linz and since 1953 in Donawitz, Austria. The revolutionary new steelmaking technology is referred to in the German literature as the LD-process, which stands for Linz-Düsenverfahren. (In the following, LD-process and BOP are used as synonyms.)

The strategies of the US steel companies regarding the diffusion of the LD-method have been scrutinised by Adams and Dirlam. They reveal the excessive caution of G.C. Tennant, General Counsel of US Steel Co., who stated that such new processes as the oxygen method (which had been described in glowing terms by engineers appearing before the Kefauver Committee of the US Senate and by State Department technicians writing from abroad) required 'further development' before they 'conceivably could be substituted for, or displace, existing practices'. Their 'growth potential', Mr Tennant felt, 'cannot be forecast'.[9]

As Table 16.5 shows, the US steel industry only gradually realised the potential of the LD-method, and the three largest steel corporations were among the slowest to adopt this method.

Table 16.6 shows that the process of adoption of the LD-process in selected US steel firms between 1960 and 1980 followed almost a random pattern.

Table 16.5

Distribution of LD oxygen capacity among United States steel producers, 1963

US Steel Producers' rank in the industry*	Oxygen steel capacity (tons)	Percentage of US oxygen steel capacity	Percentage of total US steel capacity*
1st, 2nd, 3rd	0	0	52.27
4th, 5th, 6th	6,550,000	50.62	14.76
9th, 10th, 12th, 15th, 19th	6,390,000	49.38	7.06
All companies	12,940,000	100	100

*Based on company ingot capacity as of 1 Jan 1960.
Source: American Iron and Steel Institute, *Iron and Steel Works Directory of the United States and Canada,* 1960; Kaiser Engineers, *L-D Process Newsletter,* 27 September 1963. Cited in Adams, W. and Dirlam, J.B., op.cit., p.183

The first American steel producer to adopt the LD-method was the twelfth largest at that time (by tonnage), McLouth, in 1954, followed by the fourth largest (Jones & Laughlin) three years later. The ninth largest, Kaiser, adopted it in 1958, the nineteenth (Acme) in 1959, the tenth largest (Colorado Fuel & Iron) in 1961, and the fifth largest (National) in 1962. Nine years after its American introduction, the process was adopted by the fifteenth largest (Pittsburgh), twentysecond largest (Allegheny-Ludlum), and the sixth largest (Armco). By the end of 1963 oxygen steelmaking capacity in the United States had been adopted, as shown in Table 16.6. One year later, in 1964, the two giants, US Steel and Bethlehem Steel, introduced the process. The third largest producer, Republic, followed in 1965. This despite the fact that, in the early 1960s, the 'Big Three' were sufficiently burdened by overaged and obsolete equipment, that they had incentive enough to be among the first to try out the new method. In addition, the intensification of the introduction of the LD process in the US steel industry occurred while the steel industry was using only 75 per cent of its existing capacity, that is, in 1964. In other words the innovation strategy of the three largest American steel companies in 1964 was merely a reaction to the increasing competition, especially from foreign steel producers, mainly the Japanese.

Schenck (1972)[10] characterised the resulting situation as follows:

Steel production has been less than domestic demand for a number of years, so it is clear that the American manufacturers are more interested in using imported steel because they seem to consider it in some ways as more attractive than domestic steel. In this way American mills are losing orders, which they themselves could easily fulfill. This in turn results in stagnating, insufficient utilisation of capacity. The chairman of US Steel, Edwin H. Gott, explained this situation as being due to inflation, high personnel costs and high expenditures for environmental protection.

Table 16.6

Percentage capacity in basic oxygen furnace by firm, 1960–80

Firm	1960	1964	1968	1970	1974	1980
All industry	3.7	13.5	42.5	56.9	69.7	85
Alan Wood	0	0	100	100	100	100
Allegheny Ludlum	0	0	100	100	100	100
Armco	0	28.6	66.1	71.8	77.4	78.4
Bethlehem	0	0	38.9	66.8	80.5	88.0
Colorado Fuel & Iron	0	42.7	50.0	56.2	100	100
Crucible	0	0	100	100	100	100
Detroit Cyclops	0	0	0	0	30.8	100
Edgewater	0	0	0	0	100	100
Granite City	0	0	100	100	100	100
Inland	0	0	35.0	37.7	62.3	77.4
Interlake-Acme	57.6	57.6	57.6	57.6	100	100
Jones-Laughlin (LTV)	11.8	45.3	75.9	75.9	84.5	100
Kaiser	62.4	62.4	65.1	65.1	65.1	100
Keystone	0	0	0	0	100	100
Lone Star	0	0	0	0	0	0
Lukens	0	0	0	0	100	100
McClouth	100	100	100	100	100	100
National	0	31.9	67.0	100	100	100
Republic	0	0	64.1	80.3	80.3	100
Sharon	0	65.1	65.1	65.1	100	100
US Steel	0	5.4	26.7	26.7	71.3	80.6
Wheeling Pittsburgh	0	39.5	68.7	68.7	100	100
Youngstown	0	0	0	33	33	57.4

Figures indicate share of capacity in BOF of total capacity in BOF + open hearth in the given year.

Source: Oster, S., 'The Diffusion of Innovation Among Steel Firms: The Basic Oxygen Furnace', *The Bell Journal of Economics,* vol.13, no.1, p.50.

The development of the steel industry in Japan has been completely different from that of the US. As a result, Japan became a leading steel producer in spite of the lack of natural resources necessary for such an industry. The attention of the world's steel industry was drawn to the Japanese steel strategy in the 1960s: the selection of special sites, the unusually large size of plants, which would be unthinkable even in the US, technological research, special infrastructure, and marketing considerations. The size of the R & D departments attracted special attention, as some of the steel corporations employ more than a thousand scientists. All these factors considered, it is not surprising that the Japanese won the contract for an oil pipeline in Alaska, where extreme specifications for stability of material, elasticity, strength, weldability at minus 60°C. etc. were required. The Japanese research laboratories simulated the local conditions, which enabled them to develop the required quality steel as specified.

A further basic innovation, which is presently gaining importance and, in some cases, may make the blast furnace obsolete, is the direct reduction method, also known as the Midrex direct reduction method. Midrex-technology is suitable for compact, small enterprises, which are able to adapt to the regional demands. The advantages of the direct reduction method are: up to 50 per cent lower investment costs, lower energy costs, independence from foundry coke. Already in 1976 a capacity of approximately 6 million tonnes based on this technology was available, which by 1990 should increase to 10 million tonnes. The possibility of using lignite, natural gas or any other gas, as well as titanium sands, waste materials and foundry dust in the reduction makes this sponge iron technology very interesting. The direct reduction process originated in the USA as a development of the Midland Ross Corporation in Cleveland, Ohio.

This innovation is connected with the name of Willi Korf, who had to endure a difficult battle with the steel giants in Germany. He built a steel mill in America (South Carolina), using the new method. His second mill – also based on the direct reduction method – was put into operation at Hamburg-Finkenwerder in 1972. An integrated steel mill is to be built in Kursk, Soviet Union, with an annual capacity of 6 million tonnes based on the direct reduction method. This is one of the largest German–Soviet industrial projects up to now. Korf, formerly a man violently opposed by the steel giants, opened new markets and new possibilities with his innovations in developing countries and is now a respected entrepreneur in the German steel industry. (See Chapter 7.)

The question arises in connection with the new steel production methods, how much did the LD-method specifically contribute to today's steel crisis, the most serious since World War II? The inclination to make rigorous use of the advantages of improved site locations, the size of enterprises and so on doubtless contributed to an improvement in the investment climate. The LD-method is used today (1984) by more than 200 steelworks worldwide. Approximately 65 per cent of the world's steel production is by the LD-process (Figure 16.2). The generally accepted opinion among experts is that today's excess capacity was caused by the oil crisis and by incorrect industrial predictions. They cannot promise improvement until 1985. The resulting situation requires an entirely different strategy from the steel industry and the steel corporations.[11] The attitude of the American steel industry *vis-à-vis* the introduction of the LD-method is quite instructive, and deserves special investigation.

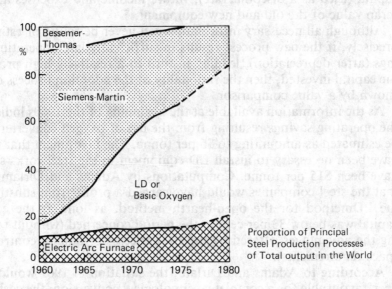

Figure 16.2
Proportion of principal steel production processes of total output in the world

Source: as for Figure 16.3

295

A critical analysis of the innovation strategies in the US steel industry with respect to the diffusion of the LD-method

The missed opportunities of the US steel industry form an instructive case of strategic behaviour of the firms in general and innovation policy in particular. In an *ex post* analysis, Adams and Dirlam (1966) tried to estimate the approximate degree of improvement in the profit level of the US steel industry. They were able to demonstrate that the application of such relatively simple decision making techniques as the net-present-value method should have influenced top management to adopt the LD-process much faster. The initial point of their ideas derives from the theory of replacement investment, according to which a new technology should be introduced if the substitution would result in an increase in the net value of the company. A comparison of present values should be based on the costs of the new process, the costs of capital (used as a discount rate), future income and expenses and net scrap value of the old and new equipment.[12]

Although all necessary information can never be found or estimated precisely, if the new process results in sufficiently large operating savings (after depreciation), leading in turn to a relatively high profit on the capital invested, then the rationality of the adoption process can be shown by a value comparison.

As the information available at the beginning of the 1960s indicated, the operating savings resulting from the use of oxygen converters may be estimated as amounting to $5 per tonne. The investment that would have been necessary to install LD-equipment in the steelworks should have been $15 per tonne. Computations by Adams and Dirlam show that the steel companies would have earned a profit from substituting the LDmethod for the open-hearth method, as long as the cost of capital was below 33 per cent in the period concerned (without regarding the advantages offered by the LD-process for quality control and space requirements).

According to Adams and Dirlam, the situation in 1961 would have been favourable for a complete technological conversion: the cash flow of the US steel industry during 1950–60 amounted to $ 16.4 billion. Only $1.3 billion or 12 per cent of the industry's actual capital expenditures of $ 11 billion would have been sufficient to install a LD-capacity of 87 million tonnes – about the same capacity that was necessary in 1960 for steel production by open-hearth. As a result of their research they come to the following conclusion: 'If 87 million tonnes of steel had been produced by the oxygen process, total operating savings of $432 million could have been realized. After-tax profits would therefore have been $ 216 million higher. Net worth could have been reduced by as much as $ 1.7 billion – the difference between the investment required for 87 million tonnes of open-hearth capacity, and the same amount of oxygen converter capacity.[13]

According to the authors the industry's computed return on net worth in 1960 would have been approximately 11.6 per cent, instead of the 7.6 per cent actually earned. This would have meant an increase of some 65 per cent in profits.

These facts show that the majority of steel companies in the US have not sufficient innovation strategies. It is also questionable whether they even had an innovation strategy at all – except a policy of 'business as usual'. The dominance of short-term thinking seems to be apparent and, as Oster (1982) has proved, the differences among firms in the rate of adoption of the LD-process can to a great extent be explained by the size of the firm: 'An increase in firm size of 1 million net tonnes reduced the probability of adoption by 1964 by about 1% and reduced the time since adoption by about 4 months. All else equal, the results suggest that a firm of the size of U.S. Steel would adopt the BOF an average of 7–8 years later than a medium size firm like Inland or National. This would seem to be quite strong support for the behavioural argument of Adams and Dirlam'.[14]

Whereas Adams' and Dirlam's explication of the actual diffusion of the LD-process in the US is monocausal, that is, it is the result of the firms' size only, Gold and Rosegger (1980), on the other hand, give us a more differentiated explanation, showing that some forecasts of the cost savings have overestimated the advantages of the LDprocess because of the economies of scale in open-hearth production. Lacking the advantages of economies of scale, the small firms had more incentive to adopt the LD-process than the leading steel producers. As Gold (1980)[15] emphasises, the net present value methodology for evaluating the relative advantages and disadvantages of adopting innovations, together with the estimated return on investment and payback-period approach, are not suitable for strategic decisions. In fact, there is a substantial lack of methods and techniques for considering 'soft facts' (individual value judgements and trend estimations). Innovation investment decisions differ from normal investments due to more uncertainty, greater risk and the difficulty of getting relevant information.[16]

Future perspectives and tasks of the world steel industry

The present steel crisis, which has existed since 1975, has caused speculation concerning the future of the steel industry. The tremendous subsidies by West European governments, and the strategy of some aggressive expanding steel giants, have fostered slogans like 'Get away from steel'. This has created further complications for the managements concerned.

Because of the above-described 'innovation-(mis)management' of the US steel industry, the importance of long-range planning and appropriate management control is now becoming generally accepted, and improvements in this field are starting to appear. Possibilities of installing early warning systems to avoid similar management failures are being discussed. In this connection future possible inventions or innovations for the steel industry are of enormous importance.

Long-range planning for the steel industry for the coming 10–15 years is based on three possible technologies:

1   the LD-process
2   electric arc furnaces combined with direct reduction method, and
3   electric arc technology on scrap basis.

The necessary energy demands are 6,600, 4,200 and 1,100 Kwh per ton respectively. Future basic innovations in steelmaking technology will probably concentrate on the reduction of the required energy consumption.

On average, 20–25 per cent of total worldwide coal production is used to produce pig iron from iron ore. (In Japan the figure is 66 per cent.) Various patented processes exist to produce process heat in nuclear power stations, replacing the traditionally used coke and oil.

As already discussed in the Midrex technology, coke was replaced by natural gas. Further patents already exist for substitution of coke through hydrogen. Thirty years ago the application of hydrogen in iron metallurgy began, and in both Minneapolis and the industrial area of Pittsburgh experiments with pilot plants have been carried out for several years. Iron ore can be reduced by the use of hydrogen to pig iron at temperatures ranging from 300° to 600° C, which is much less than the usual temperature in the conventional blast furnace process. The hydrogen reduction method has a very special advantage as regards environmental protection: its only reaction by-product is plain water. Unfortunately for this method, however, the transition from invention to innovation is expected to take place not earlier than the year 2000. The economic success of the method will of course depend on price developments for coal, coke, natural gas, oil and other energy sources in relation to production costs using hydrogen. In addition to saving coal, this new technology favours the concept of small steel mills (like the Korf works), which on account of their high efficiency were far less affected by the steel crisis of the 1970s. Thanks to Korf's management innovation, small steel mills are now feasible alternatives for developing countries. This in turn enables the development of small industries which can adapt to local demand much more easily than the huge mills of today.

In spite of all the advantages of this theoretically possible scenario, we must realise that this, and similar inventions, can only be transformed into useful innovations under specific conditions, for example the large-scale, economic production of hydrogen. It is also important to remember that, in the case of LD-innovation, the feasibility of oxygen production was a *sine qua non*, transforming the brilliant ideas of Bessemer into a successful innovation. According to Table 16.4 this took exactly 73 years. The present situation regarding production with the hydrogen method is more complicated, since the entire energy supply of the industrialised world is at stake.

As a result of cooperation between VÖEST and Korf Stahl AG, there is a new development in the technology of pig iron, called the KR (*Kohlereduktions*)-process, or 'coal reduction' (CR) method. In comparison to traditional pig iron production, the new process requires lower capital investment and makes the use of coke superfluous, thereby enabling companies to produce steel using cheap coal. The process is therefore of special interest to developing countries which have huge coal reserves, and to producers interested in building smaller but nevertheless economic steel mills.[17]

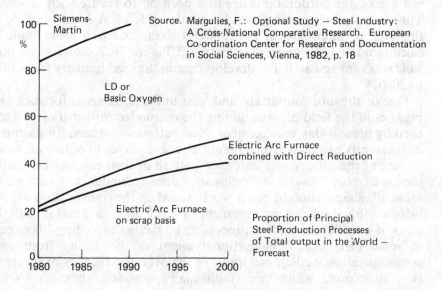

Figure 16.3
Proportion of principal steel production processes of total output in the world: forecast

Figure 16.3 shows the estimated development of steel processing methods from 1980 to 2000. It is generally expected that the openhearth process will be replaced by other steelmaking processes and will practically disappear by the end of the 1980s. Probably a dynamic expansion of electric arc technology as shown in Figures 16.2 and 16.3 will gradually diminish the rate of the LD-process from the present 65 per cent to approximately 50 per cent by the year 2000. Direct reduction combined with the electric arc furnace will apparently also expand and it may account for 10 per cent of world steel production in 2000.

While emphasising the importance of technological innovations and inventions, we should not forget the problems of innovation in the steel industry are not limited to the field of technology. On the contrary, the present uncertainty in the steel industry should accelerate the implementation of small, seemingly unimpressive piecemeal innovations which in the long run could significantly influence the effectiveness of the steel business.

Figure 16.4 shows some possibilities of piecemeal innovations in the steel industry during the 1980s. The figure implies that the steel industry needs to pay more attention to the improvement of research and development. The steel industry lacks a long-term research programme related to basic and piecemeal innovations. Because of the complexity and the financial magnitude of such longrange research programmes, not even large corporations are in a position to realise such projects. They have to be tackled on an international level. A good example of this type of international effort was a recent research programme in Europe related to continuous casting.[18] Figure 16.5 shows some possible tasks for research and development in the steel industry from 1980 to 2000.

One of the most important and least understood areas for such savings lies in the field of streamlining the outmoded information systems used by present-day management. New software systems, for example, are currently available which permit alert executives to obtain an X-ray picture of the entire communication pattern of their enterprises. Such a tool is of great benefit and helps to reduce redundancy, to cut down manual manipulation of paper work, speed up the transmission of vital information, improve management control and, in general, cut the costs of overall information processing by factors exceeding 20 per cent in some instances. Organisational measures, benefiting from such communication studies, can improve the overall efficiency of information structures, which are traditionally notorious barriers in implementing successful innovation-oriented management.

As to marketing practices in the steel industry, there are many still unused possibilities for improvement as well. Some case studies in the late 1970s and the early 1980s show the vital influence of marketing on

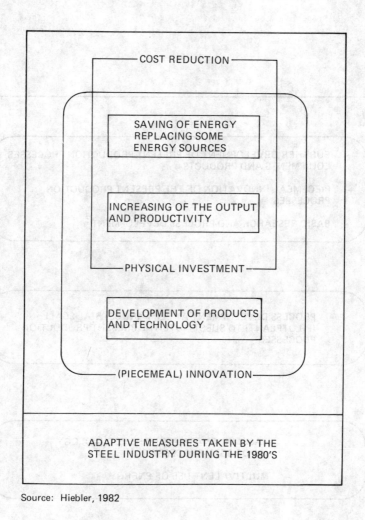

COST REDUCTION

SAVING OF ENERGY
REPLACING SOME
ENERGY SOURCES

INCREASING OF THE OUTPUT
AND PRODUCTIVITY

PHYSICAL INVESTMENT

DEVELOPMENT OF PRODUCTS
AND TECHNOLOGY

(PIECEMEAL) INNOVATION

ADAPTIVE MEASURES TAKEN BY THE
STEEL INDUSTRY DURING THE 1980'S

Source: Hiebler, 1982

Figure 16.4
Adaptive measures taken by the steel industry during the 1980s

a firm's success – even to its very survival. It is well known that international trade of steel products has changed radically since the 1960s. Traditional steel exporting countries like the United States and Great Britain have become steel importers. Some developing countries are now competitors on the international steel market, for example South Korea in Japan. The development of adequate marketing instruments and the appropriate strategies should therefore be carefully planned and implemented for the coming difficult years.

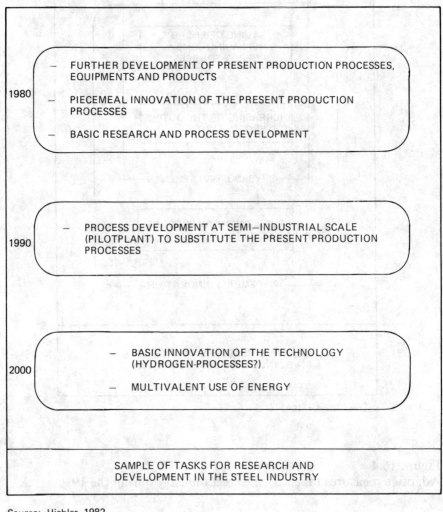

1980
- FURTHER DEVELOPMENT OF PRESENT PRODUCTION PROCESSES, EQUIPMENTS AND PRODUCTS
- PIECEMEAL INNOVATION OF THE PRESENT PRODUCTION PROCESSES
- BASIC RESEARCH AND PROCESS DEVELOPMENT

1990
- PROCESS DEVELOPMENT AT SEMI—INDUSTRIAL SCALE (PILOTPLANT) TO SUBSTITUTE THE PRESENT PRODUCTION PROCESSES

2000
- BASIC INNOVATION OF THE TECHNOLOGY (HYDROGEN-PROCESSES?)
- MULTIVALENT USE OF ENERGY

SAMPLE OF TASKS FOR RESEARCH AND DEVELOPMENT IN THE STEEL INDUSTRY

Source: Hiebler, 1982

Figure 16.5
Sample of tasks for research and development in the steel industry

Remarks and conclusions to some hypotheses and problems of innovation theory from the viewpoint of the steel industry

The following conclusions can be drawn from the developments described concerning some hypotheses and problems of innovation theory:

1   The introduction of the LD-method shows the multidisciplinary character of innovation problems. The relevant fields of industrial economics, political economics, technology, decision-making theory, economic forecasting, etc. have to be integrated into an innovation theory. The present insufficiency, incompleteness and difficult terminology of innovation theory can explain the difficulties in integrating these disciplines.

2   The success of an innovation depends on the management's ability to coordinate the relevant flow of information between the discipline concerned and to make the right decisions at the right moment. Figure 16.6 shows the various goals (targets, potential, efficiency), the different levels of managerial activity and the means of innovation management in a schematic overview.

3   Special attention should be paid to the validity of the various hypotheses of innovation theory.[19] The well known thesis by Schumpeter, whereby both the readiness for innovation and the innovation potential increase in relation to the growth of the enterprise, cannot be generally accepted in view of the facts available from the adoption of the LD and direct reduction methods.

4   The observed and described experiences in connection with the introduction of new production technologies confirm the decisive role of the promoters.[20]

5   Innovations can only be introduced successfully if the total system with its different subsystems is ready to accept the necessary adaptations. Local and worldwide effects of innovation should be considered and if necessary monitored and directed at regional, national and international levels.

6   In the future, along with the aspect of feasibility, managers will have to deal more and more with the problem of acceptance when deciding on technological innovations in industrialised countries: for example, they will have to consider environment protection laws for the steel industry, the influence of consumerism, and other new constraining features.

7   Innovations with considerable feasibility and acceptance problems require careful planning and coordination as well as co-

alitions between their promoters.

8 The quality of innovative decisions is in part determined by the quality of the information upon which it is based. The results of information economy,[21] and the relevant findings of epistemology should be paid better attention to when attempting to solve the information problems of innovation theory.

9 Since a certain amount of uncertainty and risk in making innovative decisions cannot be avoided, more attention should be paid to the problem of subjective data projection, the estimation of probabilities, and induction as a 'logic of uncertainty'.[22]

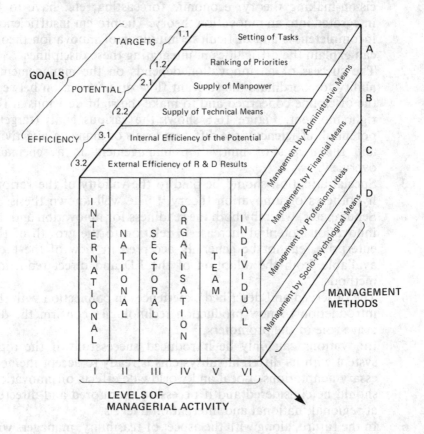

Figure 16.6
Morphological box for decision making in science and technology policy

Source: Dobrov, G. M., 'The dynamics and management of technological development as an object for applied systems analysis', Report to IIASA, Laxenburg, 1976

10   Successful innovation management is successful conflict man-
age· ment. Innovations in corporations are sources of conflicts
because they challenge the goals (targets, potentials, efficiency)
policy, products and structures of the organisation. As is generally known, innovations can, and often do, disturb the normal processes in the organisation, which generally are oriented
towards achieving short term productivity targets only, thus
neglecting long-term overall efficiency objectives. For this reason the production managers of (large) firms often perceived
innovations as undesirable disturbances of operative productivity and thus tend to raise obstacles against their introduction.

Notes

1   Institut für Arbeits- und Betriebswissenschaften der Technischen Universität Wien.
2   Chenery, H.B., 'Patterns of Industrial Growth', *American
Economic Review*, vol.5, no.4, September 1960.
3   Maizel, A., *Industrial Growth and World Trade*, Cambridge
University Press, Cambridge, 1963.
4   International Iron and Steel Institute (IISI), 1972.
5   Hinterhuber, H., *Innovationsdynamik und Unternehmens-
fuehrung*, Springer Verlag Vienna, New York, 1975.
6   Mensch, G., *Das technologische Patt*, Umschau-Verlag
Breidenstein K.G., Frankfurt am Main, 1975, p.55.
7   Adams, W., Dirlam, J.B., 'Big Steel Invention and
Innovation', *The Quarterly Journal of Economics*, vol.
LXXX, no.2, May 1966.
8   Vereinigte Oesterreichische Eisen- und Stahlwerke AG.
9   Adams W. and Dirlam J.B., op.cit., p.175.
10   Schenck, H., *Probleme der Stahlindustrie*, OEIAG, Vienna,
1972.
11   Moser, G., 'Die Konzernorganisation der VOEST-ALPINE
AG: als Antwort auf eine im Umbruch stehende Stahlwelt',
Manuscript, Linz, 1978.
12   Adams, W. and Dirlam, J.B., op.cit., p.186.
13   Adams, W. and Dirlam, J.B., op.cit., p.187.
14   Oster, S., 'The diffusion of innovation among steel firms: the
basic oxygen furnace', *The Bell Journal of Economics*, vol.13,
no.1, 1982, p.53.
15   Gold, B. *et al.*, *Evaluating Technological Innovations*, Lexing-

ton, Mass, Lexington Books, 1980.

16 Kern, W., 'Innovation und Investition', *Investitionstheorie und Investitionspolitik privater und öffentlicher Unternehmen*, Gabler, Wiesbaden, 1976.

17 *Die Presse*, 1 September 1982.

18 Hiebler, H., 'Neue Technologie in der Stahlindustrie', paper presented at the International Symposium, Automation in the Steel Industry, European Coordination Center for Research and Documentation in Social Sciences.

19 Mattessich, R., *Instrumental Reasoning and Systems Methodology*, Reidel, Dordrecht, Holland, 1978, p.24.

20 Witte, E., *Organisation für Innovationsentscheidungen – Das Promotorenmodell*, O. Schwartz, Göttingen, 1973; and 'Erfolg von Investition in neuen Technologien durch richtige Auswahl der Mitarbeiter', *Fortschrittliche Betriebsführung und Industrial Engineering*, vol.24, 1975.

21 Mattessich, R., op.cit., p.232f

22 Mattessich, R., op.cit., p.143; Kern, W., op.cit., p.293.

17 Adjustment through piecemeal innovation – the US experience[1]

Gerhard Rosegger

Introduction

Events of the past two decades have not dealt kindly with the world's traditional steel producers. The problems of the American iron and steel industry are more prototypical than unique: steadily increasing import competition; shifts in the size and composition of major output markets; relentless pressures for higher wages and fringe benefits; old technology embedded in plants whose location, layout, and scale act as barriers to major restructuring efforts; sharply rising materials input, especially energy, costs; and vacillating governmental policies.

None of the items in this depressing list of difficulties will surprise the student of developments in one of the industrialised economies' key sectors. Yet there are some elements of uniqueness to the American situation. First and foremost, there is the very size of the industry and of its markets; the latter has acted as an irresistible magnet to foreign competitors, and the former has meant that even what would be regarded as major innovative undertakings in other countries have had but marginal impact on the industry's overall performance. Second, the geographic diversity is so great that one author claimed, with some justification, 'there is no national steel industry but a group of regional steel industries which serve major industrial markets.'[2] Third, size and diversity combine to make such frequently-used concepts as 'best-practice' technique or 'industry average costs of production' practically meaningless for analytical purposes; thus, the attractiveness of markets and the ability of American firms to meet competition (on a cost basis) in these markets varies widely.[3] And fourth, the American case provides a unique illustration of the interaction between an industry's efforts to contend with the pressure of market forces and government's active, if sporadic, interest in the political and social implications of the outcome of these efforts.

A whole host of recent studies has attempted to examine the extent to which the American steel industry has been victimised by its own strategic errors as well as by wrong-headed if well-intentioned governmental policies. More important, for our purposes, are investigations of the industry's future prospects, dealing with alternative strategies for a restructuring of this key sector of the US economy's industrial base. In particular, several of these investigations[4] have been concerned with the potentials of various technological approaches to a return to full competitiveness, focusing on detailed cost comparisons and evaluating the prospects for various new investment schemes. Almost without exception, the findings have been marked by a pervasive scepticism about the industry's chances for overcoming its current disadvantages.

These pessimistic conclusions have been challenged on empirical and on conceptual grounds.[5] In arguing for the development of sounder bases for coherent national policies, Professor Gold has emphasised especially the need to improve our understanding of the results of strategies that have been implemented by individual firms and in individual plants – an understanding that cannot be gained from treating the costs and benefits of technological innovation 'on the whole and on average'.

This chapter reports the findings of a research project carried out with this objective. It deals with the results of piecemeal innovation, that is, the strategy of upgrading existing production facilities through the introduction of major process innovations. The two innovations analysed are basic oxygen steelmaking and continuous casting, both of which have been hailed as 'revolutionary' changes in steel technology; the findings are based on actual performance and cost data made available by four large, integrated firms. To the best of our knowledge, this was the first time that researchers were given access to corporate information of this type. Thus, while the sample size may fall short of the statistician's requirements, and while the firms must, of necessity, remain anonymous, the results of our case studies provide at least a highly suggestive basis for evaluating what has been the dominant strategy of American integrated steel producers during the past two decades.

Before turning to these results, however, some background data are presented on the American industry's performance and some of the broader characteristics and implications of a strategy of piecemeal innovation are considered. In a final section, some of the lessons that might be learned from the industry's efforts at structural adjustment are assessed.

308

Industry performance and structural adjustment

Imports, stagnation, and changing markets

As Figure 17.1 shows, the American steel industry enjoyed a period of vigorous expansion until approximately 1965, even though imports had begun to take an increasing share of the market since the prolonged steel strike of 1959. Initially, however, decision makers consoled themselves with the fact that these inroads occurred mostly in lowpriced, unsophisticated products, while superior process and product technology seemed to assure American firms of a continuing hold on the high-profit market segments.[6]

Expansion of imports across the whole range of products during the middle and late 1960s tended to be explained primarily by foreign governments' subsidies to their steel industries and by liberal American trade policies,[7] rather than by any concessions of problems with domestic technology and costs. Thus the assessment, 'The steel industry in the United States still enjoys an overall technological advantage. The margin of advantage has been shrinking steadily, however.'[8] Given these mis-perceptions, it was not surprising that integrated steelmakers concentrated on two strategies: the quest for 'more favourable' trade and tax policies, and the gradual introduction of new technology into existing plants.

The short-lived boom of 1973 and 1974 seemed to provide confirmation for such optimism. Euphoria[9] gave way to more sober evaluations in the aftermath of the market collapse of 1975, with industry and government agencies undertaking more searching investigations of long-term prospects, some of which we have referred to above (see note 4).

That the period since 1965 has been one of essential stagnation, and more lately of decline, is illustrated by the selective list of performance indicators presented in Table 17.1. Despite the domestic industry's efforts, it could do little more than hold its own output, with imports capturing whatever growth there may have been in the market for steel mill products. The problem was aggravated through the 'indirect' import of steel in the form of automobiles, industrial machinery, transport equipment and so on, which tended to curtail demand for the industry's output among its traditional customers. Thus, for example, the automotive market shrank from approximately one-fifth of all shipments to 14 per cent by 1980, and it has declined even further since then. Substitution of other materials also affected some markets, such as those for containers and packaging. In 1960, the steel industry's five major markets absorbed three-quarters of all shipments; by 1980, they accounted for less than 63 per cent. While it might be argued that

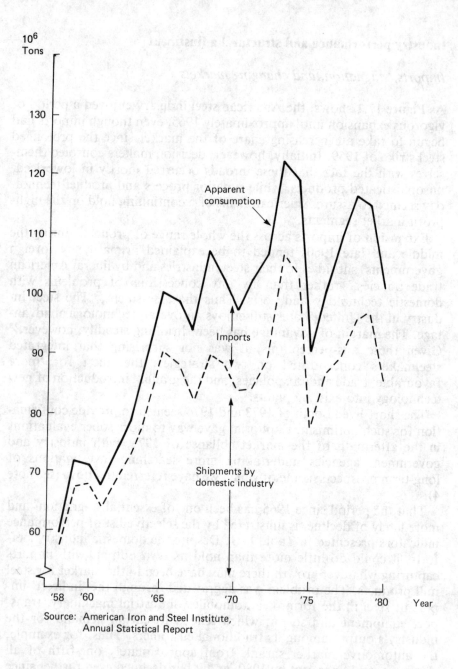

Figure 17.1
Domestic shipments, imports and apparent US consumption of steel mill products 1958–80

reducing the dependence on particular types of customers is beneficial in the long run, there are some counter-arguments: diversification of the product mix means shorter runs and increased unit costs, even if the total volume of sales were to remain constant; and a spreading about of sales and service effort will similarly affect costs adversely. In this connection it is also significant to note that, among the largest outlets, only the share of the steel service centres increased; these are essentially distributors and 'retailers' to a host of small customers.

Table 17.1

Performance indicators: US steel industry, selected years

	1960	1965	1970	1975	1980
Production					
Pig iron (10^6 tons)	66.5	88.2	91.4	79.9	68.7
Raw steel (10^6 tons)	99.3	131.5	131.5	116.6	111.8
Share of processes (%):					
Open-hearth	88.1	71.9	36.5	18.9	11.6
Basic oxygen	3.5	17.5	48.1	61.6	60.5
Electric furnace	8.4	10.6	15.4	19.5	27.9
Share of US in world raw steel (%)	26.0	26.2	20.1	16.4	14.1
Shipments (10^6 tons)	71.1	92.7	90.8	80.0	83.9
Share of major markets (%):					
Automotive	20.5	21.7	16.0	19.0	14.4
Steel service centres	17.5	17.7	17.6	15.9	19.3
Construction products	18.7	18.2	14.8	15.0	14.2
Containers & packaging	9.0	7.9	8.6	7.6	6.7
Ind. machinery & equipment	8.5	9.5	8.8	9.1	8.3
Imports (10^6 tons)	3.6	10.7	13.4	12.0	15.5
Share in apparent supply (%)	4.8	10.3	13.8	13.5	16.3
Employment					
Average number (10^3)	572	583	531	457	399
Payroll/hour (hrly empl.) ($)	3.35	3.94	5.68	10.59	18.45
Tons shipped/employee (10^3)	124	159	171	175	210
Financial					
Net income/total revenue (%)	5.7	5.9	2.8	4.8	3.0
Ret. on stockholder equity (%)	7.7	8.9	4.1	9.8	9.0
,, All US manufacturing (%)	9.2	13.0	9.3	11.6	

Sources: American Iron and Steel Institute, *Annual Statistical Report;* FTC–SEC, *Quarterly Financial Reports for Manufacturing Corporations*

The financial impact of all these developments has been discouraging: profit margins declined steadily (they experienced a brief boost in 1973 and 1974, not shown in Table 17.1); and while the return on equity may look quite favourable by the standards of steel firms in other countries, it has consistently fallen short of the returns earned by US manufacturing industry as a whole, thus effectively blocking integrated steel producers from access to the equity market.[10] To the extent, then, that capital expenditures could not be financed out of internally generated funds (depreciation and retained earnings), massive borrowings had to be resorted to. Thus the industry's longterm indebtedness grew from less than $3 billion in 1980; and the debt/equity ratio increased from the approximately 33 per cent thought appropriate in the 1960s to over 49 per cent.

Whatever the variations in profitability among firms, and these have been substantial on occasion, the industry's stagnation is reflected in these aggregate measures of performance, which are dominated by the results obtained by the large, integrated companies.

Aspects of structural adjustment. Piecemeal innovation has been the dominant strategy pursued by American integrated steelmakers over the past two decades. It has not, however, been the only factor affecting the industry's overall structure. Therefore, it is necessary to place this strategy into perspective by considering other significant changes.

Among these, the following have had significant impact:

1 The abandonment of submarginal plants by multi-plant firms. Plant closings have occurred especially in the historical steelmaking region of Pittsburgh-Youngstown, with substantial impacts on regional employment.
2 The abandonment of specific, unprofitable product lines (and of the relevant facilities) in surviving plants and firms. This trend has meant an increasing specialisation among major firms, offset only in part by an expansion into new products, such as oil country goods.
3 The exit of submarginal firms from the industry. One single-plant, integrated firm shut down its operations permanently. The bankruptcy of another, dependent almost entirely on the automobile industry for its sales, has recently been reported.
4 The construction of entirely new, integrated mills (greenfield plants) by existing firms. In sharp contrast to Japan, and even to some Western European countries, only one such plant was built by an American firm in the last two decades. Plans for a second, modern greenfield plant, announced five years ago, have since been abandoned.

5   Diversification of the large, integrated steel firms into different industries and product lines.

6   The entry of new firms into the industry. This last is one of the more interesting features of structural change since 1960. Despite the generally discouraging conditions, well over 40 firms have started up (non-integrated) steelmaking operations. Based on regional input (scrap and energy) availabilities and on a relatively narrow range of regionally marketable products, the commercial success of these so-called mini-mills has been remarkable. Though only ten of the mills had an annual raw steel capacity of over 500,000 tons, together these new entrants into the industry account for approximately 18 million tons, or 12 per cent, of the total industry capacity. Their steelmaking capability is based on electric furnace technology, which largely explains the rapid growth in the share of this process in total output (Table 17.1) and well over half of them rely on continuous casting for the production of their, predominantly billet and bar, products.[11]

The combined impact, then, of imports and of the growth of mini-mills has been to capture roughly 28 per cent of the American market for steel mill products – from their point of view, at the expense of the traditional, integrated producers.

Piecemeal innovation in integrated plants – general observations

The industry continues to be committed to the 'modernisation' of existing facilities as the primary strategy for restoring its competitive position.[12] The great bulk of the capital expenditures shown in Table 17.2 has been devoted to the piecemeal upgrading, rounding out, and expansion of integrated mills.

Although estimates of the 'tons of modernised capacity' obtained through these expenditures have been made, such calculations are tenuous at best and thoroughly misleading at worst.[13] As we have shown elsewhere.[14] the determinants of the costs and benefits of investment in process innovations vary greatly a) from innovation to innovation, depending on its functional characteristics, b) over the period of diffusion, as the innovation is modified, refined, and varied in its technical aspects, and c) from plant to plant, because of the vintage, technological specificity, and scale of existing facilities. In assessing the implications of structural adjustment through piecemeal innovation, one must therefore take account of this variety. The taxonomy proposed in the following is no more than illustrative, but it serves at least as a guide to

Table 17.2

Capital expenditures on steelmaking facilities

Year	Capital expenditures on productive steelmaking facilities* (in million '78 dollars)	Capital expenditures per ton of shipments ('78 dollars/ton)
1960	2,675	43.01
1961	1,698	29.37
1962	1,600	25.96
1963	1,811	27.39
1964	2,760	37.07
1965	3,107	38.83
1966	3,405	43.45
1967	3,367	46.42
1968	3,576	44.66
1969	2,869	36.52
1970	2,214	29.37
1971	1,705	23.10
1972	1,265	16.28
1973	1,630	17.16
1974	2,163	23.54
1975	2,684	40.48
1976	2,599	34.98
1977	2,054	27.72
1978	1,706	21.67

* Total capital expenditures less environmental control expenditures and estimated non-steel capital expenditures.

Source: Office of Technology Assessment op.cit., p.123.

such evaluations and it places the results of our case studies, to be presented subsequently, into proper focus. More particularly, it also constitutes an antidote to the typical, abstract engineering estimates of costs and benefits, which tend to be notoriously optimistic in terms of their assumptions.

Figure 17.2 shows a greatly simplified outline of the major processing steps and facilities included in an integrated steel mill, that is, at one plant site. Substantial differences among mills exist with respect to virtually every one of the production system's components, and these become especially pronounced at the 'finishing end' of plants, where the final product mix is determined. Nevertheless, the diagram is useful for our discussion.

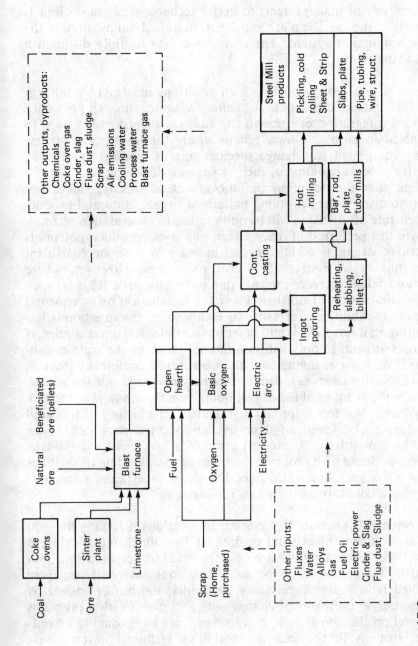

Figure 17.2
Steelmaking process steps in integrated mills
Source: Adapted from Rosegger (1974)

Functional characteristics of innovations

The receptivity of managements to major technological innovations is influenced by the specific place and function of such innovations in the existing production system. Therefore, one may usefully distinguish the following types.

Additions to available capacity. Such additions may involve innovations which change capacity in a *qualitative* sense, through backward or forward integration or through the addition of processing steps to the established product flows. Examples are the installation of new finishing equipment (to change product mix or quality) or of new, alternative process equipment, such as vacuum degassing facilities. To the extent that the adoption of innovations of this type does not threaten to disrupt the prevailing pattern of organisation and production, their rate of diffusion will be quite rapid. In general, new technologies will be preferred if they offer improved product potentials (competitive strength) at little or no increase in cost, or equivalent product quality at even modest cost reductions. However, where innovations might increase capacity in a purely *quantitative* way, such as in the elimination of bottlenecks or the installation of incremental production capabilities at certain processing steps, the situation is less clear-cut: if an innovation is still beset by technological uncertainties, if it promises advantage only within a limited range of the scale continuum, or if it increases managerial and operational complexity (that is, requires intensive learning), decision-makers may well opt for adding facilities embodying established technologies rather than take the risk of innovating. Not too surprisingly, therefore, the history of this kind of rounding-out of American plants over the past two decades has been marked by a mixture of strategies: some of it involved tried-and-true technologies, some of it required the innovative upgrading of facilities (such as the oxygen lancing of open-hearth furnaces), and some of it represented a commitment to major innovations.[15]

Displacement of functioning facilities. Innovations involving the outright abandonment of existing equipment face much more stringent criteria. These criteria involve not only the usual cost comparisons, but innovative changes of this type must also confront resistance from entrenched labour and supervisory personnel, the hurdles posed by (often excessive) estimates of change-over and modification problems in the total production system, and the tendency to rationalise a 'wait-and-see' strategy in the face of continuing technical uncertainties. Since our empirical findings, presented below, concern displacing innovations, we shall deal with their characteristics in greater detail in that context. Here it only remains to be mentioned that efforts at the comprehensive introduction of computerisation in American steel mills provide an outstanding example of both, the promises and the difficulties faced by displacing innovations.[16] The industry has argued persistently that current tax laws, especially those concerning the writing-off of displaced facilities and the depreciation periods allowed for new equipment, militate against a more rapid adoption of major technological innovations.

Replacement of capacity. Where older facilities, embodying traditional technology, have become physically worn out or competitively submarginal, managements confront the decision whether simply to abandon a facility; whether to replace it, in the same location, with equipment up to current technological standards; or whether to accompany abandonment with the addition of capacity at another plant. In the face of stagnating markets for most of their products, integrated producers have not found the wholesale replacement of old plants attractive; indeed, increasingly stringent environmental control regulations may have hastened decisions in favour of outright abandonment. In some instances, this was accompanied by (partial) additions to capacity in other, newer plants; in others, abandonment meant a reduction in total productive capacity.

Technological specificity and scale of existing plants. Which plants and firms are, at a given point in time, candidates for the adoption of major technological innovations is an empirical question; it cannot be prejudged by forcing observations into some preconceived, long-term pattern of diffusion. Although it has been fashionable to upbraid American integrated steelmakers for their alleged laggardness in adopting new technologies, such judgements typically have been based on an oversimplified view of the diffusion process. It may well be that perceptions of an unassailable position of technological, and market, leadership in the 1950s and early 1960s engendered attitudes of complacency and inertia among decisionmakers; but such a sweeping indictment could be made only with the wisdom of twenty-twenty hindsight and would be difficult to support by the objective, empirical evidence.

As we have discussed at length elsewhere,[17] the population of (realistic) potential adopters of innovations changes over time. And these changes are due not so much to the 'innovativeness' of decision-makers and their subjective assessments of situations, but to the specific conditions confronted by them, in terms of the technical setting of their production systems and of the innovation's development.

When we apply these evaluative criteria to the American steel

317

industry's performance over the past two decades, we must take into account at least the following considerations:

1   There were many plants for which the specific advantages of a given innovation appeared less than persuasive. This was so, in the first instance, because a) the technology embedded in existing facilities prevented the full realisation of the innovation's (abstract) benefits, short of incurring massive modification costs; b) the innovation affected a production step that was not, under current conditions, a binding constraint on the achievement of higher output rates or lower costs; and c) alternative commitments of scarce financial resources promised higher expected returns.[18]

2   Every major process innovation in iron and steel has undergone changes in the nature of equipment, in suitability for various types of products, and in compatibility with other technologies. Thus, the potential for adoption changes over time, as these transformations create new variants of the innovation. For example, basic oxygen steelmaking was restricted to carbon steels during the first ten years of its application; in the early 1960s, its capabilities were extended to the production of a variety of alloy steels, where it accounted for over one-third of all output within five years.[19] Similarly, it was only after the capabilities of continuous casting had been extended from bars and billets to slabs that integrated mills, whose major products are flat-rolled, could become serious candidates for adoption.

3   The single most important change over time is in the size and capacity of innovative process equipment and thus in its compatibility with various scales of operation. Again, the development of the basic oxygen furnace provides an illustration. Heat capacities of initial installations in the United States were in the 60 to 80 ton range – attractive for relatively small plants but not for the large, integrated mills. It was not until vessel sizes had been pushed beyond the 200 ton/heat level that the innovation became economically attractive for these latter facilities. At the same time, however, this development shows the limitations of piecemeal innovation, as against the construction of greenfield plants: further increases in equipment size, together with the growth of three converter (compared with the original, two-converter) plants, meant that the full economic benefits of large-scale production could be reaped in facilities of six to eight million tons of annual capacity, such as those constructed in Japan and in certain European locations. None of the existing integrated mills in the United

States comes even close to this size range, with the result that the largest American basic oxygen furnace installation has a capacity of 4.4 million tons, and most plants are considerably smaller than that.[20]

In the light of all of these factors, it is not surprising that innovations met with greatly differing evaluations by individual adopters, led to different patterns of adjustment, and brought results that were strongly influenced by the unique conditions prevailing in existing plants and firms.

Empirical findings – oxygen steelmaking

The adoption record of the basic oxygen process is summarised in Table 17.3. Newly installed capacity came on-stream very rapidly, after an initial period during which technical uncertainties were removed and standard operating practices developed.[21] No additional basic oxygen steelmaking plants were built after 1974. At this time, however, the new technology had completely displaced traditional processes in only 17 of the United States' 43 integrated steel mills; eight mills continued with open-hearth shops only; seven concurrently used open-hearth and basic oxygen facilities; and the rest used various combinations of the above processes with electric furnace shops, where the latter, however, were not necessarily utilised to produce carbon steels.

Our survey sample includes five basic oxygen furnace (BOF) shops of four major, multi-plant firms. In four of the cases, open-hearth shops were run concurrently, thus providing a direct basis for performance and cost comparisons. In each case, we were provided with complete information on the period immediately following start-up ('early'), on a phase during learning ('middle'), and on a period when managements considered the learning process essentially complete ('late').

The basic question we sought to answer was, 'How much better did the new technology perform than the old, and what accounts for the differences?' Thus, our comparisons are restricted to relative performance *within* each plant and firm. This approach had the additional merit of enabling us to work within the context of each firm's accounting system, assuring consistency of results.

Characteristics and expected effects of the innovation.

In the BOF process, molten iron and steel scrap – the latter to a maximum of roughly 35 per cent of the total metallic charge – are placed

Table 17.3

Installation of Basic Oxygen Steelmaking Facilities in the US 1954–74

Year	Percent of Raw Steel Output			Number of New Plants Started Up	Capacity of Largest New Plant (million tons)	Size of Largest Converter (tons)	Total New Capacity Added (million tons)	Cumulative BOF Capacity Installed[a] (million tons)
	Total	Carbon	Alloy					
1954	b	b	–	1	0.54	60	0.54	0.54
1955	0.3	0.3	–	–	–	–	–	0.54
1956	0.4	0.5	–	–	–	–	–	0.54
1957	0.5	0.6	–	1	0.95	85	0.95	1.49
1958	1.6	1.7	–	2	1.44c	110	2.64	4.13
1959	2.0	2.2	–	1	0.73	75	0.73	4.86
1960	3.4	3.7	b	–	–	–	–	4.86
1961	4.0	4.4	b	2	1.9	230	2.9	7.76
1962	5.6	6.2	0.6	1	2.5	300	2.5	10.28
1963	7.8	8.6	0.6	2	1.5	150	2.9	13.16
1964	12.2	13.1	3.5	4	4.7c	290	9.4	22.56
1965	17.4	18.7	8.2	4	3.5c	245	8.5	31.06
1966	25.3	26.9	14.8	4	2.6	255	7.3	38.36
1967	32.6	34.3	20.9	2	3.4	335	5.6	43.96
1968	37.1	39.3	22.6	2	3.0c	250	7.65	51.61
1969	42.6	44.8	29.0	2	3.0c	200	5.0	56.61
1970	48.2	50.2	34.6	4	3.0	265	7.5	64.11
1971	53.1	55.6	36.7	1	2.8	220	2.8	66.91
1972	56.0	58.8	38.3	1	2.3	220	2.3	69.21
1973	55.2	57.9	39.5	–	–	–	–	69.21
1974	56.0	58.9	41.4	2	2.2	210	3.2	72.41

Sources: American Iron and Steel Institute, *Annual Statistical Report*, various volumes; Kaiser Engineers Division of Kaiser Industries, Inc., *L-D Process Newsletter*, various issues.

aThis column represents the sum of initial rated capacities of plants. Because of subsequent upratings, actual total BOF capacity was higher
bLess than 0.1 percent.
cThree-converter plants (all others are two-converter plants).

into a pear-shaped, refractory-lined vessel; high-purity oxygen is blown onto this charge through a retractable, water-cooled lance. The molten steel is then discharged into a ladle. The typical BOF shop incorporates two or three vessels. In a two-converter shop, one vessel is in operation while other one is being repaired and re-lined; however, as learning and improvements in refractories' extended campaign lives, some plants manage to operate both vessels simultaneously for short periods. In three-converter shops, two vessels are in operation while the third is being serviced.

Three technical characteristics of the BOF mainly influence its expected economic performance, as compared with traditional process technology:

1   Relying on the reaction of oxygen with the molten iron charge, the process requires no external source of energy. This saving must be balanced against the energy consumed in the manufacture of bulk oxygen. (But since oxygen is generally purchased from outside suppliers on a pay-as-you-go basis, steel firms do not have to invest in oxygen facilities.)

2   Heat times are in the range of 35 to 55 minutes, compared to 8–10 hours for the open-hearth process and 4–5 hours for electric furnaces.

3   While electric furnaces operate with a cold metal (scrap) charge and open-hearth furnaces are entirely flexible with respect to the proportions of hot and cold metal charged, the BOF is limited to scrap charges no higher than the above-mentioned 35 per cent.

In combination, these characteristics carried with them another economic implication: given the differences among techniques in input composition and operating modes, and considering the fact that no American firm contemplated investment in a new open-hearth facilitiy after the 1950s, the decision confronting managements involved only the rate at which the open-hearth would be displaced in existing plants by the BOF – a decision that would be strongly influenced by such factors as blast furnace capacity and the amount of home scrap generated in vertically integrated mills.

Given these considerations, a comparison of investment costs per unit of capacity between old and new technology was immaterial. However, in an industry plagued by chronic capital shortages, scale economies with respect to unit investment costs were of great interest. From the outset, there was little doubt that a (hypothetical) comparison of these costs would yield a 33–55 per cent advantage for the BOF. More important, unit investment costs were expected to be roughly 25 per cent lower for a 1 million ton plant than for one half that size; about 15 per cent lower for a further scale-up to 1.5 million tons; and yet 5 to 9 per cent less for increases beyond 2 million tons. Thus, expected scale economies might justify the construction of BOF facilities temporarily out of balance with the rest of the plant or the provision of structures and ancillary facilities (materials handling, charging, etc.) large enough to permit a future upgrading of capacity, either through the installation of bigger vessels or through conversion from two-converter to three-converter operations.

The main focus of expectations, however, was on comparative operating costs. With materials inputs ('metallics') making up the bulk of these costs, and with their impact depending mostly on plant-specific supply conditions, estimates varied widely, depending on assumptions about these conditions. Under most circumstances, total direct costs for oxygen steelmaking were expected to range between 69 and 90 per cent of those in comparable open-hearth facilities. However, more significantly, the bulk of these savings was estimated to result from 'costs above metallics' (operating costs other than iron inputs), which were expected to be approximately 55 to 60 per cent of open-hearth costs.

Taken together, these expectations (together with others concerning the yield from the new process and operating benefits not directly measurable) provided the basis on which individual managements had to decide whether, and when, to adopt the innovation. Needless to say, the plant-specific considerations we discussed in the preceding section played an overriding role in the nature and timing of these decisions.

Comparisons of actual performance

Table 17.4 shows the comparative cost per ton for each of the three survey periods in the five BOF shops. As these figures indicate, none of the BOF installations reached the lower limit of expected advantages, and three out of five tended toward the high end of these estimates. Furthermore, as will be suggested below, a substantial portion of the apparent cost savings must be attributed to the fact that managerial adaptations to the new process led, on balance, to relatively higher unit costs for the traditional technology. The results are generally, though not overwhelmingly, in favour of the BOF. The evidence of the effects of learning is ambiguous, precisely because of the adaptations mentioned: in absolute terms, open-hearth costs rose more rapidly than BOF costs, during the survey period.

In Table 17.5 we summarise the main determinants of comparative performance; these data enable us briefly to probe into the reasons for the variations in total unit costs[22] First, it will be seen that in the four

Table 17.4

Total cost per ton of BOF raw steel, as percentage of cost per ton of open-hearth steel, in comparable plants

Plant	Period after adoption		
	Early	Middle	Late
A	82.8	80.4	71.7
B	(86)*	84.5	77.6
C	89.8	85.8	88.9
D	(97)*	95.9	95.7
E	114.1	88.3	n/a

* Estimates from accounting records inconsistent with rest of data base; they should be regarded as best approximations only.

Sources: Calculation from plant records.

mills where open-hearth and BOF facilities co-existed, both continued to play important roles in total raw steel production, even during the 'late' period. But an examination of the composition of metallic inputs shows different reasons for this condition: in plants A and B, the open-hearth shops were assigned the task of working up home scrap that could not be accommodated by the BOF (see p.above); both these plants suffered from inadequate blast furnace capacity. As a result, the open-hearth furnaces suffered the operating cost (mainly energy cost) penalties of very high scrap charges, and the seemingly greater cost advantage of the BOF facilities in these two plants derives in good part from this difference in charge composition. In plants C and D the open-hearth shops served to round out total steelmaking capacity, without having to operate with very high scrap charges. The difference is reflected most clearly in relative oxygen and energy costs per ton: the apparent advantage of the BOF is much greater in plants A and B than in the others. As far as the costs of metallic inputs themselves are concerned, the new process did not bring any notable cost benefits.

Table 17.5

BOF vs. open-hearth – main determinants of comparative cost performance

			Plant				
			A	B	C	D	E
Percentage of raw steel output (late period):	O.H.		43	16	21	48	n.a.
	BOF		57	84	79	52	–
Composition of ferrous charge, per cent:	O.H.	–molten iron	13	18	83	63	–
		–scrap	87	82	17	37	–
	BOF	–molten iron	71	72	80	76	–
		–scrap	29	28	20	24	–
BOF metallic costs/ton as per cent of OH costs (ranges)			83–104	85– 95	100–101	102	98–115
BOF oxygen costs/ton as per cent of OH costs (ranges)			119–164	157–165	147–161	140–144	108–116
BOF oxygen + energy costs/ton as per cent of OH (late)			30	17	59	45	52
BOF total labour costs/ton as per cent of OH (ranges)			22– 30	21– 28	62– 64	75– 84	33– 34
Operating labour costs/ton as per cent of O.H.			14– 17	10– 12	32– 41	68– 83	37– 41
Repair and maintenance labour costs/ton as per cent of O.H.			27– 45	37– 39	80– 95	93– 96	16– 25
BOF other direct costs/ton as per cent of OH			22– 23	81– 85	51– 61	60– 67	52– 57

Source: Calculation from survey date.

Several observations on the other cost findings in Table 17.5 are appropriate:

1　The BOF's disadvantage with respect to oxygen costs was much less than had been assumed in most theoretical analyses. These ignored a typical 'defensive' innovation, the oxygen-lancing of open-hearth furnaces, which increased output rates of the older equipment, though of course at the price of incurring charges for oxygen in addition to the usual fuel costs. All open-hearth facilities in our survey sample had been modified in this fashion.

2　Savings in labour costs appear substantial, but as will be shown below, these costs account for but a small proportion of total costs. Again, the burdens placed on open-hearth operations in plants A and B on account of input composition are apparent from the BOF's greater cost advantage.

3　All other direct costs show great variability among plants, but they do favour the BOF. Detailed investigation showed that the absolute amounts of these costs depend very much on intra-firm accounting practices. But for the mill as a whole, and for total costs per unit of finished product, these practices make little difference: items not charged or credited at the BOF or open-hearth stage of production will appear elsewhere in the system.[23]

The degree of leverage exerted on total costs by savings in specific cost categories depends, of course, on the composition of these costs. The relevant data are shown in Table 17.6. The outstanding fact, as pointed out earlier, is the domination of materials costs over all other types of cost. Net metallic costs vary between 60 and 80 per cent of all costs for the open-hearth process, and between 69 and 90 per cent of BOF costs. The variability in the latter case may appear puzzling, given the relatively rigid input mix requirements of the new technology, unless one considers: transfer pricing practices for molten iron; the internal pricing of home scrap vs. the market prices of various types of purchased scrap; and the fact that these are 'net' costs, that is, costs after credit is given for scrap generated and recycled elsewhere in the production system.

In any event, even our small sample of mills suggests how misleading 'industry average costs' might be when used to evaluate the attractiveness and the effects of innovations for individual member firms of an industry. Compared to the share of metallics, none of the other cost components is very large. Thus, even spectacular success in reducing the unit cost of such inputs as fuel and oxygen, operating labour, or

Table 17.6

Composition of total costs (in per cent), open-hearth and BOF shops (late period)

Plant	Net Metallics	Of This		Cost Above Metallics	Fuel and Oxygen	Of This		Other Costs	BOF as Percent of Open-Hearth Total Cost
		Hot	Cold			Operating Labor and Salaries	Repair, Maintenance Materials		
Open Hearth A	76.3	8.4	67.9	23.7	4.2	4.1	2.9	12.5	
BOF A	90.0	58.1	31.9	10.0	1.8	1.3	3.2	3.7	71.5
Open Hearth B	59.7	8.9	50.8	40.3	5.0	11.3	6.2	17.8	
BOF B	68.7	46.2	22.5	31.3	1.1	3.5	2.3	19.4	77.6
Open Hearth C	70.6	56.0	14.6	29.4	3.1	6.5	4.3	15.5	
BOF C	79.8	60.8	19.0	20.2	1.8	4.7	3.5	10.2	88.9
Open Hearth D	79.9	50.6	29.3	20.1	2.0	5.5	1.9	10.7	
BOF D	86.4	75.7	10.7	13.6	1.0	4.3	1.4	6.9	95.8
BOF E	89.1	63.7	25.4	10.9	0.8	2.1	1.4	6.6	88.3[a]

Source: Calculation from survey data.

[a] Cost comparison with open hearth D (same accounting system).

repair and maintenance work, will affect the total cost of production only to a relatively small degree.

The picture which emerges, then, tends to support the findings of previous empirical work:[24]

a) that even major innovations, when placed into an existing operating system, frequently do not generate enough leverage for revolution ary cost savings, because the bulk of all costs is determined by factors outside the locus of the innovation proper; and

b) that management's predilection for minor and incremental improve ments in performance and costs reflects a proper assessment of the realities of technological decision-making in large materials dominated production systems.

Empirical findings – continuous slab casting

The continuous casting of steel must be thought of in terms of two, quite distinct, processes: the casting of billets and bars and the casting of slabs.[25] The former posed fewer technical problems and was readily adaptable even in small-scale production; therefore, its diffusion since the start-up of the first facility, in 1963, has been rapid. By the end of the 1970s, billet casting accounted for approximately 50 per cent of the non-integrated producers' (mini-mills') carbon steel output.[26] On the other hand, slab-casting proved to be a technically difficult process, involving heavy and expensive equipment, major adjustments in preceding and succeeding production steps, and delicate operating problems.

As Table 17.7 shows, the first slab-casting plant in an integrated mill was started up in 1967. Installations in speciality and alloy plants followed quite rapidly but involved substantially lower capacities. We were able to investigate the performance of four of the first seven plants in integrated mills; since none of these could be said to have reached a state of smooth integration into production at the time of our survey, our observations concern only two periods, one immediately following start-up ('Period I') and one just preceding the collection of data ('Period II').

Characteristics and expected effects

The processing of the industry's key intermediate product, raw steel (hot metal), traditionally involved a batch process. The hot metal is poured from steelmaking furnaces into a ladle and then cast into ingot

Table 17.7

Year of start-up and rated capacities, slab-casting plants

Year of start-up	Rated capacity (1000 tons)		Continuous casting capacity as percentage of raw steel capacity	Source of raw steel[b]
	Carbon-steel plants	Specialty and alloy plants		
1967	2100		25	BOF
1968	2500		92	BOF
	1500		40	BOF
		450	95	EF
	250[a]		100	OH
1969		400	n.a.	EF
1970		600	45	EF
		600	100	EF
1971	800		40	BOF
		300	30	EF
1972		330	50	EF
	1300		30	BOF
1973		600	100	EF
1974	1500		30	BOF
		150	50	EF
1975	—	—		

[a]This is an installation in a small carbon-steel plant serving a regional market. Its reliance on the open hearth process for raw steel suggests its exceptional character among the adopters of slab casting.

[b]BOF = Basic Oxygen Furnace; EF = Electric Furnace; OH = Open Hearth Furnace.
Source: Rosegger (1979).

moulds, in which the metal cools and hardens. The ingots are then stripped from the moulds, placed into inventory, or transported directly to soaking pits, reheating furnaces which bring the ingots to uniform temperature before they are rolled into the standard shapes. While billets and bars involve cross-sectional dimensions of 4–8″ (10–20 cm), slabs (and blooms) are 5–12″ (12–30 cm) thick and 30–90″ (75–225 cm) wide.

As Figure 17.2 shows, continuous casting bypasses these several steps. Hot metal is taken from the furnaces to a casting machine and drained into the tundish (a refractory-lined funnel), through which the raw steel runs into the mould proper. There, the cooling process begins, and, with its outside hardened, the resulting strand of steel is withdrawn from the mould by a series of water-cooled rollers. It emerges as a slab, which is then cut to workable lengths. As usual in the early stages of innovation, casting equipment appeared in many designs and variants. Among these a system whereby casting and hardening occurs in a vertical section, but the slab is then turned into a horizontal direction through a curved mould, became more or less standard for the production of slabs.

Even this brief description suggests that continuous casting should enjoy a considerable cost advantage over the traditional processing technique. In particular, this advantage is to be derived from the following:[27]

a) Savings in fixed-capital investment of from 25 to 40 per cent, as compared to the traditional, more circuitous process technology.

b) Reductions in required plant space of over 35 per cent. This factor weighs heavily in older American plants, many of which were built in constricted areas.

c) Reductions in energy costs of 30 to 75 per cent, mainly through the elimination of cooling-reheating cycles.

d) Reductions in other operating costs as a result of the elimination of expenses connected with materials handling, in-plant transport, and ingot preparation.

e) Improvements in yield, that is, the ratio of slab output to raw steel input, from 80–85 per cent for the ingot rose to 90–95 per cent. Such reductions in home scrap would provide secondary benefits in situations where the BOF's limited capability to use scrap acted as a constraint on output.

f) Improvements in the quality of the product, especially through the reduction of surface imperfections in flat-rolled steels.

Taken *in abstracto*, these benefits would appear to be persuasive. However, these advantages depend to a large extent on the assumption that the new technology is a *replacement* for the old one. At the stage of its technical development during our survey period, total replacement of ingot facilities was not an option; and, in the American setting, the question whether slab casting should be relied on as the sole technology in greenfield plants was moot. The adoption of slab casting on an *incremental* basis or as a *partial displacement* of ingot capacity would not only eliminate many of its potential cost advantages but might even lead to increases in the costs of coordinating two types of production flow.

Nevertheless, the acquisition of incremental capacity via slab casting would prove attractive in situations where increments remained well below the minimum efficient size of traditional equipment, especially slabbing mills. In addition, partial displacement provided an opportunity for direct learning with the new technology, with the old facilities serving as a backstop in case of trouble.[28] Even under these conditions, the realisation of any advantages would still depend on such plantspecific factors as the size of production runs with particular (crosssectional) dimensions and metallurgical specifications, and the fit of the new facilities in terms of plant layout and operating practices.

Under these conditions it is not surprising that identical advance evaluations of slab-casting's technical and economic potential would lead to conflicting decisions about adoption among individual firms. And although most top managers seemed to agree that this was 'the technology of the future', many of them also had a perfect rationale for a 'wait-and-see' attitude. As the results of our investigation suggest, the experience of early adopters most likely provided confirmation for this attitude.

Comparisons of actual performance

In contrast to the BOF, continuous casting was in the early stages of development during our survey. No consistent accounting practices had been agreed upon, and only comparisons of physical performance across plants are meaningful. An effort to reduce all but the most basic cost categories to a common standard would have resulted in averages yielding few empirical insights; therefore, the cost data presented should be interpreted as reflecting individual case studies, even though they are gathered in tabular form.

Table 17.8

Slab-casting: shares of output and capacity, yield (Period II)

	Plant			
	A	B	C	E
Share of mill capacity (per cent)	23.5	31.8	30.2	26.4
Share of mill output (per cent)	11.2	28.8	34.7	18.9
Capacity utilisation (per cent)	37.6	51.5	86.6	69.7
Yield (per ton of hot metal)	.85	.93	.96	.91
Yield, ingot technology	.87	.81	.86	.83
Output/manhour (wage recipients) (tons)	1.72	–	–	2.03
Output/manhour, ingot technology	4.37	–	–	6.94

Source: Calculation from survey data.

The incremental or displacing nature of these installations is indicated by their share in total mill capacity. The difficulties experienced in production can be surmised from slab-casting's share in actual output

Table 17.9

Comparative total costs, metallic costs, and costs above metallics, cost proportions

	[Period]				[Period II]			
	A	B	C	E	A	B	C	E
Cont. cast. cost/unit as % of trad. tech.	126	114	105	144	95	94	91	–
Metallic costs/unit as % of trad. tech.	97	110	95	–	93	93	93	–
Cost above metallics/unit as % of trad. tech.	334	123	160	142	143	97	79	–
Of this: labour cost/unit as % of trad. tech.	689	180	233	200	264	114	142	–
materials and supplies	500	228	353	527	58	112	262	–
Metallic costs, cont. casting (share in %)	67	75	77	–	82	80	89	–
Metallic costs, traditional technology	87	77	85	–	91	80	88	–
Labour costs, cont. casting (share in %)	14	9	13	4	6	7	6	–
Labour costs, traditional technology	3	6	6	3	2	6	4	–
Mat. & supplies, cont. casting (share in %)	11	11	8	7	5	7	2	–
Mat. & supplies, traditional technology	3	5	2	2	3	6	2	–
Energy cost/unit as % of trad. technology*	62	65	91	89	25	33	37	–

* Calculations based on separate data set; not fully comparable with other relative costs.

Source: Calculations from survey data

and capacity utilisation rates. In plants C and E, managements declared that most of the basic technical problems had been worked out and that they regarded operations as 'reasonably smooth'. No such claims were made for plants A and B.

The yield data show a clear superiority for the innnovation, entirely in line with advance estimates. *When* the new technology was functioning, it brought substantial improvements; however, the impact on total mill performance was obviously limited by its low shares in output.[29]

Output per manhour statistics were available for only two plants and reflect a clear disadvantage for the innovation. Three factors account for this: low utilisation rates, crew sizes inflated by the need for standby personnel in case of difficulties, and the need for a careful monitoring of every aspect of the operation in a complex facility. Whether the high labour intensity of the process will continue once learning is over remains to be seen.

We attempted calculations of the extent to which expectations of energy savings were met; however, these had to be based on a number of assumptions that put the accuracy of the results in question. What little consistent evidence we could muster from the plant statistics suggests that the tonnage produce per unit of energy input was in the range of six to ten times higher than for the traditional technology. The immediate success of the innovation in this respect is not surprising: energy is on demand only during actual operations and thus, in contrast to labour, its specific use is not affected by the frequent downtimes.

Table 17.9 summarises our findings with respect to comparative costs. It will be seen that an advantage in total unit costs emerges for the innovation during Period II. However, this advantage is far from convincing. The bulk of it is due to a lowering of metallic costs per unit of output, which make up by far the largest proportion of total cost. This, in turn, is the result of two factors: the technology's improved yield and the practice of charging the continuous casting operations lower transfer prices for hot metal than are charged to the traditional processing operations. Efforts to clear up the reasons for this practice brought no satisfactory responses. The true economics of continuous casting is further beclouded by the fact that no charges are made for rerouted hot metal in case of breakdowns or difficulties, although such events clearly increase actual costs for the mill as a whole.

Labour costs were substantially higher for the new technology but showed improvement from Period I to Period II. With crew sizes more or less fixed, higher capacity utilisation rates account for most of this improvement. But labour costs amount to only a small fraction of total costs. These figures support one other general observation: once well established operations, such as the tradition ingot technology, have been brought to a higher level of efficiency through continual improvements, their greater roundaboutness is no great handicap in terms of labour requirements. Conversely, a major innovation still undergoing development and learning will be handicapped by high labour requirements (and costs) for prolonged periods.

The higher relative materials and supply costs for continuous casting reflect in part inherent characteristics of the process, such as requirements for lubricants and mould powder, and in part the effects of frequent breakdowns and repairs. Operators expected better performance as operations became smoother. They pointed out that, on the other hand, the achievement of longer casting sequences (that is, increases in the number of heats cast consecutively without shutdowns) would mean greater wear and tear on linings, refractories and lubricants, thus perhaps offsetting the above cost savings. In general, of course, greater economic benefits can be derived from reducing the time spent on repairs and maintenance than from savings in materials and supplies.

Because of the tenuousness of some of our calculations, relative energy costs are shown last in the table. They do suggest a clear-cut advantage for continuous casting even during Period I. However, since they amounted to no more than two per cent of total unit costs in the worst case, their impact is negligible. More precise answers could be gained only from the construction of complete energy balance for the two technologies.

The proposition that the innovation's emerging advantage is still governed primarily by learning seems supported by our cost findings. The most important traceable sources of advantage lie in the new technology's higher yield and in energy savings; of these, the former is clearly the more significant. Whether a cost advantage of, say, five to ten per cent after several years is sufficient to justify the investment remains an open question. Managers whom we interviewed about prospects for improvement still seemed more concerned with solving problems of physical performance than with costs. They emphasised that the innovation is 'paying off' in many other respects as well: as a marketing tool, in terms of better product quality, and by keeping personnel in touch with a technology that is expected ultimately to dominate the production of flat-rolled steel. Correct or not, all of them thought that the costs of not being among the early adopters would have been higher in the long run.

Concluding observations

Technological innovation is but one of a number of strategies leading to structural adjustment in the iron and steel industry. As Gold[30] has suggested, it may well fall short of the objective of returning the industry to full international competitiveness, unless it is accompanied by other, complementary strategies. There can be little doubt, however, that decision-makers in the United States viewed the piecemeal introduction of major innovations into existing, integrated mills as a major device for improving efficiency and costs. This predisposition was probably reinforced by attitudes of the capital market and by the fact that neither regulatory nor other governmental policies have had more than haphazard, and often contradictory, effects on the industry's performance.

Our findings point out the limitations of piecemeal innovation. We have seen that increasing proportions of the total market for carbon steel have been captured by imports and by non-integrated producers. And we have adduced technical and economical reasons why the introduction of new process technology into integrated mills may have to a) overcome evaluative hurdles considerably higher than those suggested by abstract estimates of the potential benefits of major innovations, and b) console itself with economic results that appear relatively meagre when measured against the magnitude of the industry's competitive problems.

Our case studies show, in any event, that these results have so far failed to revolutionise costs *at the locus of innovation* in the production system and that they have involved modification costs and managerial adaptations elsewhere. In an established, materials-dominated industry such as iron and steel, revolutions are hard to come by – unless one were to construct a number of competitively-scaled plants *de novo* , an option clearly not available to American firms under the constraints operative over the last two decades.

Two final observations are in order:

1   It would be a mistake to conclude from our empirical results that the overall effects of piecemeal innovation were limited to the rather discouraging economic benefits. Clearly, *not* to have followed at least this essentially defensive strategy would have had even more serious consequences for the industry. And furthermore, non measurable and qualitative, system-wide effects surely left firms in a better position to face the technological and competitive challenges ahead.

2   Seen from this perspective, piecemeal innovation obviously constituted a necessary, but not sufficient, condition for the

upgrading of America's traditional steel sector. At the very least it created a base from which to pursue longer-run, and perhaps even more aggressive, strategies. Whether the industry will commit itself to such strategies remains to be seen.

Notes

1   Part of the research for this chapter was carried out under a National Science Foundation grant (7518861). However, none of the findings and interpretations necessarily represents the views of the Foundation. The author is indebted to Professor Bela Gold for his criticisms and suggestions.

2   Boyce, R.L., 'The Structure of Steel Markets in the United States', paper presented at the Twelfth Atlantic Summit Conferences, New York, 8–11 October 1981 (mimeo).

3   For enlightening testimony on these variations, see US Congress, Subcommittee on Trade of the Committee on Ways and Means, *Administration's Comprehensive Program for the Steel Industry* , Washington DC, US Govt Printing Office, 1978, esp. pp131–6.

4   See, for example, US Council on Wage and Price Stability, *Report to the President on Prices and Costs in the United States Steel Industry* , US Government Printing Office, Washington DC, 1979; Mueller, H. and Kawahito, K., *Steel Industry Economics: A Comparative Analysis of Structure, Conduct and Performance* , Japan Steel Information Center, New York 1978; Office of Technology Assessment, US Congress, *Technology and Steel Industry Competitiveness* , US Government Printing Office, Washington DC, 1980; Crandall, R.W., *The US Steel Industry in Recurrent Crisis: Policy Options in a Competitive World* , Brookings, Washington DC, 1981.

5   Gold, B., 'Analyzing the Structure of Technological Impacts: From Physical to Economic Effects', Gold, B. (ed.), *Productivity, Technology and Capital* , D.C. Heath-Lexington Books, Lexington Mass., 1979.

6   American Iron and Steel Institute, *The Competitive Challenge to Steel* , New York, 1961.

7   In fact, the negotiation of 'voluntary' quotas between the US and Japan in the late 1960s no doubt encouraged and hastened import penetration into the high-priced segments of the market, including alloy and speciality steels.

8   American Iron and Steel Institute, *Steel Imports – A National Concern* , Washington DC, 1970.

9   See for example, Foy, L.W., 'We Can Do It', presentation at

the Annual Meeting, American Iron and Steel Institute, New York, 21 May, 1975 (mimeo).

10 On the basis of a number of heroic assumptions, Crandall op.cit. 1981 calculated that firms would face a cost for equity capital of 19.7 per cent. Gold and Boylan (1975) have argued persuasively against the relevance of such measures in the *ex ante* assessment of major, strategic investments in an industry like iron and steel. Gold (1977) systematically elaborated the argument against capital budgeting approaches to such investments.

11 Rosegger, G., 'Diffusion and Technological Specificity: The Case of Continuous Casting', *Journal of Industrial Economics*, September 1979.

12 American Iron and Steel Institute, *Steel at the Crossroads: The American Steel Industry in the 1980s*, Washington DC, 1980.

13 Crandall op.cit. 1981, estimates the cost per ton of capacity as low as $500, whereas the Office of Technology Assessment op.cit., 1980, bases its so-called 'replacement rate' on a cost of $1,100 per ton of shipments (with no indication of the rate of capacity utilisation assumed)!

14 Gold, B., *et al*. 'Diffusion of Major Technological Innovations in U.S. Iron and Steel Manufacturing', *Journal of Industrial Economics*, July 1970.

15 A major innovation of this type, although off-site from integrated mills, must be mentioned: it is the beneficiation and pelletising of low-grade iron ore (taconite), which not only vastly extended America's domestic resource base but also greatly increased blast furnace productivity (Peirce, W.S., 'The Effects of Technological Change: Exploring Successive Ripples', Gold., B. (ed.), *Technological Change: Economics, Management and Environment*, Pergamon, Oxford, 1975.)

16 See Gold, B., 'Factors Stimulating the Technological Progress of Japanese Industries: The Case of Computerization in Steel', *Quarterly Review of Economics and Business*, Winter 1978; and Gold, B. *et al*., *Evaluating Technological Innovations: Methods, Expectations and Findings*, D.C. Heath-Lexington Books, Lexington Mass., 1980 esp. Chapter 13.

17 Rosegger, G., 'Diffusion Research in the Industrial Setting: Some Conceptual Classifications', *Technological Forecasting and Social Change*, March 1976; Gold, B., 'The Framework of Decisions for Major Technological Innovations', Baier, K. and Rescher, N. (eds), *Values and the Future*, Free Press, New York, 1969; and Gold, B., 'Long-run Prospects for the World Steel Industry: Some Internal Potentials for Improvement',

The Steel Industry in the Eighties , Metals Society, London 1980.

18 A managerial preoccupation with evaluating major innovations in a capital-budgeting framework has rightly been criticised (Gold, B., 'On the Shaky Foundations of Capital Budgeting', *California Management Review* , Spring, 1977.). It is also clear, however, that institutional factors, leading among them the attitudes of lenders and of the capital markets in general, have fostered this preoccupation.

19 Gold *et al* ., op.cit., 1980, p.137.

20 Detailed statistics on worldwide installation of basic oxygen facilities, start-up years, and capacities can be found in Kaiser Engineers *L-D Process Newsletter* (occasional).

21 After 1964, annual output in each year exceeded the total capacity in existence during the preceding year, suggesting the strong preference of managements for the new process and the reduction of start-up problems.

22 A more detailed analysis of expectations, variations in operating conditions, and results can be found in Gold *et al* ., op.cit., 1980, chapter 6-8.

23 This observation is not meant to make light of the fact that methods of cost allocation to particular operating centres may have a strong influence on managerial behaviour at various process stages.

24 For example, Dilley, D.R. and McBride, D.L., 'Oxygen Steelmaking - Fact vs. Folklore', *Iron and Steel Engineer* , October 1977; Gold, B., *Exploratins in Managerial Economics: Productivity, Costs, Technology and Growth* , Basic Books, New York, 1971; Gold, B., op.cit. 1977.

25 See Rosegger, G., op. cit. 1979.

26 OTA op. cit., p.290.

27 Detailed estimates, under various assumptions, can be found in United Nations (1968) Gott (1970), Morton *et al.*, (1973), Liestmann (1975), Willim (1975), Domröse and Koch (1976); see Bibliography.

28 McManns, G.J., 'Slab Casting: Caution Gives Way to Action', *Iron Age* 6 February, 1967.

29 This observation must be further tempered: unexpected difficulties mean that hot metal scheduled for the caster has to be rerouted to the ingot facilities, which act as a backstop, with a concomitant drop in overall efficiency.

30 Gold, B., op.cit. 1980.

PART V
REGIONAL AND
ENVIRONMENTAL ASPECTS
OF BIG STEEL INDUSTRY

18   Introduction

The emergence of regional and environmental problems

The integrated steel industry is 'big' industry: like shipbuilding, production of automobiles and aircraft, it is labour- and capital-intensive. Its plants are large employers. Thus the industry, its enterprises and plants generally have a greater impact on their surrounding environment than smaller-scale industries. Steel's bigness also means that the plants need access to an infrastructure with characteristics specific to steel, for example, access to cheap bulk transportation. An integrated plant with an output of 5 million tonnes of raw steel per year will as a rule need bulk raw material inputs in the range of 30 million tonnes per year or even more.

Unlike all the above-mentioned industries, iron and steel production burdens the physical environment in the form of air or even water pollution. Iron and steelmaking processes use large quantities of energy and thus produce exhausts of different types.

Because iron and steelmaking traditionally has been located near raw material deposits – because of the already mentioned large quantities needed for the production of iron and steel – the industry is often concentrated in regions containing large deposits of ore or coal or, in some cases, of both. Steelmaking then not only implies the existence of large plants, employing thousands of (mainly) men, but also a concentrated dominance of, sometimes, a large number of plants in a region. As steelmaking means a good deal of heavy, noisy and dirty work, the steel industry usually has to pay above average wages (up to 75 per cent above the average male worker's pay in the US manufacturing industry, approximately 50 per cent above the same average in Japan, and between 10 and 25 per cent above in Europe). This implies that, if other industries want to hire employees in a steel-dominated location or region, they must be in a position to pay above average wages not necessarily exactly matching the levels paid by the steel industry, but not with too large a difference.

A consequence of this is that steel-dominated locations and regions tend to become 'monostructured': industries unable to match the wage levels given by the steel industry tend to leave or keep at a reasonable distance from dominant steel. Small- and medium-size firms will also tend to stay away. Steel firms often will have to build up their own 'infrastructure' with regard to auxiliary production and services, which are needed directly or indirectly for the operation and maintenance of steel plants, and which elsewhere would have been performed by local trade and small- and medium-size firms.

Thus old steel regions not infrequently lack important elements of economic infrastructure available in locations and regions not dominated by big industry, or maintain an infrastructure which is highly steel-oriented or steel-dependent. Big steel not only requires a particular infrastructure, directly and indirectly; it exercises influence on the long-term development of the infrastructural characteristics in its environment. Being such large employers, steel plants usually also have a variety of influences on the local and regional conditions and infrastructure. Large employers and important taxpayers usually become very influential in local and regional affairs. Their wishes or requests are given due attention, and, in good times, they usually also reward the community in different ways, ranging from supporting sports to building meeting halls, or placing their facilities at the community's disposal.

Big industry usually can mobilise strong lobbies, like the 'steel caucus' in Washington DC. Not infrequently steel employers and steel labour unions have a common understanding on issues important to the industry, despite the regular quarrels over wages, conditions and fringe benefits they both have and loudly voice. All this contributes to the stronger local, regional, state or federal influence exercised by the steel industry and indeed other big industry.

As long as business is as usual, all the points made above may be accommodated – including the sometimes rather costly, but most necessary, pollution control. When clouds loom over the horizon, the situation may shift. Soon after World War II, and for some decades to come, the US steel industry was blamed for its inflation-pushing pricing behaviour, to mention a negative example. The steel industries of several European countries in dual-structured coal and steel regions smoothly rescued the ailing coalmining industry, when oil had become so cheap that it drove coal out of the market. It was good to have big steel plants nearby, to solve the unemployment problems for displaced miners.

However, bigger clouds soon covered the sky. It may have been one of the basic policies of MITI (Ministry of Industrial Trade and Industry of Japan) which caused trouble, not only to the steel industry, but also, and sometimes to a greater extent, to traditional steel sites and regions. Before the turn of the century, MITI's predecessor, the Ministry of Agriculture and Commerce, had already recognised the advantage of siting bulk transportation-intensive steel industry at deep-sea harbour locations. Japan lacked iron ore and thus had to cater for its transportation by sea. The country's first big steel plant was consequently sited at a coastal location, becoming operational in 1901. One of the central post-World War II MITI policies was to cater for low-cost sea transportation because of the island empire's strategic dependence on imports of all kinds of goods. The policy implied the development of a highly efficient large-scale shipbuilding industry (as well as the establishment of steelworks to supply the shipyards). Japan, or its trade houses, the Zaibatsus, could afford to produce ships at little or no profit at all: by means of cheap and efficient large tonnage, freight rates would (and did) plummet. Japan – or its Zaibatsus – greatly gained from a transportation cost revolution that they had started themselves – and also gained from declining costs of energy.

One of the first consequences was the collapse of shipyards in traditional shipbuilding countries. The European and US steel industries also suddenly lost one of their major traditional advantages, viz. access to high quality ore and/or fuel nearby, requiring low transportation costs. The transportation cost 'revolution' made it possible for Japan to quickly exploit new ore and energy deposits and consequently to drastically reduce ore and coke prices, thus making its steel industry cost-competitive in a rather short time.

By the middle of the 1960s it became suddenly evident to many of the big steel firms that, in the long (perhaps even medium) run, they could only survive against their Japanese competitors by quickly adopting two strategies which had to be merged into one: producing steel at large scale and at low cost-of-transportation sites.

From the mid-1960s decisions (by individual firms or, in France, at the initiative of the state, in Great Britain by nationalising the large steel industry) were taken a) to close small or old plants; b) to move iron and steel production (at least the furnaces) to coastal (or equivalent) locations; and c) to establish large-scale integrated plants.

Three additional important changes to the steel industry occurred in the second half of the 1960s.

First, growing awareness of environmental qualities, their uniqueness and their deterioration, came to articulate requirements against pollutants of different types and sizes, amongst them the steel industry. The industry, which felt severely threatened by the new Japanese com-

343

petition, initially reacted by rejecting all claims. Later, when environmental regulation was introduced, the industry had to adapt, often reluctantly, in a variety of ways which, to a major extent, depended on the situation in which the firms found themselves at different points in time. In the late 1960s competition was rather tough; the profits of many firms were quite meagre and the willingness to comply with the environmental claims was therefore rather low. During the first half of the 1970s the industry became more and more busy in trying to meet the booming market. Protective measures taken disrupted production, which had to be run at higher and higher levels of capacity utilisation. When the boom was over, the lean years came, but also a better time to adapt to more stringent regulations.

The second change did not have as much impact upon the industry in the short or medium term as environmental regulation: as an alternative to the rapidly increasing scale of integrated steel refining, scrap and electric arc furnace small-scale 'mini-mills' emerged, choosing their sites near steel consumption centres. The mini-mills gained from several new features, which were often disadvantageous to the large integrated steel mills:

1   Scrap is a highly refined, energy-intensive raw material base, often available in regions, where steel consumption is also high.
2   Mini-mills were therefore much less constrained by the low bulk transportation cost criterion which had become vitally important for big steel.
3   The energy intensity of scrap made the mini-mills not nearly as vulnerable to high energy costs as one would assume, given that they consume the highest quality energy – electricity.
4   Mini-mills are low or nil, pollutants, and did not therefore have to invest in costly 'non-productive' pollution reduction devices.
5   Mini-mills cost only about one-tenth per tonne of annual capacity to invest, compared to big steel. This gives great advantages in flexibility and, during the years of high interest rates on foreign capital, a quite unique capital cost advantage compared to big steel.
6   Mini-mills need no site-specific infrastructural arrangements worth mentioning nor do they influence existing infrastructure.
7   Mini-mills are generally not affected by nationalisation (for example, as in Great Britain) nor by other regulations (as, for example, the Davignon plan for the restructuring of the European Community's steel industry).

The growth of the mini-mill sector has been remarkable. In the USA it now exceeds steel imports covering 20 per cent of apparent consumption in 1984 and expected to grow to 25 per cent by 1990, at the expense of big steel. The large integrated producers nevertheless have not yet reacted strategically to this substantive loss of market shares. Although they have not yet been regarded as a major threat to big steel, mini-mills are expected to grow further, at the expense of big steel.

The third change, however, is generally recognised as implying a major threat to big steel: growing competition from newly industrialising countries, and not only in low quality carbon steel. The main cost advantages of newly industrialised countries are low labour cost and, often, low cost for environmental protection. These advantages make it possible to operate plants not representing the forefront of technology, but rather the so-called 'appropriate' technology, for example, open-hearth furnaces and other low capital cost production equipment.

The changes in the competitive and regulative situation briefly described above have had far-reaching consequences for the large-scale integrated steel industry.

One set of consequences is taken up in an account of the cost of compliance with regulations superimposed upon the US steel industry in order to protect and/or restore its physical environment (air, water, soil) from the consequences of steelmaking related pollution. The account concentrates upon the situation facing US big steel industry and does not compare that particular situation and regulation to those existing elsewhere. It is evident however that the extent and enforcement of environmental regulation with regard to big steel is fairly similar in Japan (where many new steel plants operate in the densely populated coastal region near Tokyo and other large cities where environmental regulation is therefore rather rigid) or in Western Europe (which certainly includes the north, north-west, and major parts of the south). There is no reason to doubt that environmental regulation and enforcement is fairly uniform in the regions mentioned.

Cost advantages due to low environmental protection obviously apply to a rather wide range of newly industrialising countries and less developed countries, although with some remarkable exceptions.

The other complex of relevance to big steel lie in the sphere of the regional impact of steel mills or steel industry. There are at least three major issues at stake:

a) big steel's influence on local and regional industrial and economic structure;

b) big steel's influence on the local and regional infrastructure (including the physical environment);

c)   big steel's influence on local and regional labour markets.

As mentioned above, the relevance of the regional issue has to do with the mere size of integrated steel plants in the first place.

A second set of reasons of growing importance for tensions, or even conflicts, between steel producers and local and regional authorities etc. emerges from the changes in the competitive situation of the industry at large. As the competitiveness of big steel has undergone changes of increasing severity, from the middle of the 1960s, after the end of the 1973–74 boom, and since 1979, when it finally became obvious that the industry is facing its severest structural crisis to date, its stagnation, but, still more importantly, its strategic decisions, have come to cause severe concern for the local and regional, later even the national but also supra-national decision bodies.

1   As briefly described above, big industry in general, and steel mills in particular, exercise an influence in the long term on the industrial and economic structure of steel locations and regions where they operate. They tend to become 'monostructured', that is, to be vitally linked to the fate of the dominant steel mill, or firm. If the firm has to take decisions crucial to its continued viability or survival, such decisions will necessarily have an impact on the region. If the firm decides to terminate its operations, the municipality, town or region loses its economic backbone. It is left with a structure which has developed slowly over decades, and, by being the domicile for a big firm, has lost its vitally important safety net – a diversified economic structure. Unfortunately, losing a dominant big firm can practically never be overcome by bringing another big firm into the lost one's place.

Developing a desirable differentiated industrial/economic structure takes a long time, to be counted in decades rather than in years. The disadvantage of being both monostructured and deserted means a negative stigma of incredible magnitude, serving to repel firms searching for new locations, either to start new plants or to extend the operations of existing ones.

2   Big steel not only structures its environment and economic as well as political infrastructure. It regularly also expects or requests firm- or plant-related measures of different types, with or without participating in the cost of the investment and operation of the arrangement in question: streets and other 'arteries of mass transportation' like waterways, railways and so on. If the firm makes large incremental changes in the scale of operations, either upwards or downwards, the infrastructural

346

arrangements become obsolete, may have to be consequently adapted or changed, or become idle.

Big firms – and big steel seems to be a relevant example – both have an impact on land use patterns (often reflecting different phases of the firm's development) and may also appear as dominant land owners. If times are good, land may be acquired for further expansion purposes or as a kind of financial reserve. When times are bad, firms may have to sell off their land in order to cover incurred losses. All this contributes to the distortion of the organic growth and development of the location or region.

The links between the dominant firm(s) and the town or region thus go far beyond the industrial and economic structure. The older the firm, the deeper often the influence. Changes in the firm's central strategies thus frequently have a tangible impact on the local or regional infrastructure.

3   Big firms exercise a substantive impact on local and regional labour markets. The market grows and adapts to the needs of the dominant employer(s), both in quantity and quality. The local or regional labour market also draws many advantages from the presence of a large employer, for example, in providing opportunities for apprenticeships and other types of education/further education.

Steel industry influences the labour market also through its wage level, which generally is higher than that of the average manufacturing industry.

Certain types of employers will leave the region, since they cannot compete for the labour they need. Groups or individuals having employment criteria which do not fit the dominant firm, or who simply do not want to work in a steel mill, will have to look for employment elsewhere; eventually they may have to leave the region.

Steel thus forms its local and regional labour market perhaps more than other big industry does. If the industry changes its local operations – upwards or downwards – by large increments, the local or regional labour market will only adapt rather slowly. This applies in particular if operations are closed down. People are not 'perfectly mobile assets', in particular if they have families, or if they possess homes of their own.

The above summary accounts for the regional impact of big steel, and the problems and conflicts of interest. The short outline of the consequences to the region of change, and in particular of drastic change, is by no means complete or exhaustive. There is, however, sufficient evidence to further scrutinise the local and regional complex around big steel, as it is necessary both for the industry, the firm and its management and the political and administrative planners and decision-makers to improve their problem awareness, as well as their problem-solving horizon. Knowledge about local and regional problems connected with big firms is pertinent for many decisions and is by no means confined to dealing with steel.

Two approaches to the same problem region, the West German Ruhr District, are presented as examples of regional concerns. One is a rather thorough analysis of 15 years of labour market adaptation to a long sequence of strategic corporate decisions (or rather perhaps, to stress the argument, of non-decisions). It starts in the latter part of the 1960s, under the impact of the 'move to the seaside' wave, following it through a merger and its dissolution, to a nearby new merger and its cooling off (cf. also Chapter 7).

In Chapter 20, the consequences of the ups and downs of Hoesch's Dortmund operations are analysed as to their impact – real or potential on Dortmund's labour market as well as that of the region. Comparisons are made with development in the Liverpool (England) situation. In its analytical, low-key approach, the chapter represents a thought-provoking challenge to both corporate management as well as politicians, labour market and regional planners and decision makers.

Chapter 21, also on the Ruhr District, represents a rather different approach to the problems of regional and local economic, industrial and environmental structure and planning. The key chosen by the author is political–provocative rather than analytical, in comparison to the chapter on the Ruhr District labour market. The author's style may be characterised as being representative of the new intellectual environmental movement (*Die Grünen* – the Green Party) which has become a new political force in several European countries, including the Federal Republic of Germany. The approach is by no means 'anti-industry'-oriented. Industry is a vital and natural element of a region. But industry must learn to think and act in more responsible terms, to internalise the environmental and regional problems it causes rather than strictly to externalise them. The politicians, administrative planners and decision makers in the region must also learn to take new stances. Those in power during recent decades have let the region move into the trap of monostructural dominance. The author offers a range of options to escape from the trap. But all of them will take a long time, because regions grow and adapt slowly – even if firms may decide quickly. Since regions develop mainly by organic growth and adaptation, greater, more articulate long-term-oriented thinking and acting is requested from all parties involved.

The key used in the chapter may not please many readers at first glance, but it leaves behind many points and thoughts of great importance.

The issues of regional impact, as far as both the environment and employment are concerned, are of more general interest than those covered in general in this book, as they do not constitute a specific steel industry problem. The dilemma of internalising versus externalising the employment effects of technological change today concerns many industries as well as regions.

In order to survive in a highly competitive, frequently cut-throat economy firms must, on the one hand, keep productivity in the production plants at top levels (for example, comparable to Japanese top performance producing 1,800 tonnes of steel per worker per year), since running plants at lower efficiency means jeopardising their continued existence. On the other hand, the internalisation of unemployment created by rationalisation, by technological change, by change of scale or product portfolio, that is, the continued employment of people made idle or jobless because of the measures taken, means that those people will be occupied with tasks which can rarely be performed at high levels of productivity. The firm thus finds itself in a dilemma, employing people who, at often only nominal differences in pay, are supposed to perform at very different levels of productivity. This creates morale problems of great magnitude and consequence.

However, if unemployment is externalised, the morale problem is not solved: if the employed know that they will not be taken care of if they lose their jobs, they may want to reduce productivity, in order to make existing jobs 'last longer'. Firms' management must not forget that, if the logical consequence of necessary rationalisation and productivity improvement is the threat of unemployment to the workers, they will resist the measures taken. Neither can the unions continue to concentrate on the welfare of the lucky ones, who can keep their jobs, only to neglect the sufferings of their ex-members who were dismissed.[1]

As there is now considerable experience available in Europe of bringing new industries into the sites of collapsing firms, it unfortunately must be concluded that such attempts have not been overly successful. The experience in Great Britain, in particular, is not very encouraging. On top of this pessimistic conclusion, one should remember that such moves were often undertaken under conditions of better growth than we have been experiencing during the last six or seven years and which we most likely will still have to cope with for a considerable time to come.

The European approach to problems of redundancy and regional decline is stressed in this volume for the particular reason that there is comparatively, at least compared to European concern about such problems, little attention being paid to, or policies for, redundant personnel and redundant regions in the United States.

Note

1   Swedish unions by tradition have to cover the unemployment relief to be paid during the first year of unemployment to their dismissed members.

19 US environmental laws and their impact on American steel

Stanley V. Margolin

Arthur D. Little, Inc. (ADL) was commissioned by the American Iron and Steel Institute (AISI) to obtain objective industry-wide data and analysis in two areas critical to the steel industry's environmental management. Specifically, the study was designed to:

1 analyse the performance of the US steel industry in meeting regulations stemming from the Clean Air Act of 1970 (Public Law 91-604), the Federal Water Pollution Control Act of 1972 (Public Law 92-500), and the 1977 amendments to both these laws;
2 assess the overall economic consequences to the industry and the public of meeting these regulations in the future.

Other important environmental problems, such as control of hazardous pollutants and solid wastes under the Resource Conservation and Recovery Act, were not studied because regulations had not been developed in those areas when the study was initiated. Accordingly, costs emanating from such regulations were not included in the report.

There were two reasons for the study. First, the American steel industry is concerned about a number of factors it believes have combined to reduce capital formation below the level needed to modernise its facilities, to expand, and to satisfy mandated requirements so it can remain competitive in its domestic markets.[1] Among these are *de facto* price controls, earnings near the bottom of the 40 major manufacturing industries, growing imports of foreign steel often allegedly sold at 'dumping' prices, escalation of steelmaking costs, inadequate depreciation allowances, and increasing government regulations with consequent accompanying demand on available capital. The industry estimates that if it is to remain viable and meet domestic requirements, it will have to more than double its commitment of capital from the present level of approximately $4.3 billion per year to about $8.7 billion per year. (All dollar figures in this chapter are in 1980 dollars unless specifically stated otherwise.)

Second, the steel industry believed the time was then appropriate for an objective review of environmental requirements, both as they affect the industry and as they relate to overall environmental objectives and accomplishments. Almost a decade had elapsed since the US Congress enacted the basic environmental protection acts. During that time, concern had been growing about the economic consequences of the environmental regulations. Moreover, there have been advances in technical knowledge and significant improvements in reducing air and water pollution. Furthermore, the American economy has undergone a basic structural shift since the environmental laws were passed. US productivity increases are at low levels. The US balance of trade has run an average annual deficit of at least $30 billion over the last three years. The 1978 oil embargo and the resultant energy crisis and increase in energy costs have added to inflation and the trade deficit and caused pressure on the American dollar. The dollar has been devalued more than 20 per cent against several other major currencies. Inflation is high and the unemployment rate is causing widespread public and private concern.

Approach

In performing the study, Arthur D. Little, Inc. recognised its reports would, if the findings warranted, be used by the industry to support its efforts to obtain a review of environmental legislation. Therefore, the study team analysed the structure of the industry and evaluated the validity of the various issues raised by the industry in connection with environmental issues in this context. The team took special care to distinguish between industry position and study findings. Our approach is summarised below.

Arthur D. Little first characterised the US steel industry on the basis of size, complexity, plant technology, and competitiveness. With this background and with data supplied by a broad cross-section of AISI member companies, Arthur D. Little evaluated the US steel industry's performance in controlling its air emissions and water discharges. It then estimated the economic impact on the industry of: a) meeting current environmental requirements, and b) having to meet projected future environmental requirements.

The economic impact for air was calculated from capacity factors for major point sources and for fugitive emissions and then costed accordingly to the control devices considered.

The economic impact for water was estimated as follows: an environmental control profile for meeting current requirements for water was established. This profile was based on the US Environmental Protection Agency-proposed 'Effluent Limitation Guidelines for BATEA' (Best Available Technology Economically Achieveable). A contractor report to EPA was used to project future requirements.

Arthur D. Little identified the technology and the equipment to comply with total environmental requirements. Then, using an ADL/AISI computer model, the company estimated the capital and operating costs of meeting requirements on a plant-by-plant basis.

Using these costs and considering the capital available to the steel industry under present US government policies, Arthur D. Little projected those steel operations that, because of high operating costs, have the potential to be shut down regardless of environmental control costs; those that may shut down *in part* because of environmental control costs; and those that may shut down *primarily* because of environmental control costs. The tonnage lost through these potential shutdowns was estimated; the tonnage figures were converted to manpower equivalents, and the manpower equivalents converted to potential job losses in the steel industry. Through use of a standard multiplier factor, potential job losses or dislocations in steel-related industries were also estimated. Finally, the effect of these circumstances on imports was estimated.

The cost numbers stated in this report for air and water pollution control were reviewed with the US Environmental Protection Agency (EPA). The Agency stated that the incremental cost numbers associated with their immediate programmes seemed to be correct, but they have no plans for implementing such things as secondary air emission control on existing point sources, control of thermal pollution and storm run-off, and zero discharge of pollutants in water. It is recognised that the Agency may not include all these areas in its rulemaking, but it is the opinion of Arthur D. Little that, either through state or regional initiation, or by action of other federal agencies or other parties, these controls could be imposed under the present Clean Air and Clean Water Acts.

Industry characteristics

The US steel industry is large, consisting of about 80 companies with 400 plants in 37 states. These companies employ 450,000 people and indirectly provide jobs for another 1,350,000 people in many industries. The industry's manufacturing operations are complex. It employs as many as 29 unit processes and a wide variety of equipment to make thousands of products that are used at all levels of the economy. In the United States, the steel industry does not suffer from a lack of raw material, infrastructure, markets, labour, or technical knowhow.

It does, however, operate many facilities that might be considered as old plants. Although the industry has upgraded the equipment within many of these plants, the plants – because of their age and site limitations – cannot take full advantage of the modern equipment and materials flow.

Despite overall restraints, the US steel industry employs almost the same number of manhours per ton of steel as Japanese producers. Over the past decade, its financial performance (return on capital) was superior to that of the European and Japanese steel industries, and it has remained competitive in its major market areas.

The Clean Air and Clean Water Acts

The Clean Air Act establishes two principal objectives. The first, related to primary air quality standards, provides for attainment by controlling certain 'criteria' pollutants – sulphur oxides, total suspended particulates, carbon monoxide, photochemical oxidants, hydrocarbons, and nitrogen dioxide – and, in addition, hazardous pollutants which may have 'adverse effects on public health'. The second objective of the Clean Air Act is to attain secondary air quality standards to protect the 'public welfare'; that is, standards that are not necessary to protect public health.

The Clean Water Act establishes two objectives. The first is aimed at controlling designated pollutants such as 'conventional' pollutants (total suspended solids, oil and grease) and 'other' pollutants (ammonia, phenol, cyanide, and designated toxic chemicals). The second objective of the Clean Water Act is the total elimination of effluents (zero discharge).

Steel industry environmental compliance

The US steel industry has in place $8.5 billion of environmental control equipment for the control of air and water pollution sources. About $5.0 billion of this is for equipment to control air pollution towards meeting primary or health-related standards established by the US EPA. The remainder – $3.5 billion – is equipment to control water pollutants towards meeting water discharge standards.

In steel plants, the principal ('criteria') air pollutant being controlled is total suspended particulates (TSP). As a result of installing control systems at coke plants, sinter plants, blast furnaces and steelmaking furnaces, the industry has substantially improved its capability to control TSP emissions. TSP control efficiency capability (percentage of pollutants removed) of process emissions increased from 77 per cent in 1971 to 95 per cent in 1979. The industry estimates that despite very sizeable additional expenditures, TSP control efficiency will increase only one percentage point – to 96 per cent – by the end of 1982.

As a result of the industry's efforts to comply with existing regulations, the ambient air quality in several steelmaking areas has already reached primary standard 'attainment', which means that it complies with the National Ambient Air Quality Standards established by the US EPA. Most steelmaking areas, however, are not in attainment even though comparable environmental control techniques are utilised. These areas are usually in large urban locations where other industries and 'non-traditional' sources' (road dust, construction activities, wind erosion, agricultural activities, and so on) contribute significantly to total suspended particulates. A recent report[2] estimates that emissions from 'non-traditional sources' in urban areas exceed those from point sources and 'controlling' such sources would be more cost-effective in reducing TSP concentrations than additional control of point sources.

The industry also has improved and continues to improve its control of water pollutants through the installation of controls on coke plants, blast furnaces and rolling mills through various types of recycle and wastewater treatment systems. Control efficiencies for water conventional pollutants (total suspended solids, oil and grease) were 92 per cent in 1977, today are 97 per cent and, with EPA's new BPT, will be 97 per cent and, with EPA's new BAT, will be 99 per cent. For other pollutants (ammonia, phenol, cyanide, designated toxic chemicals), in 1977 the control efficiencies were 82 per cent; today they are 88 per cent and, with EPA's new BPT, will be 94 per cent and, with EPA's new BAT, will be 99 per cent. On an arithmetical average basis, in 1977 the industry was controlling 87 per cent of its water pollutants; today it is controlling 91 per cent, and with EPA's new BPT, will be controlling 95.5 per cent, and again with EPA's new BAT, will be controlling 99 per cent. (BAT = best available technology, BPT = best practicable technology).

Cost of compliance

As previously noted, the US steel industry has in place $ 8.5 billion in environmental equipment to meet air and water quality requirements. Arthur D. Little, Inc. estimates that the industry in the next four years will spend at least another $2.3 billion to comply with the current air and water environmental requirements, primarily those designated for the protection of health in attaining primary air quality standards and for achieving best available technology for control of water pollutants.

To comply with projected future environmental requirements (that is, to meet the primary and secondary on welfare standards for air and water – a worst case scenario), the US steel industry would have to incur an additional $7.7 billion in costs:

	[1981-1984]	[1985-1989]
In place	$8.5 billion	$10.8 billion
Incremental capital costs	2.3 billion	7.7 billion
Total capital costs	10.8 billion	18.5 billion
Total annualised operating costs*	$3.19 billion	$6.84 billion

* Includes operating and maintenance cost and capital recovery charges based on 10 years with interest at 10 per cent.

Therefore, by 1989, the US steel industry, under the worst-case scenario authorised in these statutes, would have a total capital cost of $18.5 billion and total annualised operating costs of $6.84 billion for environmental controls, the majority of which would be required to comply with meeting non-health related standards.

Impact of current and projected future environmental requirements

The impacts described below are broad estimates, since the accuracy of the analysis on which they are based range from [+ -]15 per cent to [+ -] 35 per cent. Nevertheless, they are indications of the seriousness of the problem.

Based on the assumption that the capital available for steel facilities would be at approximately the same level as generated in 1978 by the companies' steel segments, the US steel industry faces an annual capital shortfall of $3.9 billion, excluding capital expenditures for environmental control. Meeting current environmental control requirements increased the shortfall by at least $0.7 billion per year through 1984; meeting projected future environmental control requirements increases the annual $3.9 billion shortfall by at least $1.3 billion per year from 1985 to 1989. Thus, the US steel industry faces a minimum annual capital shortfall of $4.6–5.2 billion through 1989 which would result in reduction of shipment capability for the industry. (See Table 19.1.)

The potential effects of this capital shortfall are shown in Tables 19.2, 19.3 and 19.4. By 1989, as marginal facilities are forced by lack of capital availability or profitability to shut down, steel industry shipments could decline by 8 to 20 million tons per year, with the largest decline depending principally on the environmental requirements the industry may have to meet. Some marginal facilities will close for reasons other than the cost of additional environmental control – approximately 6 million tons of shipment capability. If the industry meets current environmental requirements, shipments could decline by 12 million tons by 1989. If the industry meets projected future environmental requirements, the decline in shipments would reach 20 million tons by 1989.

These declines would translate into substantial job losses or dislocation in both the steel industry itself and in steel-related industries. Steel industry job losses could range from 52,000 to 84,000, of which 25,000– 57,000 could be attributed primarily to environmental requirements. Job losses or dislocation in steel related industries could range from 156,000 to 252,000.

Table 19.1

Annual capital requirements of the American steel industry
(billions of 1980 dollars)

	Current requirements through 1984	Projected future requirements through 1989
Capital expenditures		
Replacement and modernisation of present steel capacity	5.5	6.1
Additional steel production capacity[a]		
Non-steel[a]	0.9	0.9
Industrial health[a]	0.1	0.1
Environmental control[b]	0.6	1.5
Total capital expenditures	7.1	8.6
Other capital requirements		
Increases in working capital[a]	0.1	0.1
Average annual capital requirements	7.2	8.7
Capital formation[c]	4.3	5.6
	(approx.)	(approx.)
Capital shortfall*	2.9	3.1
	(approx.)	(approx.)

* Including capital expenditures for environmental control.

Sources: a) AISI, *Steel at the Crossroads: The American Steel Industry in the 1980's,* January 1980. (Assumes raw steel capacity at 168 million tons and 20 million tons of imports of finished products in 1990 – Arthur D. Little projection.)
b) Arthur D. Little, Inc. estimates.
c) Estimated capital generation assuming no major changes in government environmental, tax and trade policies.

Table 19.2

Calculated future domestic steel shipments based on varying environmental requirements
(millions of tons of finished products)

Cases	1981	1982	1983	1984	1985	1986	1987	1988	1989	1990	Changes in tonnage Total	Due to environmental requirements*
With no capital shortfall	92	98	105	106	108	109	111	112	114	116		
Estimated capital shortfall excluding environmental (baseline)	92	98	99	100	100	101	101	102	103	105	11	
Tonnage lost	0	1	5	0	2	0	2	0	1	0		0
1 Industry meets current environmental requirements	92	98	98	98	98	99	99	100	101	103		
Tonnage lost because of environmental control costs	0	0	1	1	0	0	0	0	0	0	13	2
2 Industry meets current and projected future environmental requirements	92	98	98	98	97	97	96	95	95	96		
Tonnage lost because of environmental control costs	0	0	1	1	1	1	1	2	1	1	20	9
Projected consumption*	106	113	122	124	126	128	130	132	134	136		

Note: This table does not take into consideration economic fluctuations.

* Recognises anticipated actual market conditions in 1981–82; trendline growth (from 1979 base thereafter).

Source: Arthur D. Little, Inc. estimates.

Table 19.3

Effect of plant closures due to impact of environmental requirements on steel and steel-related jobs by 1990 (excluding community losses)

Cases*	Steel shipment loss (millions of net tons)	Steelworker job losses	Steel-related worker potential job losses or dislocations	Total potential job losses or dislocations
Baseline	0	0	0	0
1	2	9,000	27,000	36,000
2	9	40,500	121,500	162,000

* Case definition
Baseline – no requirements for additional retrofitting of existing facilities
1 Meeting current environmental requirements.
2 Meeting all projected future environmental requirements.

Table 19.4

Effects of plant closures on imports by 1990 (millions of finished tons)

Cases*	Total domestic consumption	Total shipments†	Imported steel requirements	% of domestic consumption
No shortfall	136	114	22	16
Baseline	136	103	33	24
1	136	101	35	26
2	136	94	42	31

* Case definition:
Baseline – no requirements for additional retrofitting of existing facilities
1. Meeting current environmental requirements.
2. Meeting all projected future environmental requirements.
† Excluding exports amounting to 2 million tons

Domestic consumption of steel is projected to increase from 117 million tons in 1979 to 136 million tons in 1989. With the decline in shipments projected for the above environmental cases, imports would necessarily increase from the current level of 17 million tons per year to 50–58 million tons (37–43 per cent of domestic consumption).

Impact of policy change

Since this report is limited to assessing the impact of regulating air and water environmental emissions, it does not include the potential changes relating to prices, depreciation, capital formation, taxes, international trade, employment, and national security. These considerations must be assessed by others.

The issues raised by the iron and steel industry with respect to the Clean Air Act and the Clean Water Act have far-reaching economic consequences. The position of the industry is that it expects to comply with all public health-related regulations. Therefore, the essential question is once having attained the required degree of health-related control efficiency, what future policies should the government adopt with respect to more stringent welfare or secondary standards? Thus, this is an appropriate time for US Government's consideration of the question. In fact, the President's Executive order 12291 recognises the need to achieve legislative goals efficiently rather than to set new standards which impose unnecessary burdens on the economy, individual, public and private organisations, or on state and local governments.

Notes

1   This was spelled out in American Iron and Steel Institute, *Steel at the Crossroads: the American Steel Industry in the 1980's* , Washington DC, January 1980; and American Steel Institute, *Steel at the Crossroads: One Year Later* , June 1981
2   Report prepared for the US EPA, entitled, 'Setting Priorities for Control of Fugitive Particulate Emissions for Open Sources', Harvard University School of Public Health 1980.

20 'Steelworks now!' The conflicting character of modernisation: a case study of Hoesch in Dortmund

Lutz Schröter

Introduction

Is industrial modernisation policy an unquestionable strategy for the restructuring of an old industrial region? The debate about economic policy in general and structural policy in particular in the Federal Republic of Germany has been dominated for years by the objective of modernising the economy in order to gain new competitive strength on the world market. With respect to the strong export orientation of many industries and firms, this seems to be a selfevident goal. In the context of this debate it is important that regional crises like the ones of the Ruhr or the Saar must be defined, in their core, as being crises of dominant economic sectors. After the crisis of the coalmining industry in the 1960s and early 1970s, both regions have been facing crises for more than eight years now in their second key industry, iron and steel, with all the signs of, first, stagnation and, now, heavy decline. Very generally speaking, two sets of causes can be defined: one is the problem of overcapacity in all western, and parts of newly industrialised, countries, which today is regarded as a long-term problem; the other is the existence of different levels of productivity and technological modernity of various firms' production sites and their economic stamina to withstand the crisis.

It seems to be a predominant condition for success in inter-firm as well as inter-regional competition to modernise the mode of production to the most advanced technology and production economy.

As far as the two steel regions mentioned are concerned, this thesis seems to be correct. Whereas, after the first big regional and industrial crisis after the war, a state-supported restructuring of the steel industry was started in the Ruhr area in 1966–67 and accompanied by a general modernisation policy for the infrastructure, there was no such thor-

361

ough development in the steel industry of the Saarland. What was achieved in the Ruhr – concentration of ownership and production to modernised plants, and of locational patterns, could not be realised in the other region.[1] Mainly because of scattered ownership, which was very stable, it was impossible to do in the Saarland what had been feasible in the Ruhr.

The consequences of the failure to concentrate did not come to the surface in times of an extraordinary boom in steel demand, but rather when demand plummeted. In 1978 the steel companies in the Saarland were in danger of a real collapse. To avoid this, it was necessary to raise 1.3 billion DM security from the state and to cut the region's steel industry employment by nearly 50 per cent.

Against the Saar, the Ruhr District seemed to be a positive example and a proof of the concentration policies' adequacy. But looking at the social and economic profile of the region, it becomes evident that there are problems. Unemployment rates differ again from the lower level of the Federal Republic of Germany after 1971, and rather drastically after 1975 (see Figure 20.1). In cities like Dortmund and Duisburg, the

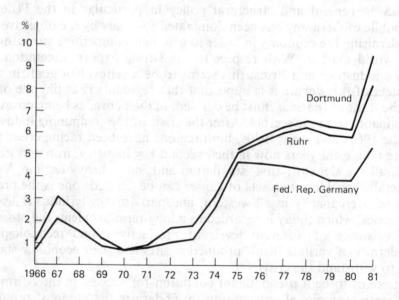

Figure 20.1
Unemployment rates (as reported by federal and regional employment offices)

Note: The data in the figure are average data over the year. The mentioned 11 per cent and 10 per cent for Dortmund and Duisburg, respectively, represent end of year — not average figures.

362

two main locations of steel activities, unemployment now stands at more than 14 per cent. There is the danger of a self-sustained process of decline when the reduction of jobs, especially in the steel industry, continues or accelerates. Whereas, in 1970, Hoesch's Dortmund steelworks employed about 27,500 people (excluding apprentices), the figure fell to 17,093 in February 1982. The last official decision of late 1981 was to reduce to about 13,000 by 1987, but in early 1982 rumours, and even statements by members of the board, were heard about less than 10,000 remaining jobs. Unemployment rates of more than 20 per cent seem about to become a reality, and Dortmund's survival as a steel production centre is becoming very uncertain.

Against this background and perspective one may raise the question what social and economic goals should a crisis policy for such a monostructured region have. It seems also to be pertinent to question the coupling of not only a dominant industry, but also a dominant firm, and the goals and aims of regional development policy.

In nearby Duisburg, with a much better location from the point of view of bulk transportation, Thyssen, the leader in German steel production, pursued a much tougher modernisation policy during the late 1960s and 1970s. Is it true that what is good for Hoesch-Thyssen is also good for the Ruhr – or are there basic and strong contradictions? The question is, of course, not whether or not to modernise, but rather who should carry the cost and who gain the profits of structural reorganisation.

It is not possible to structure this debate simply by means of a discourse between forward and backward looking observers. This becomes obvious when analysing the recent history of the Dortmund struggle. Taking the board's official point of view from 1970 until 1982, there seems to be no evidence to the contrary, or doubt, that Dortmund should survive as a steel-producing location. However, the workers soon developed a clear awareness that this could be realised only if Hoesch did not fall behind the general technological standard, so they claimed that their works should be modernised. (Nevertheless, they did not become very active in the boom period immediately after the first negative decisions were made and also remained relatively quiet on this question until 1978–79.)

Today, when asking how a firm's and a region's interests combine in the modernisation process, it is necessary to emphasise that this cannot be answered by simply pointing to overcapacities. The construction of a new plant at Dortmund was refused at a time of massive investment in new plants elsewhere. No, a new steel plant in Dortmund would have created overcapacity, although, as a matter of fact, it was the other way round. As we will see in the next two sections, overcapacity resulted from those ideas and policies which rejected investment in

Dortmund, but which shared the run to coastal sites all over Europe. The answer of other producers was to centralise and build up bigger capacities in places like Duisburg. This was the process which created overcapacities.

Regions were expected to participate in this competition either by accepting the relative, or absolute, leaving of an area (as Klöckner did, when it moved to the North Sea) or by organising the infrastructure, the labour market etc., as a consequence of this competition. Dortmund was expected to accept both: the (relative) walk-out of the region by Hoesch, and the development of a special infrastructure policy according to Hoesch's needs (see Figure 20.2). But as we will soon see, the regions have to bear an enormous burden which can be seen primarily in the central role played by the key sectors, like the steel industry, not only in their own economic life but also in that of other people who, at first sight, appear to have no connection with steel. This theme will be developed with the support of employment statistics relating to the last ten years. The burden mentioned will be defined in more detail when looking at the latest development of unemployment in typical steel regions. Finally we will return to the principal question of how to cope with modernisation in a regional context.

The history and frame of reference of the Hoesch case

The year 1966 is important in three ways:

1   The year brought the first serious economic crisis in the Federal Republic of Germany after World War II. It brought a combination of a general cyclical and a special structural crisis (in the coal mining industry) which struck the Ruhr District and weakened its economic and social dynamics. We will return to this problem when looking for lessons to be learnt from the relation between the development of firms and that of a region. It was the initial year of more direct, explicit Keynesian state intervention in economic affairs in Western Germany.

2   It was the year of the merger between the old Hoesch Company and the Dortmund-Hörder-Hüttenunion Company, of which the Dutch steel company Koninklijke Nederlandsche Hoogovens en Staal fabrieken NV owned 43 per cent of the shares. This was a major step towards the later merger between Hoesch and Hoogovens to build the Estel NV in 1972 and gives the framework for our case.

3   It was the year that the entire West German steel industry responded to the development of stronger competition on the

364

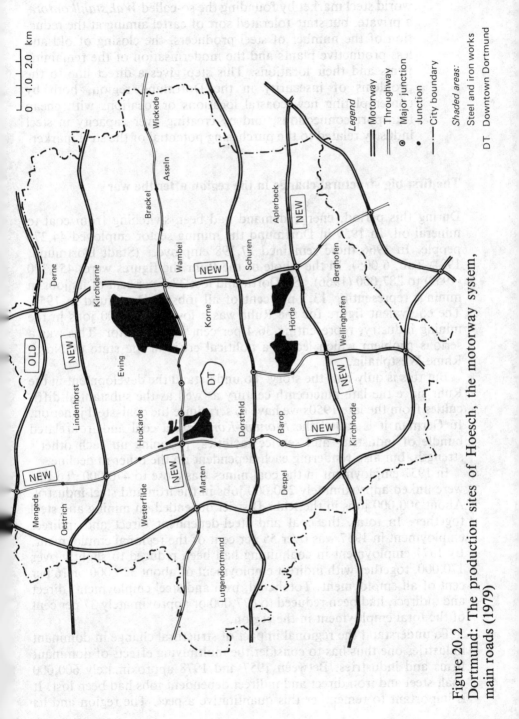

Figure 20.2
Dortmund: The production sites of Hoesch, the motorway system,
main roads (1979)

Legend

—— Motorway
═══ Throughway
⊙ Major junction
• Junction
—·—·— City boundary

Shaded areas:
■ Steel and iron works
DT Downtown Dortmund

365

world steel market by founding the so-called *Walzstahlkontore,* a private, but state-tolerated sort of cartel aiming at the reduction of the number of steel producers, the closing of old and less productive plants and the modernisation of the remaining firms and their locations. This step gives a direct line to the problems of insecurity in the Dortmund region, both by contemplating new coastal locations or locations with cheap transport connections, and by creating over capacity in steel industry related to the purchasing potential of the steel market.

The first big structural change in the region after the war

During this period, energy demand had been switching from coal to mineral oil. In 1959 in Dortmund the mining sector employed 44,270 people. In 1966 there remained 27,198 employees (Stadt Dortmund, 1981, Tab. 6,005). In the whole of the Ruhr the figures were: 451,300 (1959) to 287,000 (1966).[2] So Dortmund lost 38.6 per cent of its jobs in mining, representing 13.7 per cent of all jobs in Dortmund in 1959. The equivalent figure for the Ruhr was a loss of 164,300 jobs in the mining industry, representing 36.4 per cent in this sector. This was a serious problem which led to a political crisis in the state of North Rhine-Westphalia.

But this is only half the story. To understand the development in the Ruhr since the late nineteenth century as well as the substantial difficulties from the late 1950s we have to scrutinise the coal–steel junction. In German it is called the *Montan-Komplex,* a coal and ironrelated bundle of industries and service facilities, propping up each other's strength, but also rendering each dependent on the other in decline.

In 1957 employment in the coal mines was close to 470,000. To this were linked approximately 200,000 jobs in the iron and steel industry. About 500,000 jobs in the Ruhr District depended on mining and steel together. In total, the coal and steel-dependent direct and indirect employment in 1957 was over 55 per cent of the regional employment. By 1971 employment in coalmining had been reduced to slightly over 170,000, together with indirect employment of about 320,000 or 16 per cent of all employment. Total coal, iron and steel employment, direct and indirect, had been reduced to 677,000 or approximately 37 per cent of the total employment in the region.[3]

To understand the regional impact of structural change in dominant industries one thus has to consider the multiplying effects of dominant firms and industries. Between 1957 and 1978 approximately 600,000 coal, steel and iron direct and indirect dependent jobs had been lost. It is important to remember this quantitative aspect. The region and its industry therefore had to cope with a major structural change in the period immediately preceding the advent of the second crisis. Nevertheless, the joint coal, iron and steel complex still dominates the region, and, because of this traditional dominance, it has been very difficult to attract other industries to the region.

The first structural change after the post-war reconstruction took place under conditions of unprecedented economic growth in the Federal Republic of Germany. This favourable condition made it possible to pay for a major share of the social cost caused by the structural change and also made possible the migration of both people and jobs. However, many more people moved to jobs outside the region than jobs moved into the region. Today, the steel industry has become the heart of the regional economy and its multiplier effect on jobs in the region is stronger than that of the latter-day mining industry. And today there is no growth to speak of in Germany. These are two reasons why the social and economic impacts of the second threat to the Ruhr District will be more difficult to cope with. Today's problems are much more severe than those of the 1960s, and external conditions are much worse.

The development of the relationship between Hoesch and Hoogovens in the 1960s

To intensify the possibilities of cooperation and rationalisation the two Dortmund steel firms Hoesch and Dortmund-Hörder-Hüttenunion (DHHU) wanted to go beyond their practice, applied since 1959, of jointly employing their (different) free capacities by special contracts. To go further in this direction it became necessary to found a company to coordinate investment and to realise joint rationalisation effects gained at the Dortmund sites (see Figure 20.2). But, because the major shareholder of the DHHU was Hoogovens, owning 43 per cent of the capital, the merger was not only triggered by the opportunities to achieve rationalisation *inside* Dortmund, but also by the idea of a combination of two firms, one running plants at coastal sites and the other in the heart of a big steel market. So, before concluding the merger in October 1966, in January of the same year, a contract between Hoesch and Hoogovens was agreed upon with the aim to develop joint plants. It was clearly stated that the background of this cooperation was their mutual conviction that the centre for expanding the production of crude iron and steel in Europe should be in large and modern plants at the coast. It was considered that Hoogovens' plans of building new steelworks at the Dutch North Sea coast would fit into their common view of their companies' future.[4] From this policy derives the tenor of the struggle of the following 15 years.

After this merger Hoogovens owned 14.5 per cent of the capital of the enlarged Hoesch Company. In 1969 they concluded a contract to build a joint steel plant near Rotterdam, a plan which never materialised because they were granted no planning permission by the Dutch Government. The ecological burden on the lower Rhine region had grown to dimensions which precluded the establishment of further major industrial pollutants there.

The merger and the plans around it clearly reflected a general tendency shared by major firms in the West European and West German steel industry.

The response to new competition on the world steel market

The growth of the Japanese steel industry in the 1960s, and its aim to overcome the then leading US and West European competitors, set new standards. The answer was to modernise the technological, organisational and locational patterns of the West German steel producers. This meant trimming plants to higher capacity, employing new technologies, and searching for locations which would allow drastic cuts in transportation costs to and from overseas supply and demand markets. To realise these shifts it was necessary to reduce the existing number of producers, close old mills and organise new firms and production units to reach the financial capacity needed for new investment. To do this, the aforementioned *Walzstahlkontore* (rolled steel offices) were founded by groupings of firms which cooperated in all aspects of rationalisation and distribution. They were thought to be 'training centres' for the preparation of real mergers. They were pretty successful in closing old plants. From 1966 to 1972 18 blast furnaces, 26 Bessemer steel converters, 42 open-hearth furnaces and 80 rolling mills were closed in the Federal Republic of Germany's steel industry.

The number of employees decreased by about 37,272.[5] In 1971 socalled *Rationalisierungsgruppen* (rationalisation groups) were established. Everything was set for a big concentration and centralisation move in the German steel industry, as shown in Table 20.1.

While most of the West European countries modernised by building new plants at the coast, the West German companies – except Hoesch and Klöckner – went the opposite way. They chose to modernise the best of the existing plants, in the Ruhr District, mainly in its western part and at the Rhine at Duisburg, to correspond to the newly discovered importance of cheap transportation to the sea. In this way they achieved a very quick renewal of their technological base. But as is shown in Table 20.1 this modernisation was combined with a capacity

Table 20.1

Major mergers in the iron and steel industry in the Federal Republic of Germany in the 1970s

1970	Thyssen-Mannesmann Ringtausch (tube & pipe; rolled steel)
1971	Stahlwerke Salzgitter-Ilseder-Hütte 'Stahlwerke Peine-Salzgitter'
1971	ARBED-Röchlingsche Eisen und Stahlwerke (50–50 partnership)
1971/72	Hoesch-Hoogovens ESTEL
1973/74	Thyssen-Rheinstahl
1974	Mannesmann-DEMAG
1974	Deutsche Edelstahlwerke (Thyssen), Edelstahl Witten (alloy steel)
1974	Krupp-Stahlwerke Südwestfalen
1977	Klöckner-Maxhütte
1977	ARBED-Röchling-Neunkirchner Eisenwerke ('The big Saar Solution')

Combined with the reorganisation of the ownership there was a *strong growth of new capacities* (see Table 20.2).

Table 20.2

Crude steel capacities 1965–75 in the Federal Republic of Germany (million metric tonnes)

1965		1973		1975	
FRG	NRW	FRG	NRW	FRG	NRW
47.5	33.4	58.8	37.8	63.2	39.9

Source: Bömer op.cit. 1977, p.85

FRG = Federal Republic of Germany
NRW = North-Rhine-Westphalia

increase. (A detailed example is the case of Krupp — see Table A.1 in the appendix to this chapter.) The results of this process are evident from Table 20.3.

Table 20.3

Key data of the iron and steel industry of the Federal Republic of Germany, 1970–79 (quantities in million metric tonnes; Employment in 1,000s; DM figures in billions)

	1970	1971	1972	1973	1974	1975	1976	1977	1978	1979	1980
1 Crude steel production employment	45.0	40.3	43.7	49.5	53.2	40.4	42.4	39.0	41.3	46.1	43.8
2 Steel enterprises only	295	280	272	279	284	273	271	259	251	251	243
3 Steel and dependent enterprises	374	355	340	345	344	331	324	308	300	297	288
4 Raw steel capacity	53.1	57.9	57.0	58.8	60.4	62.9	65.8	67.7	68.9	69.0	69.2
5 Capacity utilisation	84.8	69.2	76.7	84.2	88.1	64.3	64.4	57.6	59.9	66.7	63.3
6 Physical investment (bn DM, at 1970 prices)	2.4	3.1	3.1	2.0	1.7	2.2	2.7	1.6	1.1	1.4	
7 Change in %	–	29.8	–0.3	–35.3	–14.4	25.5	23.8	–39.8	–31.8	28.6	
8 Nominal net production capacity	14.8	15.6	16.4	17.0	17.1	17.3	17.6	17.8	17.6	17.3	
9 Index (1970 = 100)	100	105.0	110.9	114.5	115.6	116.6	118.9	120.2	118.7	116.8	
10 Export	7.1	7.2	7.8	10.0	13.6	8.8	8.0	8.3	9.6	10.0	
11 Export ratio	26.4	30.6	33.8	34.5	40.0	33.7	31.8	38.5	40.3	39.0	
Change in producers' prices (% as against previous year)											
12 Iron & steel	8.6	4.4	2.9	8.1	18.0	–5.8	1.0	–4.0	0.1	3.0	
13 All industries	4.9	4.4	2.6	6.7	13.3	4.7	3.7	2.7	1.1	4.9	

Sources: 1–4 Statistisches Bundesamt, Fachservice 4, Reihe 8.1, Eisen und Stahl
5–9 Krengel u.a., Produktionsvolumen und Potential...22. Folge 1970–1979.
10–11 IFO – Datenbank, Branchenservice Eisenschaffende Industrie, Ausgabe Juni 1980.
12–13 Statistisches Bundesamt, Lange Reihen zur Wirtschaftsentwicklung 1980.

Source of table: Memorandum 1981, p.212

What are the main aspects of this development for our investigation?

The main point is development of capacity, until 1980. From 1965 to 1974 (the absolute peak year) there was an increase of 27.2 per cent. But there is growth also from 1974 to 1980, the six crisis years, by 14.6 per cent. But capacity utilisation has been continuously declining except during the peak years 1973–74. Even the record figure of 1974 was export-triggered: the export share rose from 26.4 per cent in 1970 to 40.0 per cent in 1974 (cf. also Figures A.3 and A.4 in the appendix to this chapter). This means, the period after 1966 with its strong orientation towards modernisation and reorganisation of the steel industry led to overcapacity, thus jeopardising the possibilities for a stable development of the region. The single firm found, in one or the other case, ways to solve or minimise the problems by diversifying into other industries or, in some cases, by moving out of regions which did not correspond to the new bulk transportation cost criteria (neither the Ruhr nor the Saar do). But the region was left behind. The chance to 'change the sector', that is to diversify into other industries, was rather small under conditions of general economic decline.

Inter-regional migration is a way out of this dilemma, but it is only open to some, and it makes the situation worse for those who, for various reasons, cannot escape. This all led to the question, asked more and more frequently in the regions, why modernisation of industry, which is thought to be a positive process, could not produce positive consequences for the regions and their inhabitants as well.

It is often said that overcapacity is imported from abroad by producing overcapacity in foreign countries which pay 'wrong' subsidies etc. This causes the bad situation for the regions. There is no doubt that a similar process of overinvestment appeared on the international level, with substantive capacity increases in a great number of countries, which then did not know where to sell the resultant products. However, it is not possible to blame only the international development for the steel market's problems. The rationale of some German firms' policy was not only one of a reaction to changing international circumstances; it also implied strong *actions* by themselves to change the international competitive situation. Like other companies, they tried to achieve positions of greater strength in relation to both their national and international competitors, by gaining higher productivity and capacity – a process by which the potential additional demand in the long term is claimed several times over by different producers worldwide in the medium term. This competition by growth is one of the most important reasons for pinpointing the contradiction between the interests of firms and those of a region from the point of view of regional analysis and regional policy.[6]

371

A similar point may be made by looking at the run to the coast. Dortmund was hit twice by this tendency. The first blow was the transfer of most of the Ruhr steel industry to the west, the Rhine (only Klöckner went to Bremen on the North Sea coast). As a result, most of the competitors gained transport cost advantages, which was the reason why Hoesch merged with Hoogovens. But it did not solve Dortmund's problems. The struggle went on inside the new firm about long-term strategy in relation to the two locations. What could have been a useful move, that is, uniting two locations with different advantages and disadvantages into one company in order to give more stability to both, became the cause of a fundamental struggle which has been going on in the Dortmund region for more than ten years.

Not many other industrial problems relate so explicitly to regional problems and thus accentuate the contradiction between the interests of a firm and those of a region. The question of how to obtain secure jobs dominated local policy from late 1980 to mid-1981 in a very extreme way, but discussions and actions failed to influence or change the economic and industrial policy of both state and industry.

The struggle about modernisation at Hoesch: 1970–1982

The historical account a few pages earlier ended with the conclusion of the 1969 contract to build a joint steel plant at the coast near Rotterdam, a plan which failed because the Dutch authorities refused to give the necessary public planning permission.

The continuation of the struggle, now concentrating on the issue of where to build a new steel plant, is only given by means of rough key points.

1966–1970 . There were discussions about the optimal steel product portfolio policy inside the new company, mainly forced by F. Harders, president of the old DHHU, vice-president of the enlarged Hoesch until 1968, president 1969–73 (that is, until his death). His vision was to compete on the world market by building the 'optimal steel plant' of 10 million metric tonnes capacity a year. Because of transportation cost advantages, he argued, this plant had to be built at the coast. So a *real* merger of Hoesch and Hoogovens seemed to be the best way. Although Hoesch owned some important advanced steel refining process plants, Harders tried to make Hoesch one of the biggest bulk steel producers on the international level.

Opposition came from two quarters: the workforce; and parts of the old Hoesch management. Both asked to build a second basic oxygen steel plant in Dortmund. (The existing one was built in 1963.)

1970 . W. Ochel, the chairman of the supervisory board (*Aufsichtsrat*; its members are elected by shareholders and – because of the German Codetermination Act – by the workforce) resigned because of the abovementioned differences of opinion over Hoesch steel policy. Ochel was the old Hoesch president, then president of the enlarged company until 1969, when he changed to the supervisory board. He asked to build the second oxygen steel plant in Dortmund. There were some controversies about the structure of the coming joint company as well.

His successor, the president of the Deutsche Bank, the biggest merchant bank of West Germany, H.J. Abs, became a strong supporter of Harders' policy. This change decided the debate among the capital owners and management fraction in favour of Harders' option.

During the negotiations about the merger, the workforce was threatening to launch a strike if no clear decision was made to develop the Dortmund site: after that, and because they needed the consent of the employees' representatives, a double compromise was reached:

1   to build a new oxygen steel plant at the Dortmund site and
2   to guarantee a future production capacity of 7.2 million metric tonnes per year in Dortmund, despite the aim to implement the investment at the coast.

1971 . However, the first part of this deal was made dependent on the financial possibilities, and this was used as a permanent excuse to postpone its realisation. The foundation stone for the new steel plant was laid at the end of 1971. However the realisation was later nullified.

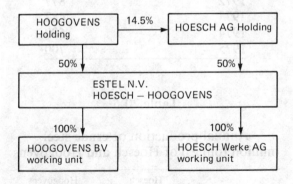

Figure 20.3
The Hoesch – Hoogovens merger of 1972

1972 . Hoesch and Hoogovens merged to form Estel NV (see Figure 20.3). Tables 20.4, 20.5 and 20.6 give a rough idea of the development between the two location complexes of the new company. In the late 1960s Hoesch was twice as big as Hoogovens.

Table 20.4

Comparison of Hoesch–Hoogovens characteristics

		Hoesch	Hoogovens
1967–68	crude steel production (mill.tonnes)	6.1	3.4
1968–69	crude steel production (mill.tonnes)	6.66	4.34
	turnover (mill.DM/hfl)	3.89 DM	1.75 hfl
	employees	52,400	21,900
	original stock (mill.DM/hfl)	569 DM	206 hfl

Table 20.5

Estel total investment 1969–75

Year	mill. hfl
1969	663
1970	784
1971	1,007
1972	877
1973	482
1974	589
1975	700

Table 20.6

Annual production of crude steel
(million tonnes) of Hoesch and Hoogovens

	Hoesch	Hoogovens
1974	6.7	5.4
1977	4.8	4.6
1978	5.1	5.3
1979	6.0	5.5

374

Relations were changed by the investment policy of the following years, most of it before the official merger was implemented in 1972. In that year, Hoogovens had reached a capacity of 6.6 million tonnes. The plan was to produce by 1980 7 million tonnes per year at Dortmund and 8 million tonnes per year at Ijmuiden (the Netherlands). There also existed a plan to let the capacity of the Ijmuiden site grow to 12 million tonnes per year.

The original plan of the late 1960s, to specialise in steel production at Ijmuiden and to develop the processing part in Dortmund, was undermined by the policy to build a new integrated plant in Holland.

1973 . A new middle range strategy was formulated, consisting of two points. The first was a compromise to modernise and extend the existing oxygen steel plant in Dortmund from 280,000 tonnes a month to a capacity of between 350,000 and 400,000 tonnes a month. The second point was that the existing open-hearth furnaces should remain in use up to 1990. Earlier, they were supposed to be shut down at the beginning of the 1980s. Later approximately 100 million DM were invested in dust extraction and desulphurisation devices, partly financed by the Federal Department of Science and Technology within the framework of a pollution abatement programme. This was by no means a step to modernise the Dortmund operations in the long run. The share of open-hearth furnace technology in 1973 in the Federal Republic of Germany was 18 per cent, in Dortmund about 39 per cent. Only during the short boom period was this no matter of concern because they reached a high load factor, far above the break-even point.

The new strategy implied dropping the 1970 proposal to build a new steel plant. The aims were:

a) no extension investments in steel production;
b) maintenance of open-hearth furnaces as a reserve capacity for boom periods;
c) reduction of steel-producing capacities. After 1990 the Dortmund Phoenix plant would be closed.

1974–75 . During the big steel boom the idea was launched to build an entirely new steel complex on a Hoesch-owned free site in the northwest of Dortmund to completely replace one of the old plants after 1990. The building of a second oxygen steel plant was rejected as being totally unrealistic and died during the ongoing steel crisis.

1976 . In the second half of the 1970s three contradictory developments took place in the processing part of Hoesch:

1 Hoesch left the special steel sector, although it possessed a reasonable share of a firm in the alloy steel sector and was permitted to acquire more stock in the firm in question. The special steel capacity was sold to Krupp.
2 It closed an engineering plant acquired as late as 1974–75.
3 It bought the majority of a steel and metal processing company (Siegener AG) in the region south of Dortmund.

It would be an interesting piece of research to study more deeply the motives of Hoesch's diversification policy because this could have played an important role in a regional modernisation policy. If we look at manpower savings on account of technological progress, especially in the steel-producing processes, a regional strategy should create alternative jobs to compensate for the lost ones (as one aspect of a more comprehensive programme needed to solve manpower problems).

In 1979 Hoesch was the only big steel company in the Federal Republic of Germany producing more than 50 per cent of its turnover in steel production proper.

1978–79 . In the West German steel industry, a large-scale strike was launched, aiming at shortening working week to 35 hours. The workers and their unions had thought that, by cutting the working time for all workers, they could help to secure jobs for steelworkers, who otherwise, given productivity increases and constant working hours, would have to leave their employment. The outcome was an agreement (to be implemented in 1983) to cut the working hours by one hour for a quite small group of steelworkers.

An explicit debate about the firms' location policy did not start until 1979.

1979 . In May 1979 another new strategy was submitted to public discussion:

1 Steel production in Dortmund should stagnate. Because it is 38 DM per tonne more expensive than in Holland, only rationalisation investments and investments to improve some technological processes would be permitted. Dortmund's function would be reduced to serving the local and regional market. Export markets should be covered only by Hoogovens. Steel production in Dortmund should be concentrated in one location. Reduction of employees should be accelerated, but with-

376

out dismissals (social plans were to be developed; 'natural' reduction of workforce by retirement and natural leavers to take place).

2   Subsequent processing should be strengthened. Some processing plants ought to be closed. This strategic plan never gained the confidence of those involved.

The workers' answer was that no further agreement would be given to rationalisation and reduction of employment without a comprehensive plan. The union body inside the works demanded that no further reduction of jobs should be made without creating alternative jobs to compensate for lost workplaces.

However, because there were differences between the shop floor and the employees' representatives on the corporate board and the works council, there was no real pressure in that direction in the first place.

Autumn 1979 . The board of directors published a new planning concept revising the issue of replacing the old open-hearth furnaces on the northern site by a new oxygen steel plant. This would imply a capacity reduction from 7.3 to 6.3 million tonnes per year.

A definite decision was promised within the 'coming weeks'. The possibility of a state subsidy was mentioned by a member of the Federal Parliament.

After all the years of promises without positive decisions and real activity and, in view of advanced rationalisation programmes being in process in nearly all parts of the company (including research and development), the workforce was heavily sceptical. Leaflets were distributed summarising the haphazard policy changes and unfulfilled promises. A more radical discussion took place about the causes of the problems and the way to overcome them, which led to a stronger and wholeheartedly supported demand to reach binding decisions about the new steelworks. There was a serious discussion whether the problems of the steel industry and of Hoesch would not be better solved by nationalisation. The workers' demands were now strongly submitted to politicians. Hoesch's problems were regarded not just as an internal question of concern to the employed, but as a major issue of critical importance to the whole region. Thus they reflected the linkages inside the regional economy.

January–February 1980 . A formal deicision was reached by the different bodies of the firm to realise a concept implying in principle that a new steel plant should be built, beginning in 1983, and that a further 4,200 jobs should be abandoned.

This plan was backed by a state promise to give financial assistance by means of a cheap (low interest) loan of 240 million DM, which was calculated to render a 100 million DM real subsidy.

The reduction was to be organised without dismissals, by using the possibilities of a combination of unemployment grants, firm subsidies, early retirement and natural leavers.

Autumn 1980 . It was questioned whether the January plan could be realised under the conditions of the ongoing steel crisis. Again, the specialisation between the coastal plant and Dortmund was being stressed: there to concentrate on the production of steel, here to develop the processing part, and to concentrate on one site in Dortmund.

November 1980 . A demonstration of about 70,000 people, shouting 'steelworks now!', moved through the streets of Dortmund. The core of the argument was that, without a steel plant, even the processing part would not survive, except perhaps the cold rolling mill.

December 1980 . A standing political conference was founded by the Lord Mayor to develop solutions, but it turned out to be a great public relations show rather than a real policy body.

May 1981 . The *Aufsichtsrat* of Hoesch (board including workers' representatives) agreed on the following plan:

a) concentration on one site in Dortmund;
b) a new steel plant with 4.5 million tonnes per year replacing the existing oxygen plant rather than the old open-hearth furnaces;
c) reduction of the range and variety of products, to concentrate on plain steel;
d) reduction of employment to 13,000 by 1987;
e) the construction of one continuous annealing furnace.

1981–82 . Negotiations took place between Krupp and Hoesch to found a new joint steel company, Ruhrstahl AG. The merger between Hoesch and Hoogovens was to be dissolved. The existing plans incorporated a new steelworks, but there was strong scepticism inside and outside the factory after 12 years of ineffective 'decision making'. The reduction of jobs was to continue, in any case, by a stronger rationalisation. With the new plan, Dortmund would have to compete with Krupp's remaining steelworks in Duisburg on the Rhine. So whether or not the new plant will be built will most likely remain a question as open as it has been for a long time now. During 1982 even the merger between Krupp and Hoesch to new Ruhrstahl AG became more and more uncertain. Instead of Ruhrstahl (Krupp and Hoesch), it is most likely that Rheinstahl (Krupp and Thyssen) will become a much more attractive solution, although this will, however, leave Hoesch and Dortmund in a worse situation than could ever have been imagined. By the end of 1984 no more mergers are envisaged.

There is an ongoing debate in Dortmund as to whether the gradual decline of steel production here was a deliberate strategy of the Hoesch management. There are arguments, heavily stressed mainly by the Hoogovens management, that this would have made no sense in their concept, since, as they were interested in the large market in the Ruhr and in the processing activities of Hoesch, Dortmund would have had to have a steel supply of its own. To deliver steel at such quantities from the coast would have been uneconomical. Many reasons were given why the investment decision always was postponed, though by the end of 1984 Hoesch aimed to solve its problems by upgrading.

To take these arguments seriously, it would be argued that the investment problems were the result of a too-small merger, compared with their competitors, taking place under conditions of crisis. Estel was not capable of managing the three core points of the plan:

a) to realise the expansion and modernisation of the coastal steel production,
b) to modernise the Dortmund steel production in principle; as well as
c) to expand the processing activities there.

But whether deliberate or not, Estel's corporate policy always happened to end up in decisions which put Dortmund at a disadvantage.

It seems as if there had been no chance of marrying the best interests of the firm and the region. World market competition, overcapacity and new coastal sites led to a 'mechanism of decline' in an old industrial region.

Regional impacts of declining industry: the economic complex around the steel industry

We have mentioned the important role of the production complex, when analysing the first big crisis in the Ruhr. The decline of the mining sector led to a change in the upstream and downstream linkages: the steel industry now took the leading position because of the investment boom of the late 1960s and beginning of the 1970s. The mechanical and electrical engineering industries therefore developed with good fortune, especially in the Ruhr District (see Figure 20.4).

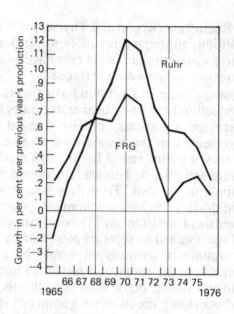

Figure 20.4
Annual growth rates in the investment goods industry, Ruhr District and Federal Republic of Germany

This development was not bad, because it gave the opportunity to gain new strength out of the existing structure. That meant to have a 'sliding shift' with the existing productive forces in the region, both with regard to physical investment resources but, more important, to the knowledge and experience of technicians and workers. From the point of view of further regional development this is important, because this could give a chance to the people living in this region with all their social and financial ties, rather than an abstract imagination of a 'modern population', which in the dreams of some politicians and planners was thought to move into the region with highly sophisticated new technology firms. But these never would have provided a sufficient number of jobs to all those people who would have become unemployed, if the new beginning had to start with an ongoing breakdown of the existing industrial pattern. We have to remember: the first university in the Ruhr District was founded in Bochum in 1963, the next one in Dortmund in 1968. Today, there are five universities in the region with some 120,000 students, but it takes a long time to transform such a change of a regional structure of production. Some results are shown in Table 20.7, which depicts the outline of the new production complex in the mid-1970s. As can be seen in the last two columns, the steel industry – despite employing fewer people than the mining sector – has a more broadly influence on jobs. In fact, 121,300 jobs in the steel industry are economically linked to an additional 21,000 jobs in the very closely related sectors of steel processing and mining and – more important – to more than 191,000 jobs in a wider defined production complex round the steel industry.

Table 20.7

The impact of the mining and iron–steel industries upon employment in the Ruhr District, 1976

	Mining and steel industry in total		Mining	Iron & steel
	in 1,000	in %*	in 1,000	
Mining and steel industry	317.30	–	129.80	142.40
Coal mining	155.00	–	126.30	18.50
Iron & steel	132.50	–	1.50	121.30
Iron, steel temper casting	9.90	–	0.55	0.90
Cold milling	10.20	–	0.30	0.50
Steel forming	9.70	–	1.10	1.20
Dependent employment†	356.20	23.5	124.90	191.80
Energy and water supplies	4.00	15.3	0.90	2.60
Stone and earth working industries	2.20	18.8	0.70	1.20
Chemical and petrochemical	2.50	4.5	0.90	1.35
Steel and aluminium construction	4.10	10.1	1.20	2.50
Machinery	13.60	19.2	6.30	5.90
Vehicle and cars	1.70	7.8	0.50	1.00
Electrotechnical	12.60	24.4	4.55	6.75
Metal working industry	4.80	5.2	2.20	2.10
Foodstuffs	8.10	28.1	2.90	4.30
Handicraft & trade	45.30	31.2	16.00	24.00
Construction industry	30.50	21.0	11.00	16.10
Commerce & distribution	51.80	18.9	16.90	28.55
Surface traffic, communication	27.90	26.8	8.20	16.80
Banking & insurance	7.70	21.4	2.65	4.20
Services, housing	51.70	29.3	18.10	27.50
Public & private organisations	70.00	34.7	25.10	37.10
'Montan'-complex total	673.50	36.7	254.70	334.20

* In per cent of the employed in the sector or industry in question
† Directly and indirectly dependent work places or jobs
Source: Brune, R. *et al., Wirtschaftstrukturelle Bestandsaufnahme für das Ruhrgebiet,* 1. Fortschreibung Essen, 1978, p.72

Table 20.8

Total employment* in the steel production complex A† (narrowly defined steel complex) in different regions 1970 and 1978–79

| | 1970 | | 1978/1979 | | Variation 1970–78/79 | | Steel job losses |
	absol.fig 1	% 2	absol.fig 3	% 4	+ -absol. fig 5	+ -% 6	% 7
FRG[a]	944,210	3.6	756,000[c]	3.0	− 188,210	− 19.9	
NRW[b]	505,030	7.2	410,250[d]	6.0	− 94,780	− 18.8	
Ruhr (KVR)[b]	255,920	11.6	204,450[d]	9.7	− 51,470	− 20.1	
Dortmund‡ I[a]	50,120	11.0	44,600[c]	10.4	− 5,520	− 11.0	
Dortmund II[b]							
(City)	31,770	11.4	28,600[d]	10.6	− 3,170	− 10.0	10.7
Duisburg (City)[b]	69,000	26.4	58,500	24.0	− 10,500	− 15.2	28.8

Notes:

*   Total employment: all employees and self-employed, helping members of the family etc.

†   Steel complex A: iron, steel, non-iron metal foundries, drawing plants, rolling mills steel forming, fitters' shops.

‡   Dortmund I: Dortmund (city), Kreis Unna, Hamm (city); Dortmund II: Dortmund city only.

a   Prognos (1980);

b   KVR (1980);

c   data for 1978;

d   data for 1979. The difference between 1978 and 1979 on the FRG-level was very small in this economic sector. The number of all employed people in units > 20 employees differed only by about − 0.8 per cent.

Column 2/4: percentage of all jobs in the regional unit

Column 7: lost jobs in the steel complex A as a percentage of all lost jobs in the regional unit.

FRG = Federal Republic of Germany

NRW = State of North-Rhine-Westphalia.

However, the positive role of such a complex will change if its leading sectors are in severe trouble – as happened in the mid-1970s. The consequences of a declining steel sector for towns like Dortmund and Duisburg play an important role in all regional and local discussions. The above figures were compiled by the economic research institute RWI on the basis of an input-output matrix. It is not possible to compile equivalent data for a city like Dortmund in order to get a better idea about the impact of massive reductions in the steel industry for

such a region. We will see that the present development of the unemployment figures signals danger to the region. But before looking at
the figures, some other data should be analysed. They will provide
some information about the steel policy in the two major steel cities,
Dortmund and Duisburg, and will be used a little later when we try to
draw some conclusions.

Table 20.8 shows the significantly higher proportion of this sector in
all parts of the Ruhr, with an extreme level in Duisburg. The difference
in Dortmund's and Duisburg's steel policies at that time is quite obvious. In Dortmund, modernisation and rationalisation were not pursued as intensely as in Duisburg (further use of open-hearth
technology, only one continuous casting plant etc).

The reduction of jobs in Dortmund was the lowest of all regions in
our table. The concept of emphasising the coastal site for Estel seemed
to be a good policy for the workforce for a short time and is one reason
for the relatively quiet time in the mid-1970s. It was nevertheless a
further threat to the maintenance of jobs in Dortmund after the mining
crisis.

Table 20.9
Total employment in Dortmund (II) (City)

1970	1979	Lost jobs	New jobs	Net change
278,300	270,600	−29,640	21,940	−7,700

Table 20.10
Total employment in Duisburg

1970	1979	Lost jobs	New jobs	Net change
261,090	243,450	−36,410	18,770	−17,640

The reduced number of jobs in the steel industry accounted for about
10.7 per cent of all lost jobs. This appears to be a rather modest
change, but it reduced the needed supply in the late 1970s. Several
other sectors lost a greater number of jobs, both by absolute figures
and by percentage. This is the point we have to scrutinise more thoroughly. Before doing this, however, some words about Duisburg.
Duisburg was hit much harder at that time, suffering a net loss of 6.8
per cent of all jobs, compared with only 2.8 per cent in Dortmund (see
Tables 20.9 and 20.10). The share of the declining steel jobs was nearly
29 per cent. Here, the strong impact of a single sector on a region can
be shown very clearly. This change can be underlined better when using
the steel complex B, which is more broadly defined (although it does

not include some related industries like electrotechnical engineering and the tinplate industry). (See Table 20.11).

Table 20.11
Total employment in the steel production complex B*
in different regions 1970 and 1978-79

	1	2	3	4	5	6	7
	1970		1978/1979		Variation 1970-78/79		
	absol.fig	%	absol.fig	%	+ -absol. fig	+ -%	%
FRGª	3,380,710	12.7	3,218,000ᶜ	12.8	− 162,710	− 4.8	
NRWᵇ	1,178,240	16.7	1,029,550ᵈ	15.0	− 148,790	− 12.6	
Ruhr (KVR)ᵇ	456,120	20.7	384,200ᵈ	18.3	− 71,920	− 15.8	
Dortmund‡ Iª	84,570	18.6	73,420ᶜ	17.1	− 11,150	− 13.2	
Dortmund II (City)ᵇ	52,920	19.0	45,200ᵈ	16.7	− 7,720	− 14.6	26.0
Duisburg (City)ᵇ	91,370	35.0	75,050	30.8	− 16,320	− 17.9	44.8

* Steel production complex B = complex A plus steel and light metal construction, mechanical engineering, motorcar production and repair, shipbuilding, aeroplane building.

All other notes and signs see Table 20.8

The regional impact of industrial change is much better visible in this complex. We have to remember that it reflects only rather direct linkages. Compared to the definition of the RWI, complex B is rather narrow. With nearly one-fifth of all jobs in Dortmund city this sector exercises quite a strong influence, in this case on stagnation and decline. The reduction of jobs is larger, and this complex's share of all lost jobs is rather alarming. In Duisburg the job reduction in this complex accounts for nearly one-half of all reductions, in Dortmund for more than one quarter. This is not merely an effect of statistical manipulation resulting from taking a wider range of industries. The definition of the sector is narrow compared to that used by RWI.

It is thus obvious that autonomous decisions taken inside the big steel companies – which often control a major proportion of those related industries – affect the cities or regions to a considerable extent.

In attempting to estimate the employment impact of the iron and steel industry crisis one may find two ratios of particular interest. One is the Ruhr District's share of all jobs in this complex in the Federal Republic of Germany in 1970, the second the equivalent share of the job fluctuation from 1970 to 1978. (See Table 20.12.)

384

While the Ruhr held 13.5 per cent of all jobs in the Federal Republic of Germany in 1970, it had to accommodate 44.20 per cent of the reduction between 1970 and 1978. This ratio for Dortmund and Duisburg is not so bad. Related to the figures for the Federal Republic of Germany, Dortmund held 1.6 per cent of all jobs in the steel complex B in 1970. By 1978 it had lost 4.7 per cent (average 4.8 for Germany): the Duisburg figures are 2.7 per cent; it lost 10.0 per cent (as against the German average of 4.8 per cent for the steel complex B).

Table 20.12

Ratio of steel-related employment and employment reduction (1970–79): Ruhr District, Dortmund, Duisburg, to FRG, in per cent (based on Tables 20.8 and 20.11)

	Steel complex A		Steel complex B	
	col.1	col.5	col.1	col.5
Ruhr District	27.1	27.3	13.5	44.2
Dortmund II	5.3	2.9	2.5	6.9
Dortmund I	3.4	1.7	1.6	4.7
Duisburg	7.3	5.6	2.7	10.0

In these differences we see the declining strength and weight of the Ruhr District in its interregional relationships, as caused by the problems of its industrial core.

To all managers, politicians and regional planners who intended to convert the Ruhr District from a steel-producing into a steel processing region it ought to be a shocking insight not only to see how little has been achieved, but how deeply the region was affected by the decline. The above ratios speak very clearly. For the narrowly defined steel complex, the data reflect the fact that the existing steel plants were concentrated in Duisburg and Dortmund whereas many other plants outside and inside the Ruhr District were closed.

The weakness of complex B shows that the strong development of the investment goods industry could not be maintained during the 1970s. Thus the stabilisation plans for the Ruhr failed. Compared with the period after the crisis in the mining sector we now lack an equivalent factor of support. This is one reason why conflicts about the policies to choose for overcoming the new social problems will become more serious. Any further massive reduction of jobs will have to be paid out of the very substance of the region, since the region now lacks the comparatively favourable conditions it enjoyed when having to solve the coal mining crisis.

The development of unemployment and the lack of jobs in the Ruhr District and some of its major steel-dependent cities

We have seen the big influence of the steel complex on the regional labour market during the last 10 years. When now analysing regional unemployment we will come to an understanding of the social impact of the steel crisis on the lives of the region's inhabitants.

The present situation is characterised by a level of unemployment which, a few years ago, no-one would have ever thought realistic for the industrial core region of the Federal Republic. We will take five towns in NRW, which are the main locations of the two steel companies Hoesch and Krupp and where there is a discernible trend towards a strong rise in unemployment. There are nevertheless some differences which we shall try to relate to the special situation in the steel industry.

The unemployment level in February 1982 reached 11.6 per cent maximum and 8.5 per cent minimum (see last column of Table 20.13). All range above the national average, all – except Siegen – above the NRW level as well.

To develop a temporal perspective we shall look back to the peak of the first big cyclical and structural crisis which hit the Ruhr region in the 1960s.

In Table 20.13 and Figure 20.5 we can see that from the point of view of 1982 the 1966–67 crisis was rather modest. In none of the mentioned regions was there more than 4.0 per cent unemployment. Bochum was worst hit by the decline of coalmining. The three Ruhr towns were in a significantly worse situation than both the German and the NRW state averages. Only Hagen and Siegen were equal to the regional and federal averages.

The boom period after that crisis gave new employment opportunities to all the areas. Unemployment was never higher than 0.6 per cent in 1970, and all areas had more or less equal unemployment figures. The boom period ended with a strong growth of unemployment which brought all parts back to the level of the mid-1950s. In the Federal Republic of Germany the number of unemployed reached more than one million. However, there were only small differences between the single regions, although the three Ruhr towns were hardest hit. After 1975, the strongest boom year the steel industry ever experienced, the gap widens, as it had already done in 1967 (cf. variation rates in Table A.2 in the appendix). In the beginning, Duisburg suffered most. In 1979 and 1980 this city[7] took the peak, mainly because of the radical manpower reduction policy then adopted by the Thyssen and Krupp companies in the late 1970s. With more than 6.0 per cent unemployment, Duisburg suffered from the rationalisation and closure moves in the dominant sector, under the condition of a generally weak

Table 20.13

Unemployment and unemployment rates in different regions
(All figures relate to the districts of the manpower services commission,
Arbeitsamtsbezirk). (Month/of year)

	9/67	9/70	9/75	9/79	9/80	9/81	2/82
Duisburg	4,120	1,188	7,905	9,493	12,988	16,984	21,709
Bochum	9,943	1,491	11,433	11,810	11,402	16,028	20,606
Dortmund	9,488	1,697	14,307	14,524	15,681	24,172	32,091
Hagen	3,680	822	9,775	8,542	9,384	14,240	19,470
Siegen	1,993	352	5,528	3,691	4,494	7,729	11,960
NRW	118,659	30,096	292,538	252,418	275,592	411,662	558,561
FRG	341,000	97,000	1,005,000	737,000	822,565	1,256,374	1,935,316
Duisburg	2.1%	0.6%	4.8%	6.3%	6.1%	8.0%	10.2%
Bochum	4.0%	0.6%	5.2%	5.6%	5.4%	7.6%	9.9%
Dortmund	3.0%	0.5%	5.1%	5.4%	5.7%	8.7%	11.6%
Hagen	1.6%	0.4%	4.5%	3.9%	4.6%	6.6%	9.0%
Siegen	1.5%	0.3%	4.0%	2.7%	3.2%	5.5%	8.5%
NRW	1.9%	0.5%	4.6%	4.0%	4.4%	6.5%	8.7%
FRG	1.6%	0.5%	4.4%	3.2%	3.5%	5.4%	8.1%

Sources: BELR 1980. Tab.2; monthly reports of the manpower service commission NRW
('Landsesarbeitsamt NRW')

economic climate, whereas in the Federal Republic of Germany and NRW (as well as in the two smaller places, Hagen and Siegen) there was a short recovery between 1975 to 1979. Since then, the speed of unemployment growth (since the last quarter of 1980) is most remarkable.

Unemployment rates rose rapidly in the entire Republic, but changed the relative position of our regions. Cities like Hagen and Siegen, which had always been in a better position, now saw higher growth rates of unemployment, partly due to the general situation, but also due to the closure of some plants and the reduction of jobs in the steel sector (see the last two columns of Table 20.13).

Dortmund, since late 1980, has been the region in the worst situation. This was caused when Hoesch started to reduce the number of employees, as was decided by its board in February of 1980. Table 20.14 shows the rapid slump in workforce: more than 2,000 people were dismissed in 1980.

This was when Hoesch tried to meet the more radical rationalisation policy of Thyssen (which we already recognised in the difference of lost jobs in Duisburg and Dortmund between 1975 and 1980), which had

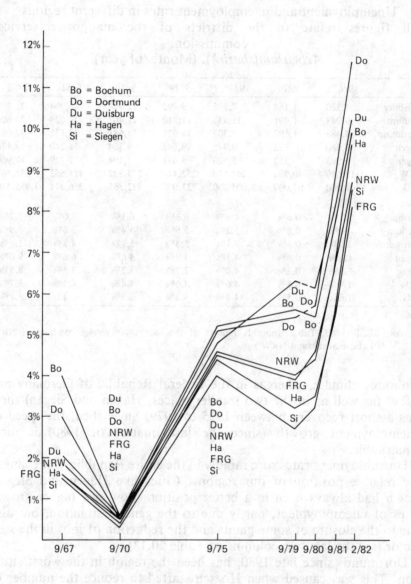

Figure 20.5
Unemployment rates 9/1967–2/1982 in specified steel centres, state of North Rhine Westphalia (NRW) and Federal Republic of Germany (FRG)

managed to achieve a higher output per capita in recent years. Again, we see the diverging interests between firm and region, which we could observe in the growth period ten years ago, but which are now in the process of reduction. In both periods the firms tried to improve or to regain their viability by a policy incompatible with the region's economic development. The strategies of both Thyssen and Hoesch, however different they were, nevertheless had the same effect on their regions. The viability of a firm is by no means a good guide when selecting a development policy for a region. Because of Hoesch's policy to close and to reduce in 1981 (− 3,455 jobs), combined with the regional interlinkages and the general economic weakness, Dortmund kept its no. 1 rank in unemployment.

Table 20.14
Employment at Dortmund Hoesch Estel-Hüttenwerke
(steelworks) 1966–82 (without apprenticeships)

Date	Number	%	+ /-
30.09.66	29,564	100	
30.09.67	27,120	91.1	− 2,444
30.09.68	26,885	90.9	− 235
30.09.69	27,098	91.7	+ 213
30.09.70	27,507	93.0	+ 409
30.09.71	27,088	91.6	+ 419
30.09.72	25,531	86.4	− 1,557
31.12.72	25,444	86.1	
31.12.73	25,782	87.2	+ 251
31.12.74	26,399	89.3	+ 617
31.12.75	25,620	86.7	− 779
31.12.76	25,327	85.7	− 293
31.12.77	24,440	82.7	− 887
31.12.78	23,615	79.9	− 825
31.12.79	23,529	79.6	− 86
31.12.80	21,419	72.5	− 2,110
31.12.81	17,964	60.8	− 3,455
28.02.82	17,093	57.8	− 871

Looking at Table 20.15 we recognise an increasing speed in unemployment growth, especially in Hagen and Siegen, since 1979 and 1980 compared to Duisburg and even Dortmund. The plans were made to concentrate many of the Hoesch and Krupp plants formerly operating in the area of Hagen and Siegen towards the Ruhr region. These plans were designed with the view to the contemplated merger of the two companies to Ruhrstahl AG and strengthened this development. These two regions have lost their good position and will continue to

Table 20.15

Unemployment September 1979 – February 1982 in specific steel centres (and variation to demonstrate the speed-up of the development) (month/of year)

	2/79	2/80	2/81	2/82	2/79 – 2/82	2/80 – 2/82
Duisburg	12,521 (8.3%)	13,678 (6.4%)	15,908 (7.5%)	21,907 (10.2%)	+75.0%	+60.2%
variation		+9.2%	+16.3%	+37.7%		
Bochum	15,644 (7.4%)	13,265 (6.3%)	14,545 (6.9%)	20,907 (9.9%)	+33.6%	+57.6%
variation		−15.2%	+9.6%	+41.7%		
Dortmund	20,047 (7.4%)	17,559 (6.3%)	21,776 (7.9%)	32,091 (11.6%)	+60.1%	+82.8%
variation		−12.4%	+24.0%	+47.4%		
Hagen	11,137 (5.1%)	10,481 (4.8%)	12,476 (5.8%)	19,470 (9.0%)	+74.8%	+85.8%
variation		−5.9%	+19.0%	+56.1%		
Siegen	6,565 (4.8%)	5,660 (4.0%)	7,979 (5.7%)	11,960 (8.5%)	+82.2%	+111.3%
variation		−13.8%	+41.0%	+49.9%		
NRW	340,360 (5.4%)	297,654 (4.8%)	379,319 (6.0%)	558,561 (8.7%)	+64.1%	+87.7%
variation		−12.5%	+27.4%	+47.3%		

Source: monthly reports of the manpower service commission NRW (*Landesarbeitsamt NRW*)

deteriorate as time goes on. The situation in all steel locations will come to an approximately equal level in the next few years.

What may the future bring to Dortmund? In this context, we have to be aware that the crisis, especially in the Dortmund region, will last and even grow. We can get an idea of its potential dimensions when analysing the results of a simulation model which was designed to forecast the development of the region from 1970 up to 1990 on the basis of past trends (M. Wegener, 1981; M. Wegener, M. Vannahme, 1981). Although the authors do not think these results are a precise forecast, but rather a test of the model, it would not be justifiable to ignore their results.

The authors took the decision of the Hoesch board from May 1981 to reduce employment from 20,500 in 1980 to 13,000 by 1987. They further assumed that another 22,500 jobs in the iron and steel industry would be lost during that period of time (Figure A.5 in the appendix shows the mentioned regions). In their model a further 90,000 jobs in the area will be affected by the initial reduction of 30,000. The wider Dortmund region (which is larger than that mentioned in Tables 20.8 and 20.11 and includes some parts west of Dortmund, like Bochum, and south of it, like Hagen) may lose up to 120,000 jobs by 1990. Figure 20.6 shows the significant difference between the model assuming a) the 'normal' development, on the basis of the generally prevailing tendencies in NRW and b) the special Hoesch-induced change. The worst situation will occur in the southern part of Dortmund, at Hörde, where a blast furnace and an integrated steel plant will be closed (subregion no. 8). Figure 20.7 shows the growth of the unemployment rates (up to 20 per cent) in the region, as derived from the two options of the simulation model. Even regarding the difficulties (validity and reliability) incurred in modelling such a process and the effects they may have on the results, we should scrutinise this scenario carefully.

Firstly, the assumed job multiplier of 1:4 seems to be high compared with that used by the RWI institute, 1:2.76. But this was employed for the entire Ruhr District. The eastern part of the District, depicted in the simulation model, however holds much larger shares of the iron and steel industry than the District as a whole. In conclusion, the assumed 1:4 multiplier seems to reflect the particular situation in the chosen regional subset quite well. Secondly, the model gives less than 10 per cent unemployment for Dortmund itself in 1982, which is too low with regard to the 1980–83 situation. Thirdly, the number of lost jobs will exceed 13,000, as has already been mentioned when referring to the history of the Hoesch case. Numbers of less than 10,000 Hoesch employment are being discussed, so we should not discard the possibility of an unemployment rate between 20 and 30 per cent in the late 1980s.

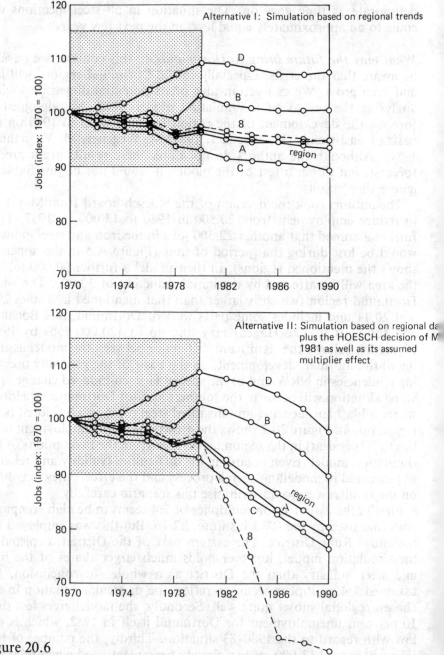

Figure 20.6
Development of employment in the eastern Ruhr until 1990
(1970 = 100)

Source: Wegener and Vannahme, 1981; a tentative result from a simulation model. No real
forecast

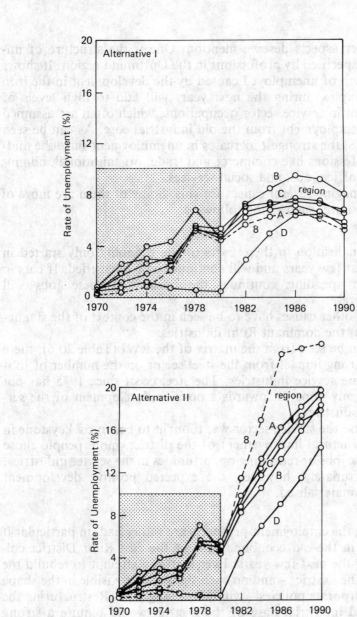

Figure 20.7
Development of the unemployment rates (%) in the eastern Ruhr until
1990. Alternatives I and II (cf. Fig 20.6)
Source: Wegener and Vannahme, 1981

Two further aspects deserve mention. One is the structure of unemployment specified by professions in the Dortmund region. It shows that the growth of unemployed caused by the development in the iron and steel complex during the next years will add to high levels of unemployment in service sector occupations, which often are assumed to absorb unemployment from the old industrial core. As can be seen by Figure 20.8, the strongest increases in unemployment since the mid-1970s hit professions like commerce and trade, organisational, administrative and office jobs, and social services.

One can imagine several causes for this. Some of them are more of an external nature, such as the following:

1   Rationalisation in these parts of the economy only started in the last few years and will continue to be intensified. If cuts in public spending continue, the decline in these jobs will accelerate.
    Some other causes have to be seen in the context of the stagnation of the dominant Ruhr industries.

2   As can be seen from the matrix of the RWI (Table 20.6), there is a strong impact from the steel sector on the number of jobs in some service industries. The steel crisis since 1975 has not given any impetus towards a positive development of the service industries.

3   Because the service sector was thought to become a keystone in the structural improvement of the district, more people chose to seek jobs or retraining opportunities in the service industries. Unfortunately, however, the expected positive development never materialised.

Obviously, the employment prospects are rather bad, in particular if employment in the old economic base of the east Ruhr District collapses during the next few years. Even a strong attempt to rebuild the economy of the District – and no such signs are yet visible in the shape of state or corporate policies – will take a long time. Restructuring the economic and industrial base of the District would require a strong general growth climate in the entire Federal Republic, which, as we saw, was one of the strategic elements present in the recovery of the Ruhr in the late 1960s and early 1970s.

The second point, supporting a rather pessimistic view regarding the employment trends, is derived from a forecast of the balance of job demand and supply in the Ruhr and in some of its towns.

The Batelle-Institute[8] claims that, by 1985 (based on the estimated peak of the growth curve of people able and willing to work), there will

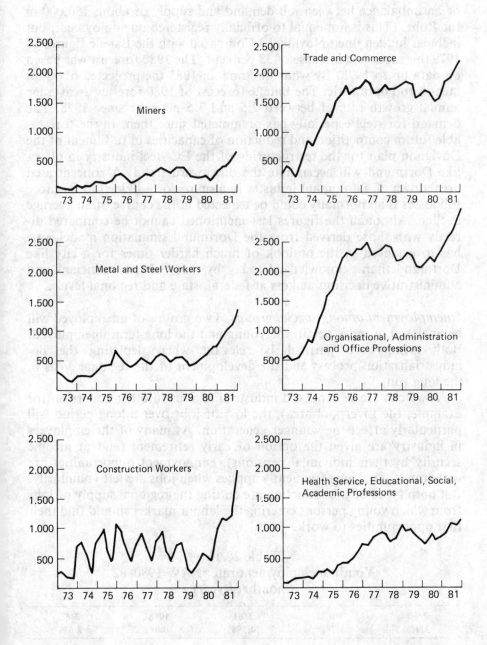

Figure 20.8
Unemployment in Dortmund 1973–81 specified by selected professions

be an imbalance between job demand and supply of about 150,000 in the Ruhr. (This is not equal to officially registered unemployment, but includes hidden unemployment.) Compared with the Batelle figures of 1979 this means an increase of 33 per cent. The 1980 forecast was based on data up to 1978. By what we know in 1983 the projected development is too optimistic. The Batelle forecast of 1980 foresaw an average annual growth rate of between 2.5 and 3.5 per cent since 1978. The demand for steel especially has plummeted since then. In the foreseeable future competition and reduction of capacities in fulfilment of the Davignon plan for the restructuring of the EC steel industry in places like Dortmund will accentuate the downward trend. A concentrated breakdown of a dominant industry in a subregion will lead to far more job losses than normally could be expected in the course of an average decline. Although the figures last mentioned cannot be compared directly with those derived from the Dortmund simulation model, one has to get used to the outlook of much harder times for a city like Dortmund than acknowledged today by most of the politicians and administrative decision makers at federal, state and regional levels.

Unemployment among special groups. Two groups of unemployed will be analysed in more detail: the young and the long-term unemployed. Both groups are particularly relevant when depicting the deindustrialisation process and the development of the regional crisis in the long run.

As we can see in some old industrial regions in other countries (for example, the Liverpool area), the loss of jobs over a long period will particularly affect the younger generation. As many of the employees in industry are given the option of early retirement (and of income security by their individual pensions), employment opportunities are withdrawn. The same frequently applies when jobs are left voluntarily. But both types of job reduction are cutting the regional supply of jobs from which young persons entering the labour market should find their first opportunities to work.

Table 20.16
Youth unemployment rate: NRW 1980–82
(month/of year)

2/80	10/80	2/81	10/81	2/82
3.9%	3.7%*	4.5%	6.0%	8.6%

Sources: monthly reports of the manpower service commission NRW (*Landesarbeitsamt* NRW)

* This figure is lower than in 1980, because of technical changes in the statistics, not because of real reduction

The reduction of entrance opportunities is depicted by the youth (under the age of 20) unemployment rate for North-Rhine-Westphalia (similar data are not available for comparable districts). (See Table 20.16.)

If we look again at our special regions we see a stronger development than in NRW and in the Federal Republic of Germany except for Bochum and (partly) Dortmund (Table 20.18). Table 20.17 shows the absolute figures.

Table 20.17

Youth unemployment 9/1978 – 2/1982

	9/1978	9/1979	9/1980	9/1981	2/1982
Duisburg	1,115	834	1,422	1,984	2,057
Bochum	1,297	987	978	1,467	1,742
Dortmund	1,516	1,057	1,435	2,121	2,240
Hagen	779	597	871	1,274	1,640
Siegen	519	323	485	997	1,196
NRW	25,663	20,352	24,519	40,302	47,257
FRG	92,030	68,593	78,792	132,811	165,071

Source: monthly reports of the manpower commission NRW *(Landesarbeitsamt NRW)*

Table 20.18

Changes in youth unemployment 9/1979–9/1981

	9/1978-9/1979 (1 year)	9/1979-9/1981 (2 years)	9/1981-2/1982 (6 months)
Duisburg	– 25.2%	+ 137.9%	+ 3.7%
Bochum	– 23.9%	+ 60.5%	+ 18.7%
Dortmund	– 30.3%	+ 100.7%	+ 5.6%
Hagen	– 23.4%	+ 113.4%	+ 28.7%
Siegen	– 37.8%	+ 208.7%	+ 20.0%
NRW	– 20.7%	+ 98.0%	+ 17.3%
FRG	– 25.5%	+ 93.6%	+ 24.3%

After a reduction between 1978 and 1979, which was partly due to the cyclical development and partly to the success of a great campaign to offer more apprenticeship opportunities mainly in the craftsman sector, we can see the development in the different regions. In some of these, it does not seem to be so dramatic compared with the total development in the Republic and in NRW. But Duisburg is remarkably higher in 1980 and 1981, which is exactly the time which we recognised as the main rationalisation period. According to the strong impact of changes in both the narrowly and the broadly defined steel complexes (see above) this is not surprising. Although the dominance of the two complexes is not as strong as in Duisburg, the development in Dortmund in the next few years will be similar, because the reduction of the workforce in the steel sector will be faster and more thorough as the general crisis continues. Cities like Hagen and Siegen will suffer from high youth unemployment, as they are more disadvantaged in employment opportunities compared with Dortmund because their service sector is much more dependent on the local purchasing power, whereas Dortmund has some more 'countervailing power' by virtue of its interregional importance.

Compared with data from regions like Liverpool, where unemployment under the age of 20 accounts for nearly a quarter of all unemployment (1979 data), the youth share in the steel region in, and south of, the Ruhr is relatively small (cf. Table 20.19). But the Liverpool figures, despite all differences in the educational and vocational training system, should be looked at as the result of a process which has been going on for years. In relation to this, regions like the eastern and southern part of the Ruhr are at the beginning of a contractive process. If it continues, the 'growth rates' of youth unemployment during the last few years point upwards further.

Table 20.19

Share of unemployed youth of all unemployed (September figures)

	9/1978	9/1979	9/1980	9/1981
Duisburg	10.1%	8.8%	10.9%	11.7%
Bochum	9.5%	7.6%	8.6%	9.2%
Dortmund	9.2%	7.3%	9.2%	8.8%
Hagen	7.8%	7.0%	8.8%	8.9%
Siegen	10.8%	8.8%	10.8%	12.9%
NRW	9.0%	8.1%	8.9%	9.8%
FRG	10.6%	9.3%	9.6%	10.6%

Source: monthly reports of the manpower commission NRW *(Landesarbeitsamt NRW)*

The second group we want to look at may as well be regarded as an indicator, and as a result, of a long-lasting process of structural unemployment which leads to a filtering process. A large number of unemployed cannot be thought of as a stable group of individuals, but as a floating mass of people made redundant and – partly – engaged again by other firms. The best members of this mass also will replace 'weaker' employees, who will be dismissed as soon as qualitatively better labour becomes available. High unemployment during a long period always tends to carry with it the danger of a growing number of long-term unemployed being kept longer and longer outside the active workforce. This group will carry the hardest social burden, and its chance of being reintegrated into the active population is very small. This development thus must be observed carefully. Here again, one needs a longer perspective to look for trends, because there is a strong relationship with cyclical developments. (The long-term unemployment share is higher during a small cyclical boom in a time of a longer crisis, because its constituents tend to remain unemployed as a kind of deposit at the very bottom of the groups of hard-to-employ persons.)

Table 20.20

Long-term unemployment (1 year [+]) and variation rates
(September figures 1978–81)

	1978	1979	1980	1981	78–79	79–80	80–81
Duisburg	3,214	3,151	3,507	3,995	− 2.0%	+ 11.3%	+ 12.9%
Bochum	3,804	3,573	3,047	4,137	− 6.1%	− 14.7%	+ 35.8%
Dortmund	3,979	3,995	3,396	4,950	+ 0.4%	− 15.0%	+ 45.8%
Hagen	2,038	1,812	1,735	2,491	− 11.1%	− 4.2%	+ 43.6%
Siegen	829	615	587	1,009	− 25.8%	− 4.6%	+ 71.9%
NRW	63,728	58,070			− 9.0%		
FRG	175,502	148,338	140,102	203,386	− 15.5%	− 5.6%	+ 45.2%

Source: monthly reports of the manpower service commission NRW (Landesarbeitsamt NRW)

At first sight it might be surprising that, except for Siegen, the increase of this group of unemployed is equal to, or even below, the national trend. But it can be argued that this is a consequence of the effects of the social plan policy[9] in the steel industry by means of which handicapped persons, for example, can leave the active workforce. Otherwise, they would join the group of long-term unemployed. The steel regions can still reduce this group at the expense of state subsidies, social security funds, and some additional money from the companies. But this policy is highly dependent on the general level of wealth and the struggle for its redistribution. Times of affluence are withering away in Germany and less money will be available even for such purposes in the next few years. The last social plans of the Hoesch Company did not meet the levels of those of the earlier years. Therefore, the further development of long-term unemployment is dependent on the level and duration of general unemployment, and the availability of reserves to be distributed amongst the victims of structural change and crisis.

The Dortmund figures contain an alarming point. Nearly 40 per cent of those unemployed in September 1981 for between one and two years (of a total of 2,984) were in the age group 19 to 24 years. Thus not only old and physically handicapped workers belong to this group. This is a dangerous perspective as it may imply the establishment of a passive workforce for many years, by means of a self-sustained process of decline for those of the unemployed who are out of work for a long time during the normally most active part of their life.

Conclusions

Positive adjustment to structural change: the case of planning the process. As we said in the beginning, it is not merely a question of saying 'yes' or 'no' to structural change, when it comes to developing new foundations for continued economic and social welfare. What has to be questioned in connection with the necessary change is:

a) whether or not the development of regional instability is caused by general gaps between supply and demand, or by the process of modernising firms and industries – particularly in highly capital demanding ones – under conditions of competition which will lead not only to overaccumulation but, in many cases, to regional stagnation or decline rather than to development;

b) the idea that structural change can be managed as a piecemeal process whilst dominant industries are being destroyed at a rapid pace;

c) that the regional population is only a residual variable, which has to adapt to regional and other changes.

One central answer to those questions is that more deliberate attention should be paid to causes and consequences of change, to make possible smoother change at considerably less suffering and losses to the concerned groups and individuals.

It is not simply a question of too little technical progress. None of the industrial strategies of Thyssen and Hoesch can be seen as a base, or even a promise, for a continued regional development.

It is not mainly a question of too old products. Even in modern industries one has to harmonise accumulation, productivity growth, market demand, and regional employment. (The production of synthetic fibres is another example of problem industries.)

A point has to be made about the time-dimension of adaptation processes, not only in firms or industries, but in regions. To modernise the latter takes much more time than the former. The reconstruction of a firm can be managed quite quickly and also in a fairly radical manner, for example, by means of inter-industry capital flow. To change occupational structures and a general mind is much more time and resource-consuming. This difference in dimensions is one important element of the contradiction between a firm and a region as to their different options and their different abilities to respond in flexible ways.

Both Liverpool and the Ruhr thought they had overcome their 'structural problems'. Up to now both have failed, though on different levels and to different degrees.

One of the dangers is that a totally unplanned process will occur, which may multiply the dimension of lost jobs compared with the 1960s and 1970s.

To use an industrial landscape as nomads do, moving firms or industries from region to region, makes no sense in the late twentieth century. Structural change in its core is addressed by the question: how to develop existing productive forces in an old industrial region.

Notes

1 Bömer, H., 'Regionale Strukturkrisen im staatsmonopolistischen Kapitalismus und marxistische Raumökonomie am Beispiel der Ruhrgebietskrise', *Marxistische Studien* , Jahrbuch des IMSF, February 1979; and Väth, W., 'Industrielle Restrukturierung und Strukturpolitik – das Beispiel Saarland', in *Probleme der Stadtpolitik in den 80er Jahren* , Essen, October 1980.

2 Lauffs, H.W. and Zühlke W., *Politische Planung im Ruhrgebiet* , Göttingen, 1976.

3 Source: Brune, R. *et al.* , 'Überlegungen zu regionalpolitischen Massnahmen für das Ruhrgebiet', *Mitteilungen des RheinischWestfälischen Instituts für Wirtschaftsforschung* , Essen, Heft 1, 1979; and Lamberts, W., 'Der Strukturwandel im Ruhrgebiet – eine Zwischenbilanz, 1. Folge: Die Bedeutung des Montan-Komplexes für die Ruhrwirtschaft', in *Mitteilungen des Rheinisch – Westfälischen Instituts für Wirtschaftsforschung* , Essen, Heft 3, 1972.

4 Mönnich, M., *Aufbruch ins Revier – Aufbruch nach Europa. Hoesch 1871–1971* , Munich, 1971, p.417.

5 Bömer, H., *Internationale Kapitalkonzentration und regionale Krisenentwicklung an Beispiel der Montanindustrie und der Montanregion der Europäischen Gemeinschaft* , Dortmund Beiträge zur Raumplanung, Bd. 5, Dortmund, 1977, p.82.

6   Bömer H. and Schröter, L., *Ursachenanalyse regionaler Krisenanfälligkeit. Zur Anwendung der Theorie der Überakkumulation/ Entwertung auf regionale Probleme* , Seminarberichte no. 10, Gesellschaft für Regionalforschung, 1974; Bömer, H. op.cit., 1979; Carney, J. *et al.* , (eds), *Regions in Crisis* , London 1980.

7   Some of the unemployment figures from 1980 onwards are not fully comparable to the earlier statistics because of a spatial reorganisation of some manpower service districts. The growing tendency is nevertheless obvious.

8   Batelle Institute, *Der Arbeitsmarkt in Nordrhein – Westfalen und im Ruhrgebiet. Analyse der Entwicklung bis zum Jahre 1995* , FrankfurtMain, 1980.

9   Planned lay-offs of groups of employed, e.g. when a factory or part of it is being closed down, require the consensus of the concerned works council. The employer is obliged, by the rules of the enterprise legislation, to submit a detailed 'social plan', showing how social security will be extended to the dismissed employees.

Appendix to Chapter 20

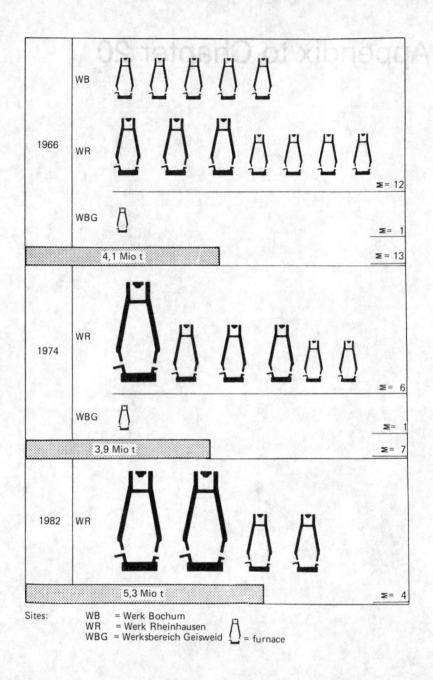

Figure A.1
Rationalisation and capacity growth of crude iron production at Krupp
1966–82
(w = plant + location, Mio t = million metric tonnes)

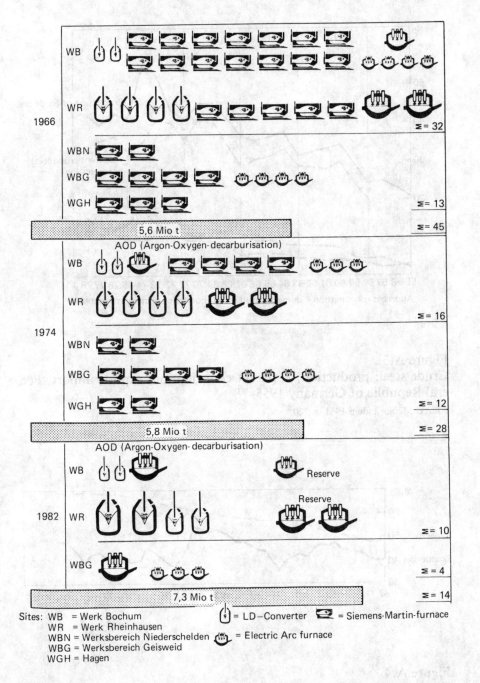

Sites: WB = Werk Bochum
       WR = Werk Rheinhausen
       WBN = Werksbereich Niederschelden
       WBG = Werksbereich Geisweid
       WGH = Hagen

$\dot{Q}$ = LD–Converter $\blacktriangleright$ = Siemens-Martin-furnace

$\text{\small (m)}$ = Electric Arc furnace

Figure A.2
Rationalisation and capacity growth of crude steel production at Krupp 1966–82

405

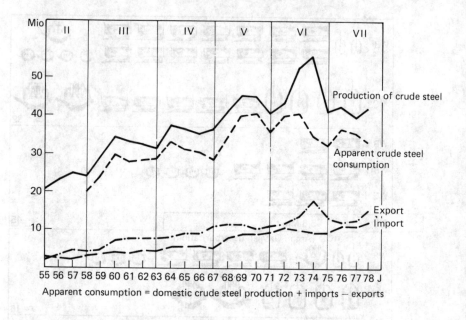

Apparent consumption = domestic crude steel production + imports − exports

Figure A.3
Crude steel: production, apparent consumption, export, import, Federal Republic of Germany 1955–78
Source: Memorandum 1981, p.220

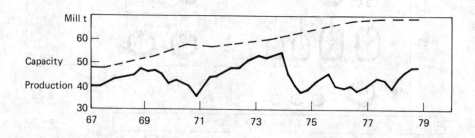

Figure A.4
Crude steel: capacity and production, Federal Republic of Germany
Source: Memorandum 1981, p.214
Production figures are seasonally adjusted

Table A.1

Employment shares, in per cent, by sectors, in North Rhine-Westphalia, the Ruhr District, Dortmund and Duisburg, 1970 and 1979*

	NRW		NRW excl. Ruhr District		Ruhr District		City of Dortmund		City of Duisburg	
	1970	1979	1970	1979	1970	1979	1970	1979	1970	1979
0 Agriculture/forestry	4.9	2.7	6.3	3.4	1.8	1.1	0.8	0.6	0.5	0.4
1 Energy and water supply, mining	4.1	3.8	1.5	1.7	9.8	8.4	8.9	7.6	5.3	3.8
2 All industry (without 1,3)	40.6	36.2	42.3	37.7	37.0	32.8	33.1	27.4	45.8	41.3
3 Construction	7.7	6.6	7.3	6.2	8.6	7.3	8.3	7.9	7.3	5.9
4 Trade, commerce	15.0	14.6	14.6	14.4	15.8	15.1	17.8	16.7	13.7	13.9
5 Transport + communication	5.1	5.7	5.0	5.5	5.5	6.4	7.1	7.5	8.8	10.3
6 Banking and Insurance	2.4	3.1	2.6	3.4	1.9	2.4	3.4	4.0	1.5	2.1
7 Other private service sector	14.2	18.2	14.1	18.0	14.5	18.5	15.4	20.6	13.4	16.2
8 Non-profit organisations	0.8	1.7	0.9	1.8	0.6	1.5	0.8	1.5	0.4	1.1
9 Public bodies, social insurance agency	5.1	7.5	5.4	8.0	4.5	6.5	4.4	6.3	3.4	5.1

* all employed people, self-employed etc. (Source: KVR, 1980)

Table A.2

Variation rates of unemployment figures

	9/67–9/70	9/70–9/75	9/75–9/79	9/79–9/80	9/80–9/81	–2/82
Duisburg	−71.2%	565.4%	20.1%	36.8%	30.8%	27.8%
Bochum	−85.0%	666.8%	3.3%	−3.5%	40.6%	28.6%
Dortmund	−82.1%	743.1%	1.5%	8.0%	54.2%	32.8%
Hagen	−77.7%	1088.9%	−12.5%	15.7%	44.1%	36.7%
Siegen	−82.3%	1470.5%	−33.2%	21.8%	72.0%	54.7%

Source: NRW— Monthly Regional Employment Statistics (monatliche Veröffentlichungen des 'Landesarbeitsamtes NRW')

407

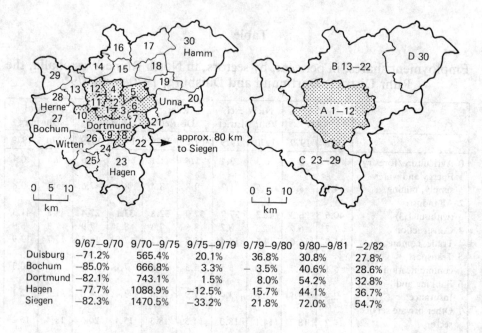

	9/67–9/70	9/70–9/75	9/75–9/79	9/79–9/80	9/80–9/81	–2/82
Duisburg	–71.2%	565.4%	20.1%	36.8%	30.8%	27.8%
Bochum	–85.0%	666.8%	3.3%	– 3.5%	40.6%	28.6%
Dortmund	–82.1%	743.1%	1.5%	8.0%	54.2%	32.8%
Hagen	–77.7%	1088.9%	–12.5%	15.7%	44.1%	36.7%
Siegen	–82.3%	1470.5%	–33.2%	21.8%	72.0%	54.7%

Figure A.5
The wider region of Dortmund

Source: NRW — Monthly Regional Employment Statistics (monatliche Veröffentlichungen des "Landesarbeitsamtes NRW")

21 Structural problems of an old industrial area: the case of the Ruhr District

Klaus R. Kunzmann

Introduction

Industrial concentration, increasing worldwide competition, technological innovation, changing demands and saturated markets have seriously threatened the iron and steel industry of the Ruhr District, a region which for decades has contributed to the wealth and affluence in the Federal Republic of Germany. The development of the district on the one hand is a pertinent example of the long-lasting crisis of western capitalism; on the other, it is just one unavoidable consequence of technological and structural change in the late twentieth century.

This polycentric urban agglomeration of more than five million inhabitants was and still is the major regional centre of iron and steel production in the Federal Republic of Germany. Since the early nineteenth century, iron and steel production has been the motor of industrial development of the Ruhr District. Coal exploitation and steel production were the major sources of wealth and strength of the region. Coal and steel industries have dominated the region's social, economic, and political life for about 160 years.[1]

Consequently, spatial and physical development in the region has always been planned to primarily, if not exclusively, meet the infrastructural demands of the coal and steel industries. Landscape and townscape, urban structures, public utilities, and transport systems in the Ruhr District reflect, to a certain extent, the monostructure of the regional economy. Today the structural crisis of the steel industry is paralleled by a structural crisis of the agglomeration as a whole. Most of the present spatial, social and economic problems of the District are consequences of regional policies which traditionally have been dominated by the interests of the steel and coal industries.

This chapter is devoted to a few structural problems of an urban region whose formerly strong economy is declining rapidly. The central thesis is that a region's spatial, social, political and environmental problems affect all the efforts of the iron and steel industry to adjust to structural change. Unfavourable social, economic, political and environmental conditions will have considerable influence on the future development of the steel industry itself. Thus these aspects should gain much more attention, and also should be continuously monitored as well as considered much more seriously by the steel industry than hitherto. It is obvious that the industry of the Ruhr District, which functions as the core of a big urban agglomeration, where such resource-related locational advantages as coal and human capital still seem to exist, cannot act as if it were in a free enterprise zone on an isolated, uninhabited island.

I will try to elaborate on this interrelationship, taking the Ruhr District as a pertinent example. Some quantitative information on the relative dominance of the steel sector in the Ruhr District will underline the extent of the problem. The case will be further illustrated by the description of the negative impacts that the steel industry has had, and still has, on the spatial structure of the region. To really understand the regional situation, a few 'complaints' of the steel industry about planning, planners and politicians are then presented. Some structural problems of the Ruhr District determining its future, spatial, social and economic development will be described in order to demonstrate the necessary extent of political action required to cope with the present crisis.

I will conclude this chapter by making recommendations to the steel industry, in whose interest it would be to carry them out, as they could well contribute to the restructuring of the region as a whole.

The dominance of the steel industry in the Ruhr District

It is undisputed that any dependency of a region, its population and its labour force on one single industrial sector and on one single source of income has rather strong effects on the overall economic performance of the region. If, in addition, this sector is highly exposed to an international competition which cannot be controlled, for example by a national authority, the relative vulnerability of such a region is obvious.

As early technological innovations at the beginning of the nineteenth century induced an extremely successful industrial development in the Ruhr District, coal and steel dominated here, unchallenged for more than 150 years.

Even by 1957, 55.2 per cent of all jobs in the Ruhr District were directly or indirectly dependent on the coal and steel sector. This figure declined to 36.7 per cent only in 1972.[2] A comparable figure for 1984 is not yet available, but will certainly be below 30 per cent. In the year 1972, 20.6 per cent of the jobs in the region (as compared to 21.1 in 1957) were directly and indirectly connected with iron and steel production, using a 2.76 multiplier of total steel employment.[3] Thus, in 1972 every fifth job in the Ruhr was dependent on iron and steel production. Even if this value has probably further declined, such figures document the factual and structural strength and power of the steel industry in the region. The steel industry surpassed the coal industry as the leading industrial sector in the Ruhr District after the coal crisis in the 1960s.

Apart from its considerable impact on the labour market, the dominance of the steel industry signifies power and influence in the economic and political life of the region as well. This relates to a variety of fields ranging from direct financial support from the regional (*Land*) government, through transport and energy subsidies to land use control and construction permits. Due to its power and dominance, the steel industry has always been able to influence regional and local transport systems in line with its needs. Roads, railway lines and canals have been aligned and continuously expanded in such a way that they primarily meet the transport requirements of the big industrial enterprises rather than the unarticulated demands and environmental aspirations of citizens living in the region. This may have been justified to a certain extent by economic reasons, but it is doubtful whether such prerogatives may be accepted today, even if job creation measures in times of economic stagnation deserve the highest possible priority.

Another feature is the fact that the District's steel industry probably owns more urban land than any other landowner in the region. This is the result of the farsighted land acquisition policies of the steel 'barons' in the nineteenth century, who bought all the agricultural property around their plants they could lay their hands on. Apart from land used for production itself, considerable areas of (derelict) land 'left over' from former coal exploitation and production is also in the hands of the steel industry today. For good financial reasons, land was not transferred to Ruhrkohle AG when the ailing coalmines were merged in 1969 in attempts to control the crisis in the coalmining industry. Considerable housing stock and fully developed land reserves are also controlled by the steel industry's own housing corporations. This is due to the fact that the supply of housing for employees has long been a paternalistic tradition in the Ruhr District. It is certainly no secret that the steel directors in charge of issues of real estate, infrastructure policy and public planning meet regularly to coordinate their respective land policies when having to deal with municipal or regional authorities. Cases are also known in which they use their bargaining

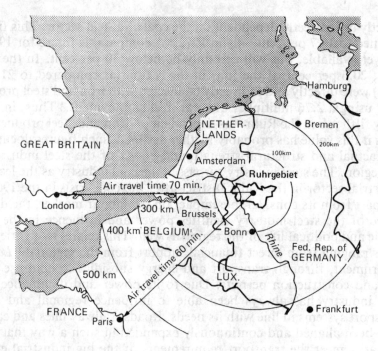

Figure 21.1
The Ruhr District in the European context
Source: Institut für Raumplanung (IRPUD) Universität Dortmund

power – sometimes supported by local politicians in order to put emphasis behind their interests when it comes to issues of developing the traffic system, the power supply or related topics. This has happened more than once and still happens because local politicians are seldom sufficiently independent to oppose a town's biggest industrial employer. Quite a few infrastructural arrangements, emphatically enforced by the industry for reasons of internal production needs – resulting from corporate transport cost calculations or more sophisticated cost–benefit analysis at a certain time – turned out to be rather shortsighted. Even if they contributed to more competitive steel prices for a few months or years, they more than once delayed necessary, more farsighted, decisions concerning technological innovations. A region in which spatial, social and economic development is monopolised by one or two industrial sectors is not independent in its decisions for future urban development. If the Lord Mayor and almost one-third of all members of a city council are in fact employees of the city's biggest employer, as once was the case in Dortmund, independent decision is hardly to be expected from this council. No decision of any importance will be taken against the interests of the all-powerful employer. Such a

412

strong corporate partnership between politics and major industries as in the Ruhr District is certainly no unique case, but rather common in monostructured industrial regions.

Given the dominance of the large industries in the Ruhr District and knowing of the close formal and informal relationships between leading industrialists, unionists and politicians in the region – we use the German colloquial expression *'Filz'* (felt) for that kind of 'relationship' – planners and administrators at low or medium level are extremely hesitant to touch any issues which could lead to political and/or personal inconveniences or would be potentially deterrent to any career prospects. It is difficult to describe this kind of syndrome. It is an amalgamation of frustration, resulting from earlier useless efforts to enforce any measures against big industry, because the industry can employ better lawyers, with the experience that compromises can hardly ever be reached on issues concerning the environment or disadvantaged groups when widely divergent attitudes with respect to social values and aims are involved.

Another indicator of the power of the steel industry in the Ruhr District is the close interrelationship of big industries with the regional chambers of commerce which serve as spokesmen for industrial and economic policy in the District. This informal coalition has eased the formulation of industrial policies which favour the big industries and neglect the problems and requirements of small and medium size firms.

Land, housing, infrastructure and environment are important areas of influence and impact. The steel industry uses its influence in all communal bargaining decision making processes related to spatial, urban, and regional development. Even in times of uncertainty, their prime interest is to further adapt existing urban settlement structures to their own needs.

The steel industry cannot be blamed for using its power and influence in these daily communal or regional bargaining processes. But internal long-term considerations and responsibility for a region should lead to a better understanding of the differing goals of any other local interest group, and of those decisions a local council has to make to balance conflicting arguments.

The steel industry's impact on the spatial structure of the Ruhr District

Apart from the economic repercussions the decline of the steel industry in the *Ruhrgebiet* has on the economic development of this old industrial area, its considerable impact on the living environment deserves mention.

In the last decade or so, non-economic aspects of human life have

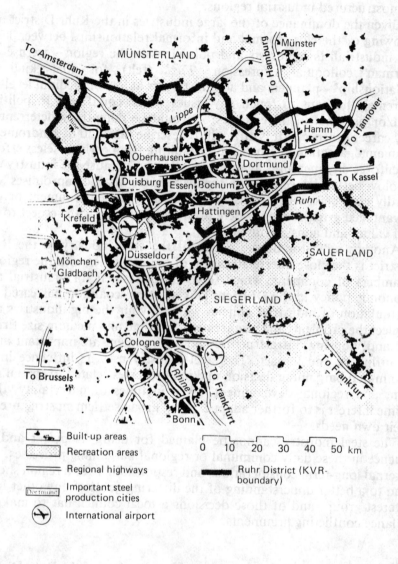

Built-up areas

Recreation areas

Regional highways

Dortmund **Important steel production cities**

✈ **International airport**

Ruhr District (KVR-boundary)

0 10 20 30 40 50 km

Figure 21.2
The Ruhr District: regional context and settlement structure

Source: Institut für Raumplanung (IRPUD) Universität Dortmund

414

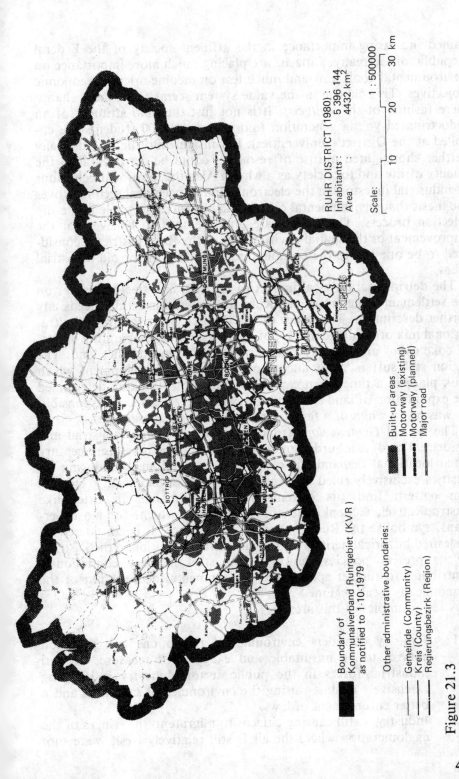

RUHR DISTRICT (1980) :
Inhabitants : 5 396 144
Area : 4432 km²

Scale:
1 : 500000

10   20   30 km

Figure 21.3
Spatial structure of the Ruhr District in 1980
Source: Institut für Raumplanung (IRPUD) Universität Dortmund

Boundary of
Kommunalverband Ruhrgebiet (KVR)
as notified to 1-10-1979

Other administrative boundaries:

Gemeinde (Community)
Kreis (County)
Regierungsbezirk (Region)

Built-up areas
Motorway (existing)
Motorway (planned)
Major road

gained increasing importance in the affluent society of the Federal Republic of Germany. Citizens are placing much more importance on environmental goals now and much less on income-oriented economic objectives. This change in the value system seems to be more than a mere fashion of the *Zeitgeist*. It is not just the 'anti-attitude' of an indoctrinated young generation (more than 100,000 students are enrolled at the District's universities), but insight into the fact that any further shortsighted misuse of resources could be detrimental to the quality of life and to society as a whole. Also the locational behaviour of industrial investors in the electronic and biotechnical sectors proves the thesis that environmental criteria play an increasing role in the site selection process. Thus, there are good reasons to believe that the improvement of the spatial environment of the Ruhr District is considered to be one key to the economic restructuring of the old industrial areas.

The detrimental physical impact of the coal and steel industries on the settlement structure of the Ruhr agglomeration hardly needs any further description. The image of the District is derived from the traditional mix of mining derricks, of coal washing and separating plants, of coke ovens and blast furnaces, of collieries and derelict land. The region still suffers from uncontrolled industrial development which took place in a time when capitalism and profit-thinking determined the exploitation of land and natural resources, when social responsibility was just a whim of a few paternalistic industrial magnates.

The typical *Gemengelage* (functional mix of urban industrial and residential land) is a burdensome heritage from the nineteenth century when locational demands of coal exploiting and steel-producing industries exclusively ruled the region's physical development.[4] This urban pattern finds its extreme factual expression in cities like CastropRauxel, Gelsenkirchen or Oberhausen which, for most Germans, symbolise the Ruhr's negative appearance. It is considered an undesired historical heritage of the industrial revolution in the District. By its present policies, however, the steel industry continuously contributes to further spatial and environmental depreciation of the economically weak region.

A few examples of this are:

1   Industry considers environmental regulations and pollution control to be unsuitable and exaggerated measures of anti industrial forces in the public sector. Industry is lobbying extensively against raising the environmental standards and a better enforcement of laws.

2   Industry is still causing citizens to migrate to the fringes of the agglomeration where the air is still relatively fresh, water not

yet polluted or extensively chemically treated, where roads are not yet jammed and children are still safe, and where land for residential purposes is available at relatively moderate prices.

3   Industry is thus encouraging further exploitation of scarce open agricultural land which is so urgently needed to balance the endangered ecosystem in the immediate hinterland of the agglomeration.

4   It necessitates additional commuter traffic, which again requires more, better and faster roads if newly given leisure time for recreation is not to be spent in traffic congestion.

5   It furthers demographic decline in urban cores, causing implications for the social sector and the social budgets of the cities.

6   Industry indirectly constrains the restructuring of the urban fabric, that is the transformation of the former uncontrolled developed urban pattern into habitable and attractive urban environments which may attract potential entrepreneurs of small and medium size firms to the District.

It should be mentioned here that the dominance of the big industries discourages small and medium size firms from settling in the region. This fact has further impacts on the migration of qualified young technicians and economists, who see much better career chances in other agglomerations of the Federal Republic of Germany.

The other side of the coin

The catalogue of complaints against the District's infrastructure and administration as brought forward by representatives of the steel industry or the regional chambers of commerce on every suitable private or public occasion is no less impressive.

The major stumbling block of the steel industry in the District – as with all other polluting industries – is stringent environmental regulations. These instruments for environmental control have been introduced by the state government to improve the quality of life in the area. One of these regulations is the so-called *Abstandserlass*, an ordinance which controls the minimum distance between industrial and other (mainly residential) land uses.[5] It is only the second best approach to protect residents in the region from noise and air pollution caused by nearby industries, as it retards the introduction of environmentally sound technologies. Nevertheless this ordinance has hindered further industrial expansion in the densely built areas at the expense of residents in their immediate neighbourhood. Big industry, however,

417

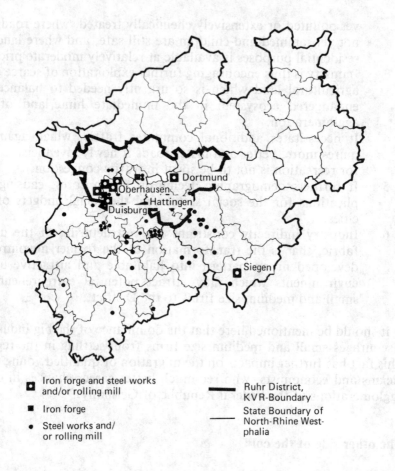

Figure 21.4
Location of iron and steel producing industry in North Rhine-Westphalia (1976)

Source: Institut für Raumplanung (IRPUD) Universität Dortmund

considers any environmentally justified argument against one or the
other expansion project as an obstacle against economic development,
as central state investment control, or as an interference in market
principles.

It is astonishing that the close interrelationship of economic development on one side and environmental quality of life on the other side –
an interrelationship which can be studied in all post-industrial western
countries by help of migration figures and the locational choice of
modern innovative industries – is not known or just neglected by the

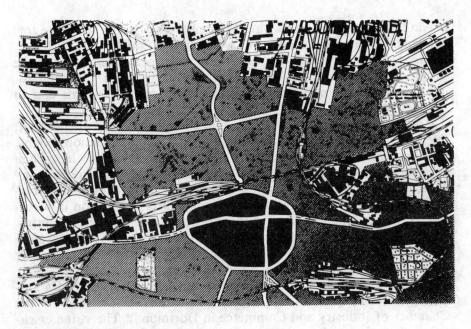

Prevailing land use:

Residential areas

Industrial areas

Open areas

City (shopping precinct)

Major urban road

Scale: 1 : 30000

0   500   1000   1500 m

Figure 21.5
The pattern of land use in the Ruhr District: the example of the inner city of Dortmund
Source: Institut für Raumplanung (IRPUD) Universität Dortmund

management boards of the big industries and by the unions' representatives on these boards.

Another steady object of complaint in the steel-producing industry is the public bureaucracy. The matter of complaint is that extensive bureaucratisation of the public sector – especially with respect to development-related approval procedures, ranging from building approval permits to physical investment control – impedes quick and spontaneous reactions to changing market forces. This, in fact, is partly true.

Efforts to speed up formal procedures and administrative decision making processes have already been undertaken, but the argument should be weighed against internal bureaucratisation which has made big industrial firms just as immobile as those in the public sector. Bureaucratisation seems not to be a question of 'public' and 'private', but to depend, *inter alia* , upon the number of persons involved in the decision making process. It is certainly a function of standard, qualification, readiness and authorisation to make decisions and to take responsibility.

Another argument put forward again and again is the evidence that, due to a number of reasons, but mainly due to militant action groups, the construction of transport and energy facilities in urban agglomerations has come to a near standstill. There are indeed many good financial reasons for stopping excessive public investment in further expansion of transport and energy facilities in a region in which the economic structure is changing and where a certain level of infrastructure has been reached. Cost–benefit calculations incorporating 'social and environmental costs' would reveal how uneconomic many of the strongly postulated new public investments in fact are.

These complaints, and some others, can be easily traced in a catalogue of essentials for future political action as published by the Chamber of Industry and Commerce in Dortmund.[6] There, ten essentials are put forward:

1 Guarantees for future steel production in Dortmund;
2 Development of the eastern Ruhr area into an energy centre;
3 utilisation of new technologies for coal refining;
4 Improvement of transport facilities;
5 Development of the service sector;
6 Adjustment of urban planning and public land policies to economic needs;
7 Increased utilisation of the regional research potential;
8 Test and application of new communication technologies;
9 Improvement of the residential and leisure qualities of the region;
10 Moderate tax and tariff policies of the communities.

This political shopping list prepared by a strong regional lobby group is typical for the region with its mixture of solely industry-related requests with assumptions, which indicate a certain understandiog of the overall structural problems of the region.

Detailed study of the elaborations under the headings reveals that, apart from well-known arguments, such as 'building more and better roads', 'providing cheaper energy', 'lifting rigid regulations' and 'streamlining public policy towards the needs of the big industries', no really new proposals are put forward to overcome the regional economic crisis. Thus, this catalogue is but a document of biased shortsighted regional lobby politics, a document of ignorance and helplessness. Such a catalogue of traditional means is not a seriously meant contribution to overcome the structural problems of an old industrial region. The catalogue is as bloodless as the overall regional policy, which seems to be totally paralysed by the extent of the problems from which it suffers.

Six billion Deutschmarks have recently been calculated as being indispensable to maintain steel production in Dortmund and to save 10,000 jobs in the sector. This huge amount of public money can certainly be much better used for the creation of at least the same number of jobs in other, more promising occupations.

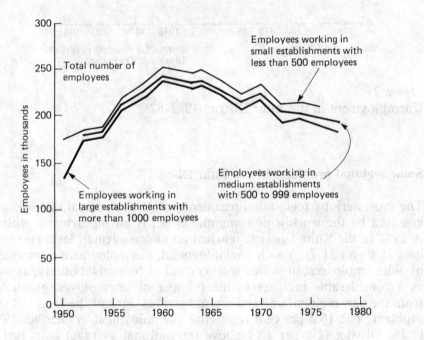

Figure 21.6
Employment in the iron and steel producing industry in North Rhine Westphalia 1950–80

Sources: ROJAHN, G. Der Einfluß von Großunternehmen auf die raum- und siedlungsstrukturelle Entwicklung im Verdichtungsraum Rhein-Ruhr, Dortmund 1983
Institut für Raumplanung (IRPUD) Universität Dortmund

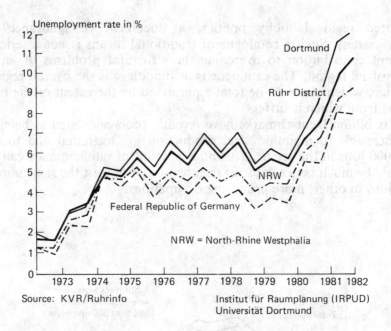

Figure 21.7
Unemployment in the Ruhr District 1973–82

Some regional problems of the Ruhr District

The most serious regional consequence of the structural change, aggravated by the worldwide economic crisis, is unemployment, which, in 1982 in the Ruhr District, reached an unprecedented level (see Figures 21.6 and 21.7). Youth unemployment, unemployment of women, of older employees, unskilled workers and of foreign labourers, as well as a considerable increase in the number of unemployed graduates from the new regional universities, are serious signs of the extent of the problem. The 10.8 per cent registered unemployment in October 1982 in the District (2.9 per cent above the national average) is in fact a tremendous challenge for politicians, economists and planners.

In an attempt to identify some internal reasons for the economic decline of a once powerful region, a number of regional problems which, in one way or the other, characterise the present situation will be pinpointed. This effort will, in a way, exculpate the regional steel industry. It will give some evidence for the fact that the relative economic decline of an old industrial area is to be attributed to numerous regional actors and power groups and their closely interrelated policies.

Physical urban structures

Uncontrolled industrial and urban development for more than a century has converted the region between Duisburg and Hamm, between the Ruhr and the Lippe rivers, into an unstructured urban agglomeration. Here industrial land use is scattered around the region wherever it was required. Residential areas, in turn, developed nearby whenever need was felt to provide nearby housing for industrial employees.[7] Today, old traditional town centres, derelict land, industrial villages and numerous modern suburbs are interconnected by an extensive road and rail network, probably the densest of any agglomeration in the Federal Republic of Germany.

During the war the Ruhr was obviously a favourite target of Allied bombs. In addition to what was destroyed during the war, the reconstruction fever of the 1950s and 1960s demolished nearly as much as the bombs did before. Due to a number of reasons, the reconstruction of the urban structure after the war was not primarily guided by the aim to create a better environment. Thus the uncontrolled spatial heritage of the nineteenth century, and its adaptation to post-war industrial needs during the 1950s and 1960s, are a major constraint to any future-oriented spatial development policy in the District.

Inferior environmental conditions

Despite considerable achievements during the 1960s and 1970s, environmental pollution is still a problem. Even if occasional smog alarms are no longer a special feature of life in parts of the Ruhr agglomeration (today Munich, Stuttgart and Frankfurt have similar problems), there is no good reason why economic recession should bring the necessary further pollution abatement measures to a standstill.

Any effort to recreate a sound environment in the District will probably contribute more to its long-term economic development than short-term action against public pollution control. This includes the rigid preservation of still existing open land, of fresh air corridors and interconnected recreational belts, but also a long-term strategy for the regional reclamation of industrial land emphasising the conversion of former industrial land into ecologically regenerated areas. Similarly, further alignments of additional transport corridors and energy transmission lines through still unbuilt areas should not be permitted as well as the location of new public and private physical investment in the attractive green areas of the south.

Wrong emphasis on energy development

For years Ruhr District politicians, arm in arm with heavy industry, have been stressing the importance of the area as a federal or regional energy region. Originally based on coal exploitation, energy production has a long tradition in the area. But bringing additional power plants and coal refineries to existing sources of environmental nuisance would further accentuate the pollution of the region without creating those jobs which might contribute to its necessary economic diversification. Proclaiming the Ruhr the future energy region of the Federal Republic of Germany would mean accepting the passive gradual depopulation of the region. Anyone who has been walking during his weekends under whistling energy transmission lines in the Ruhr will doubt whether its conversion into a new energy-producing centre is the right policy for the District. The image of an energy region may have been attractive for big basic industries, but their expansion has probably come to a standstill anyhow. Due to strong worldwide competition all major new industrial investment of some size, if there is any, will have to be located at the coast with access to deep-sea harbours, and not in an old industrial region where existing constraints form a considerable burden to new investment and its future profitability.

Availability of land

The availability of land for industrial purposes in the Ruhr is limited, even if considerable areas of under-used (derelict) industrial land appear in urban land use plans. Most of the urban land is still in the hands of the coal and iron industry and not available on the regional real estate market. The threatening competition for qualified workers, the hope for future industrial expansion, unrealistic speculative land prices, very high costs for transforming former industrial land into land which is suitable and attractive for modern small and medium size plants, are some of the deficiencies characterising the inflexible land policies of the big firms. The lack of planning permission and inflexible economic promotion agencies of the cities further restrict the real estate market. Many firms prefer to develop virgin land on the green edges of the agglomeration; often they are lured by tax-hungry communities which offer additional open or hidden incentives[8] to new industrial settlement. Among insiders, the Ruhr is known as a region where decisions for industrial investment can be implemented only after time-consuming and costly procedure. Thus, potential investors prefer to settle at other places whenever there is a choice.

The Ruhr's chances of attracting new industries and investors from outside the region are rather limited. During the last decade, the number of newly established firms in the industrial sector has declined tremendously. Their number is low compared to the number of firms being closed. Among the regions which compete for the very few young innovative investors, the Ruhr has certainly more disadvantages than advantages to offer.

Lack of innovative small and medium size firms

The Ruhr's traditional image and the unpleasant physical environment, compared, for example, with Hamburg and Munich, or that of numerous other attractive medium size cities in Germany, are reasons for the hesitancy of innovative small and medium size firms to settle in the District. Limited access to attractive recreational areas are an important locational disadvantage. The adjoining Sauerland and Münsterland can only be a second choice to the Alps or the North Sea islands.

Lack of universities and research institutes

The region still suffers from a hundred-year-old Bismarckian doctrine to settle neither universities nor military establishments in it. Until 1965 it did not have a single university in its boundaries. Whereas the technical universities of Munich, Stuttgart, Berlin and Hannover contributed considerably to the industrial development of their respective regions, the Ruhr industry had to recruit its engineers and economists from the universities of Aachen, Bonn, Münster or Cologne.

This deficiency was partly removed during the 1960s when some new universities were established. Probably due to traditional interrelationships in influential circles of regional opinion leaders and decision makers, however, the number of departments dedicated to traditional and modern technologies stayed rather low.

For a long time industry in the Ruhr employed a relatively small number of academically trained people. One reason was probably the shortage of qualified manpower available in the region. Today the Ruhr universities produce a surplus of academically trained engineers, economists, chemists and computer scientists. As the industry is suffering from economic recession, there is little chance to employ them and for the region to profit from this newly created human capital.

Housing shortage

The housing sector in old industrial areas like the Ruhr poses problems for a number of reasons. Despite a large stock of working class houses owned by the region's big coal and steel industries, and a considerable stock of housing for low-income households, built and managed by communal housing associations, the availability of low-priced housing has become a serious problem. Urban renewal and modernisation fashions have reduced the amount of low-rent accommodation in the inner areas of the cities. The steady increase of floor space per household, the absolute increase of the number of households and an extreme growth of student and one-person households have caused a serious shortage of cheap housing.

The limited land supply for residential purposes in attractive areas is similarly felt by middle and high-income households. The attractive south of the region along the river Ruhr is already overdeveloped and requires strict sprawl control, whereas the core and the north of the Ruhr District is hardly attracting any new private housing investors.

Insufficient housing conditions constrain the region's economic development, as well as intra-regional migration and job mobility. These and other, already mentioned, negative aspects make the District most unattractive to the innovators and entrepreneurs it needs so badly.

Political nepotism and inflexibility

When analysing the region's structural problems, it must be mentioned that the communal politicians in the District are a special breed. One may characterise their minds as being as monostructured as the region's economy. For decades political power in the city hall has been in the hands of the Social Democrats. Unchallenged by strong political opposition and backed by very conservative, but influential, labour unions, the Social Democratic Party dominates the public life of the area. There is very little political innovation.

The finely knit network of political nepotism and mutual favours (a sophisticated kind of corruption) has always succeeded in surviving attacks from the outside as well as social or economic disturbances. Even the 'internal' conflicts between a traditional, union-related Social Democratic Party, representing a relatively affluent working class, and the young Left, consisting mainly of radical academics at the regional universities, could not trigger any political innovation. The quarrels to the Left rather paralysed innovative actions.

Despite its political image of being 'red', the region is extremely provincial; its citizens are conservative and traditional in mind. This makes it difficult to introduce necessary innovations. Apart from the social sector, there is hardly any courage in the region to support openended experimental actions.

426

Unsuccessful administrative reforms

Administrative reforms in the early 1970s, resulting from long-lasting political bargaining processes, led to further mergers of formerly independent communities into a few big cities, a process which has been continuously going on since the late nineteenth century. Instead of improving public services, as intended, instead of avoiding *'Kirchtum'* politics (parish politics) and local chauvinism, this so-called 'functional reform' rather added further bureaucratisation and obscurity to political decisions.

Today we know that this reform impeded political decisions rather than bringing them closer to the citizens. Increasing mistrust in local politics is only one consequence.

The coordination of urban and regional development in the Ruhr District was the task of one of the world's more famous regional planning authorities, the *Siedlungsverband Ruhrkohlenbezirk* (SVR). Despite a low performance on its regional planning tasks – resulting from political competition and mutual neutralisation among the powerful Lord Mayors of the Ruhr cities – this planning authority played a major role in the development of the region and as an effective public relations and cultural promotion agency for the Ruhr.

Relieved of its planning functions and left with the responsibility for a few coordinative and intercommunal services, the *Kommunalverband Ruhr* (KVR) – as the authority was renamed a few years ago – now plays only a minor role in the development of the region. This formerly primary task of SVR has been transferred to three regional authorities, residing in Münster, Düsseldorf and Arnsberg, all of them beyond the boundaries of the District. No other region with 5 million inhabitants, and of similar economic importance to the Ruhr District, would accept its spatial development being planned, guided and controlled exclusively by outside authorities. This is an essential constraint to the management of necessary restructuring efforts for the region.

Low performance of urban and regional planning

For a number of reasons spatial (urban and regional) planning for the region is mediocre and far from exemplary. Whatever the reasons are, the fact is that urban planning in the region was never much inspired. After a short but intensive reconstruction period, when the war-damaged centres of the major cities were remodelled, urban planning was guided by compromises between functional, aesthetic and financial objectives. For political reasons spatial planning is subject to a piece-meal approach and day-to-day incrementalism. Convincing long-term strategies for a better urban and regional development of the urban

427

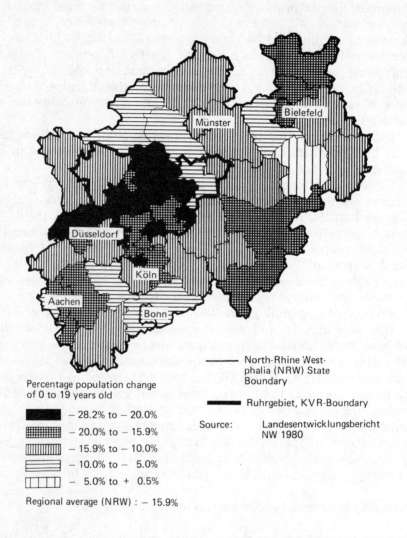

Percentage population change
of 0 to 19 years old

■	− 28.2% to − 20.0%
▦	− 20.0% to − 15.9%
▥	− 15.9% to − 10.0%
▤	− 10.0% to − 5.0%
▦	− 5.0% to + 0.5%

Regional average (NRW) : − 15.9%

———— North-Rhine West-
phalia (NRW) State
Boundary

▬▬▬ Ruhrgebiet, KVR-Boundary

Source: Landesentwicklungsbericht
NW 1980

Figure 21.8
Population development in North Rhine Westphalia 1980–95. Age
group 0–19 years

Sources: Landesentwicklungsbericht NW 1980
          Institut für Raumplanung (IRPUD) Universität Dortmund

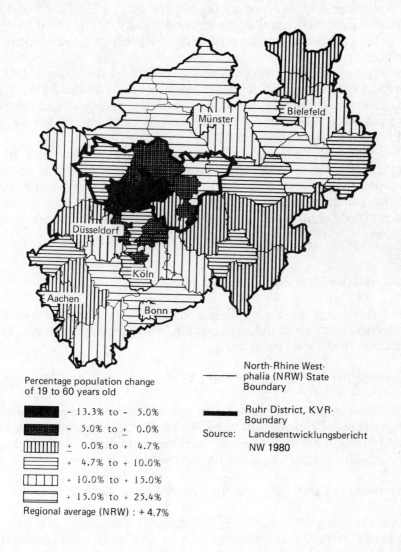

Figure 21.9
Population development in North Rhine Westphalia 1980–95. Age group 19–60 years

Sources: Landesentwicklungsbericht NW 1980
        Institut für Raumplanung (IRPUD) Universität Dortmund

regions are nonexistent. Urban and regional planning is poorly backed by politicians and attacked by the big industry, which still considers spatial development planning as an ideologically infected tool to strangle free economic development (that is, on the sole terms of capitalism).

Interesting planning tasks normally constitute a challenge and inspiration for planners. Such tasks are ubiquitous in the Ruhr. There are, however, strong reasons why outstanding, excellent, and independent planners never have been chosen and employed by the various political councils of the Ruhr cities.

The political and social climate of the Ruhr seems not to be favourable for those planners who engage in the long-term development planning of their cities beyond the day-to-day petty cash of party politics which emphasises compromises for the short-term solutions and the 'long-term' considerations of the 'maximum election period' horizon.

Some wishes addressed to the steel industry

From this quick review of the problems of an old steel and iron production-dominated industrial region, a few messages emerge, to be addressed to the steel industry.

Responsibility for the natural and physical environment of the region.

This does not always require more money. In many cases, it just requires expertise and the will to accept monitoring methods to be integrated into internal decision making processes, for example, environmental impact analyses for development projects.

Restraining the shift of social costs to the public sector.

Endeavours to shift social costs of industrial investment to the public sector are certainly one consequence of increased competition. But complaints about excessive expansion of the public sector are untrustworthy, as long as the social costs and consequences of industrial investment and production are left totally to the public to take care of. Sharing the burden through tax contributions from industry's revenue alone is insufficient. The immense liabilities incurred by the cities in the District should be a warning.

Responsibility for overall urban and regional development.

In the long run the steel industry could profit much more from the region having a good overall image than from short-term profit maximisation. The physical image of a region as a whole plays an essential role in the postindustrial society and in the interregional or international competition. Thus the better the national or international image, the better is probably the overall economic performance potential of its industries.

Explicit support of small and medium size firms.

This may be another possibility for the steel industry to contribute to the gradual restructuring of an old industrial area. Instead of further exclusively pursuing its own expansion policies, close cooperation between large industrial enterprises and small and medium size firms in the region should be sought. Such a policy of differentiation and deconcentration of production, services and research would bring back necessary flexibility. It will certainly facilitate innovative developments.

Decentralisation of non-producing affiliations.

Whereas production certainly has to be concentrated rather than dispersed, to keep internal transport costs down by the help of improved and new communication technologies, related tasks in the service sector should be regionally decentralised. Such a policy would avoid any further concentration of jobs to a few locations and contribute to a further reduction of enforced mobility and commuting.

Employment bubble.

Unavoidable rationalisation in the productive sector would certainly be more acceptable to the unions if paralleled by the creation of new jobs to keep the overall number of employment opportunities balanced. Why not create jobs in the social and recreation sectors? The changing patterns of qualification and work attitudes among industrial employees could certainly be an inducement for new arrangements of the working environment, incorporating social and leisure oriented activities, within the spheres of the industrial firms' financial responsibility.

Responsible land policy.

The large holdings of urban land still in the hands of the steel industry should no longer be looked upon either as a reserve for industrial expansion, or as an instrument to balance the firms' annual profit and loss statements. Even if the land is not to be sold, the possibility of interim use should be examined, thus not only using the land lots as deposits for scrap iron and recycling stock, but making it available for purposes aiming at the social, environmental and cultural improvement of the region as a whole.

Accepting the change of the value system.

Adjustment to structural change in a region also means accepting the new values and aspirations of a considerable number of the younger generation. Their attitudes and values are different from those of their parents, who profited from the reconstruction-oriented spirit of the post-war decades. From the planner's viewpoint already a partial accomplishment of some of the above 'wishes' could contribute to the restructuring of the region, to a sound economy and to a better future for the District.

Notes

1   Hall, P., *The World Cities* , Weidenfeld and Nicolson, London, 1966. This book contains, among other things, an excellent introduction to the development of the second biggest European agglomeration. Other good sources of further introductory information on the region are:
Hellen, J.A., 'North-Rhine-Westphalia', *Problem Regions of Europe* , Oxford University Press, 1974.
Bowden, P., *Regional Development in Action* , Anglo-German Foundation for the Study of the Industrial Society, London, 1979.

2   Brune, R. *et al.* , *Wirtschaftsstrukturelle Bestandsaufnahme für das Ruhrgebiet* , Rheinisch-Westfälisches Institut für Wirtschaftsforschung, Study, Essen, 1978 (in German).

3   Ibid.

4   Niethammer, L., *Umständliche Erläuterung der seelischen Störung eines Communalbaumeisters in Preussens grösstem Industriedorf – oder die Unfähigkeit zur Stadtentwicklung*, Syndikat, Frankfurt, 1979 (in German).

5   Runderlass des Ministers für Arbeit, Gesundheit und Soziales. Abstände zwischen Industrie bzw. Gewerbegebieten und Wohngebieten im Rahmen der Bauleitplanung (Abstandserlass)', *Ministerialblatt für das Land Nordrhein-*

Westfalen 35 (1982) , No. 67, 1984, p.1376. This ordinance is a late outflow of the ideology of separating land uses, an ideology which has ruled the thinking of German urban planners and planning for more than half a century as an easy means for a better compromise between the diverging and conflicting demands of industry and citizens. (In German.)

6   The threatening closure of the steel industry in Dortmund has certainly incited the Chamber of Commerce to produce this document. A second similar but more general report, covering the problems of the whole region was published a few months later. See Chambers of Commerce of the Ruhrgebiet, *Wirtschaftspolitik für das Ruhrgebiet*, Dortmund, 1982.

7   Compare the detailed analysis of Hübner, H., Krau, J. and Walz, M., 'Lohnabhängigkeit und Wohnabhangigkeit', in *Bauwelt 69* (1978), p.1114 to 1126 and Krau, J. and Walz, M. (1982) 'Die betriebliche Wohnungsversorgung bei Stahl- und Bergarbeitern in Ruhrgebiet', in *Bauwelt 73* (1982) pp.956 to 961.

8   The levy of the local industry tax is the major source of income of the local communities in the Federal Republic of Germany.

PART VI
STRATEGIES

22 Introduction

Strategies to cope with the steel industry crisis in different regions and countries are treated throughout the entire book. The introductory, statistical and country-regional sectors deal with strategies in a descriptive manner: what strategies have been employed, what were the outcomes. (Occasionally, for example, in Chapters 12 and 13 by Crandall and Hirschhorn on the US steel industry, pleas for strategies to be adopted are launched.)

The strategy section, which we are entering upon now, is essentially of the 'prescriptive' type: what could, should or ought to be done to cope with the crisis. The tenor of the contributions is, however, quite different.

Chapter 23 by Neumann, 'Improved competitiveness of steel-producing firms by means of diversification', is based on a micro-analysis, employing the capital asset pricing model, to examine how efficient the diversification strategy adopted by German steel producers has proved to be. The conclusions drawn are that diversifying firms are better suited to resist the crisis than undiversified firms. Firms, however, cannot diversify into 'all' kinds of business. Only the government (of a larger country) could have access to a true market portfolio, and thus be able to hedge against industrial risks. Professor Neumann feels, however, that governments, for political reasons, would not use the opportunities given by their superior portfolios in optimal ways. Thus the 'second best' strategy of enterprise diversification is actually and obviously the best one.

Messrs Takano and Horie of Nippon Kokan demonstrate, in Chapter 24, against the example of their firm, how large steel producers in Japan are adapting to the crisis and what future strategies they aim to pursue: continuous investment in the most modern production technology, close cooperation with major steel customers in order to solve their functional problems by providing adequate new steel qualities, streamlining, rationalising production capacities, productivity improvement, mechanisation, and computerisation. The authors also stress the importance of all categories of personnel and, consequently, the emphasis Japanese firms lay on employee relations.[1]

Chapter 25 by Professor Bela Gold combines his eminent experience in dealing with steel industry problems over 30 years with his respect for the dynamism of the Japanese steelmakers. It is not overstressing the conclusions drawn if they are summarised thus: a steel firm which is not willing to play the game according to the Japanese rules has no future and might as well leave the industry right away.

Notes

1 An interesting piece of side information provided by the authors are the statistics on employee attitudes towards the Japanese quality circle movement within the steel industry.

23 Improved competitiveness of steel-producing firms by means of diversification

Manfred Neumann

Mergers between steel producers and downstream firms have been considered with much concern because of their alleged anti-competitive effects in downstream industries. Notably mergers, as for example Thyssen and Rheinstahl, Mannesmann and Demag, Krupp and Buckau R. Wolf as well as Koppers, which took place in the early 1970s in West Germany and which were followed up by a series of minor acquisitions of downstream firms, have eventually induced the West German Monopoly Commission to propose an amendment to the law against restrictive practices with the intention to outlaw further acquisitions of that kind. On the other hand, the impact of diversifying mergers of steel-producing firms on the iron and steel markets has been largely disregarded. This aspect, however, seems to be of substantial interest in view of the recent crisis of the iron and steel industry in Europe. It therefore appears to be worthwhile investigating to what extent the ability of the West German steel industry to survive competitive pressures has been enhanced by past diversifying mergers.

This chapter consists of three sections. In Section I a basic theoretical model is set up. In Section II it is shown how the West German steel industry fits into the theoretical pattern. In the concluding Section III it is checked how the ability to survive competition pressures from outside has been affected by diversification.

Section I

The starting point is the capital asset pricing model.[1] According to this model maximisation requires that the present value of capital invested in firm i should be

$$V_i = \frac{\bar{p}_i q_i - C(q_i) - S \, cov(R_i, R_M)/\sigma_M}{r}$$

where

r	=	rate of interest on a risk-free asset,
$\bar{p}_i$	=	expected price of commodity i,
$C(q_i)$	=	operating costs,
S	=	market price of risk,
$cov(R_i, R_M)$	=	covariance between the returns of firm i and the return of the market portfolio,
σ_M	=	standard deviation of the return of the market portfolio.

Consequently expected profits of firm i are given by

$$\begin{aligned} E(\pi_i): &= \bar{p}_i q_i - C(q_i) - rV_i \\ &= S \, cov(R_i, R_M)/\sigma_M. \end{aligned}$$

Hence expected profits are positive or negative or zero according to whether the covariance of the returns of firm i with the return of the market portfolio is positive, negative or zero, respectively. Note that V_i is the present value of capital invested in firm i and rV_i is the opportunity cost of capital. Therefore, negative or zero expected profits do not necessarily imply that expected operating profits, i.e., $\bar{p}_i q_i - C(q_i)$, are negative or zero, respectively. Nevertheless expected operating profits can be presumed to be comparatively lower or even negative if the covariance is low or negative.

Provided the firm is run in the interest of its owners, it is operated in such a way that the present value of capital is being maximised. If that is done in a competitive environment by fixing the quantity of production, q_i, it is required that[2]

$$\bar{p}_i = C'(q_i) + S \, cov(R_i, R_{(M)})/\sigma_M q_i$$

Therefore, if the covariance is negative, the expected price must be lower than the marginal cost. The firm expands output until its marginal cost exceeds the expected price which, in a competitive environment, can be assumed to be given to the firm. Actually, however, if all firms operating in a market where a negative covariance is obtained behave in the way indicated above, the average market price will drop below marginal cost.

In a perfect capital market, capital owners would be willing to cover up losses entailed in lines of production with a negative covariance because holding such stock entails a diminution of risk of the individual portfolio. The individual investor is thus trading in expected return against a lower risk.

In fact, capital markets are far from being perfect. Since the individual investor is only imperfectly informed about the firms' operations he cannot be sure whether a loss is due to a deliberate policy in line with the rules outlined above, or whether it is due just to bad management. The individual investor will therefore normally expect a firm to be run profitably. Things are different if a line of production with a negative covariance is run by a multiproduct firm. The management of the respective line of production is controlled by the top management of the firm. Within such a firm it can be reasonably expected that the individual lines of business are run in the way indicated by the capital asset pricing model. Therefore, lines of business with a negative covariance *vis-á-vis* the market portfolio can be subsidised by profits earned in other lines of business.

Section II

It will now be investigated how the West German iron and steel industry fits into the pattern of the capital asset pricing model.

In Table 23.1, time series of returns from 1965 to 1973 for several industries including the iron and steel industry and the total of all West German industrial firms are given.

From the data assembled in Table 23.1 coefficients of correlation were computed, which are listed in Table 23.2. The correlation between the returns of the iron and steel industry and the returns of all firms, which can be taken as representing the market portfolio, is close to zero. On the other hand, the returns of the machine building industry and the construction industry are positively associated with the returns of the market portfolio. It thus appears that, in a perfect capital market, the steel industry should have lower expected profits or should even be subsidised by the other industries.

It might be noted that the construction industry fits neatly into the theoretical framework. As can be seen from Table 23.1, the average of this industry is comparatively high. That is the way it should be. Notice that, in this industry, the coefficient of variation of returns is comparatively high. As $\text{cov}(R_i, R_M)/\sigma_M = \varrho_{iM} \sigma_i$ where ϱ_{iM} is the coefficient of correlation between returns of firm i and those of the market portfolio and σ_i the standard deviation of the returns of firm i, given a certain positive correlation, the expected return should be the larger, the higher the standard deviation. This result lends additional support to the suggested interpretation based on the capital asset pricing model.

Table 23.1

Gross profits (annual surplus)[1] 1965-1976, in billions of DM

Year	Iron and steel industry	Machine building industry	Construction	All firms
1965	3.00	4.42	0.94	83.26
1966	2.20	4.13	0.92	81.42
1967	1.78	4.35	0.84	82.20
1968	1.81	3.66	0.41	72.88
1969	3.12	4.45	0.63	84.04
1970	3.52	4.42	0.86	87.05
1971	1.78	4.73	1.20	90.56
1972	1.48	4.72	1.20	98.11
1973	2.15	4.18	0.84	97.93
1974	3.92	3.98	0.63	94.21
1975	2.76	4.22	0.85	95.64
1976	2.57	5.51	0.88	118.38
Average	2.51	4.40	0.85	90.47
Coefficient of variation (per cent)	29.4	10.1	25.2	12.4
Gross profits/ Equity	0.186	0.306	0.403	0.392

1) Gross profits before taxation
2) Including reserves
Source: Deutsche Bundesbank, Jahresabschlüsse, 1978.

Table 23.2

Correlation between gross profits, 1965–76

	Machine building industry	Construction	All firms
Iron and steel industry	-0.093	–0.373	0.080
Machine building industry		0.578	0.756
Construction			0.368

Source: Computed from data of Table 23.1

Section III

The above reasoning suggests that, within the diversified firms which emerged from the acquisitions of the early 1970s in the West German steel industry, an internal subsidisation takes place. This is completely rational from the point of view of the capital asset pricing model. Consequently, the ability to survive the competitive pressures from outside to which steel producing has been exposed has increased substantially. It can be presumed that the more diversified firms perform better than the less diversified ones. This presumption seems to be well in conformity with the actual performance of the various steel-producing firms in West Germany.

Actually, the potential of internal subsidisation is limited because the scope of diversification is limited, too. Hence, the financial pool from which subsidies can be drawn is less than the actual market portfolio. Therefore, in a perfect capital market, where each investor holds a portfolio which duplicates the properties of the market portfolio, cross-subsidisation is based on a broader spectrum of assets than those available to the present steelmaking firms in Germany. Although diversification is apt to decrease the degree of imperfection of the capital market, a substantial degree of imperfections remains in the German market.

On the other hand, the broadest possible pool for cross-subsidisation can be provided by the government's taxation power, which encompasses the true market portfolio. Thus the rationale for government subsidisation of steelmaking would be a simulation of the working of a perfect capital market. Obviously, in view of the broad coverage of government's taxation power, such a scheme would be vastly superior to any private scheme of diversification.

However, it cannot be taken for granted that governments would strictly adhere to economic principles. Inevitably, purely political pressures will arise in determining the direction and the scope of subsidisation. Serious distortions can therefore be expected to result, deteriorating allocative efficiency. Hence, private schemes of diversification by appropriate mergers seem to be more advantageous from an economic point of view.

Notes

1   See, for example, Fama, E.F. and Miller, M.H., *Theory of Finance* , Holt, Rinehart and Winston Inc., New York, 1972.

2   See Neumann, M., '"Predatory Pricing" by a Quantity Setting Multi-product Firm', *American Economic Review* , vol.72, 1982.

24 Responsibilities of the steel industry of industrialised countries: the case of Japan

Hiroshi Takano and Shigeyashi Horie

Meeting the challenge of steel

Over the past ten years, the steel industry of industrialised countries has been confronted with various problems and challenges. We would like to discuss some of the major challenges facing the industry and how those challenges can be met.

In the case of industrialised countries, we know that, as the industrial base is expanded, there is a relative decline in the position of the steel industry within the overall structure.

To overcome difficulties caused by such decline, we would suggest six basic measures (see also Figure 24.1):

1 Ensuring a healthy business footing by gaining and holding a stable position in the domestic market.
2 Maintaining competitiveness in the world market through modernisation and rationalisation of operations.
3 Expanding business by developing new markets through production of advanced and speciality steels.
4 Expanding business through further diversification into related industrial fields.
5 Promoting revitalisation of steel companies by recruiting high calibre personnel and enhancing employee morale. To this end, one would, for example, have to increase the attractiveness of the company by improving the working environment and by taking the steps described in 3 and 4 above.
6 Requesting government support for an industrial policy based on the long-range perspective.

We would like now to briefly discuss the principles of these points, citing a few examples of the actions we in Japan have taken in these directions.

```
┌─────────────────────────────────────────────────────────────┐
│  Basic Measures for Steel Industry in Industrialized Countries │
├─────────────────────────────────────────────────────────────┤
│                                                               │
│   1.   Assurance & capture of stable domestic market          │
│                                                               │
│   2.   Modernization and rationalization                      │
│                                                               │
│   3.   Emphasis on advanced and specialty steels              │
│                                                               │
│   4.   Diversification of business activities                 │
│                                                               │
│   5.   Sustained vitality of an enterprise                    │
│                                                               │
│   6.   Governmental support from a long-range perspective     │
│                                                               │
└─────────────────────────────────────────────────────────────┘
```

Figure 24.1
Basic measures for steel industry in industrialised countries

Assuming a stable domestic market

Generally, there exists a large domestic market for steel in industrial countries. To ensure a healthy business base, the steel producers must gain and hold their domestic market.

In this sense, a major advantage of domestic producers is that their mills are located in proximity to domestic customers and this facilitates gathering information about the customers and also providing them with many services. Furthermore, from the standpoint of customers, the steel industry is often an important buyer of goods and services.

The domestic steel industry, by fully utilising these advantages, must gain and hold the domestic market.

As one example, we would cite one of the unique services Japan's steel industry provides for Japanese automobile companies with respect to deliveries of steel coils. This type is also extended to other major customers such as shipbuilding companies.

Auto manufacturers place orders for steel coils about two months in advance. The placing of the orders is based on their *preliminary production plans* for the press lines. Based on the orders, the steel mills

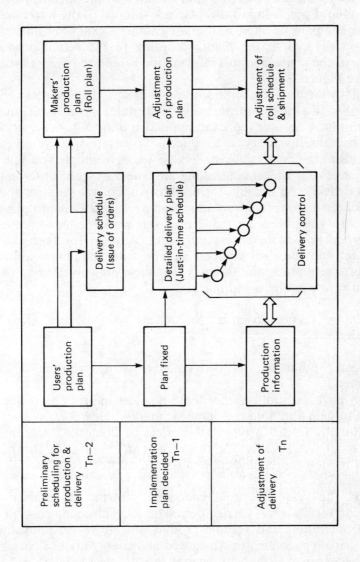

Figure 24.2
Delivery control of steels to automotive sector in Japan

447

set up *their* preliminary production plans to include production and operating schedules for each manufacturing process. One month before the particular coils are to be used, the detailed delivery schedules are established according to the specific production plan of the auto maker. Based on this, the steel mill finalises its production plan. The coils produced according to this plan are shipped to the warehouses of the steel company located near the customer's plant. From here the coils are delivered to the plant according to the actual production schedule of the plant, and this information is reported daily to the steel mill. (See Figure 24.2.)

This daily information on the auto maker's production plans enables us to realise fine-tuning of our rolling schedules. As a result, our mill operations have become more rationalised and the in-process stocks, as well as product inventory, are reduced. To ensure speedy and precise operations of the timely delivery service, we have adopted on-line computer systems for use in controlling all processes from order entry to product delivery. As a result, the inventory of coils at the steel mills has been decreased from five weeks to three weeks. At the auto plant, the coil inventory has been decreased from one week to almost nothing.

This reduction of inventory constitutes savings for both the auto companies and the steel companies. The auto companies appreciate our efforts very much and they say that these are contributing to their competitiveness in the world market.

Maintaining competitiveness by means of modernisation and rationalisation

The second point we wish to discuss is the rationalisation of operations.

As we have mentioned, the domestic steel industry is in the best position to gain and hold the domestic market. However, as steel is an internationally traded product, the prices of domestic steels must be internationally competitive. In this respect, there must be constant efforts for modernisation and rationalisation of production facilities and their operation.

During the years of rapid economic growth, modernisation of production facilities was carried out in Japan. This made possible both rational operations and volume production. However, following the first oil crisis, economic growth slowed and energy prices skyrocketed. These changes made it imperative that Japan's steel industry be further rationalised.

With the stagnation in steel production, two of the main actions the Japanese steel industry took were as follows. One, they concentrated production in the more modern and most efficient production facilities. Two, with respect to capital investment, they stressed rationalisation for realising improved saving of materials, energy and manpower. Outstanding examples of these developments are up-grading the continuity of processes by use of continuous casting and continuous annealing technologies, and saving energy, mainly heavy oil, through recovery of waste heat including use of CD, and effective uses of gases generated at various shops. These energy-saving efforts have been made to cut energy consumption, resulting in a constant decline of energy consumed per ton of crude steel. (See Figure 24.3.)

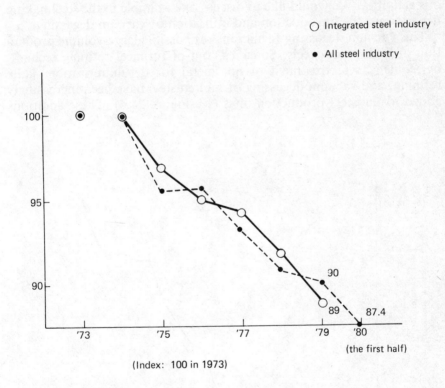

○ Integrated steel industry

● All steel industry

(Index: 100 in 1973)

Figure 24.3
Trend of energy consumption per ton of crude steel in Japan

Expansion of business through production of more sophisticated steel and steel products

Together with the advances in the industrial structure, there has been an increase in the demand for high grade and speciality steels. In the energy development field, there is an increased demand for steels capable of withstanding severe service conditions. Also in the sector of energy saving and manpower saving, there is increased demand for steel having specialised properties. Production of these steels requires advanced technologies which are available only in advanced countries. Thus, the steel industry of advanced countries must supply requirements for both domestic and foreign markets. The increased efforts towards production of such steel contributes to improving steelmaking technology and helps to vitalise the company's operations.

Concerning this matter, Japanese steelmakers have conducted various activities. We would like to discuss one example in the steelmaking department, the introduction and utilisation of vacuum degassing.

This vacuum degassing facility makes possible large-volume production of high-grade steels. So-called 'out-of-furnace' refining technologies such as pretreatment of hot metal for desulphurisation, ladle refining, and vacuum degassing of molten steel have been increasingly added to the steel production lines (see Figure 24.4). These additions

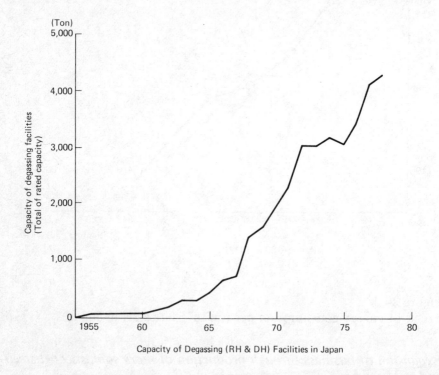

Capacity of Degassing (RH & DH) Facilities in Japan

Figure 24.4
Capacity of degassing (RH and DH) facilities in Japan

450

have been made without disrupting the conventional production line consisting of blast furnace, basic oxygen converter, and continuous casters. Thus, there have been great improvements in steel quality, for example, by reducing the sulphur content of steel used for manufacturing line pipe.

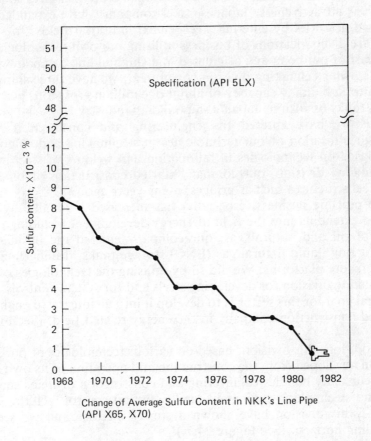

Change of Average Sulfur Content in NKK's Line Pipe
(API X65, X70)

Figure 24.5
Change of average sulphur content in NKK's line pipe

The use of technologies such as we have just mentioned, has made it possible to realise large-volume production of super-low-sulphur steel with a sulphur content of less than 10 ppm (see Figure 24.5). This was never thought possible in the past. Now, steel capable of being used in an environment of gas containing high hydrogen sulphide can be produced in large volumes.

Expansion of business through further diversification

No remarkable increase in steel demand in developed countries can be expected any longer. An increase in value-added products by stressing the more sophisticated qualities of steel may not be sufficient to fill the gap. Thus, steel companies may have to take other measures to maintain their attractiveness. Japanese steel companies have expanded and vitalised activities by entering steel-related industrial fields. In short, these are diversifications of business utilising our own technologies.

A part of our company originated in a shipbuilding company which had its own steelmaking division. At present, we have three shipyards with a total building capacity of about one million gross tons per year, which ranks us fifth in Japan's shipbuilding industry.

Thus we have entered the engineering and construction fields through utilisation of our technologies in steelmaking and shipbuilding. Based on technologies in fabrication and welding of steel in our shipbuilding division, our company started construction of pipelines and steel structures such as bridges, many years ago.

The pipeline business, especially, has increased gradually. With a view to extending into the field of energy development including treatment of oil and natural gas, our company entered the business of constructing liquid natural gas (LNG) storage tanks, drilling facilities and offshore platforms. We did so by utilising the technologies of our steelmaking division for developing steels and for system controls. Our eventual goal for this sector is to develop it into an integrated engineering and construction company in the energy-related field. (See Figure 24.6.)

In addition, this division, based on various technologies of our steel division developed from the construction and operation of its own steel mills, currently handles engineering and construction of items such as iron and steelmaking equipment, and pollution control facilities. The sales of this division have shown a steady increase and we see an expanding horizon. (See Figure 24.7.)

Maintaining the vitality of the company

The next point is maintaining the vitality of the company. The most important factor in the growth of a company is its personnel. This means the securing of able personnel and improving their morale. In order to secure able personnel, the company must be attractive and have expectations of continued growth, and give personnel places to work to match their accomplishments.

Japanese steelmakers make efforts to attract engineers, by promot-

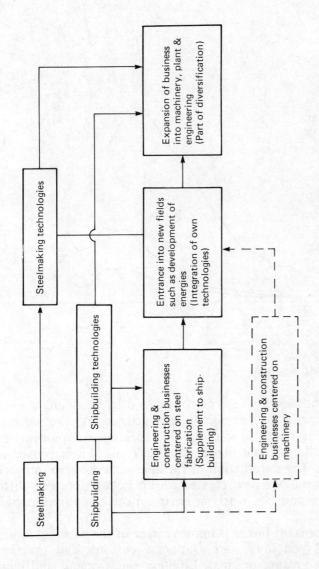

Figure 24.6
Steps in diversification of NKK's business into engineering and construction fields

453

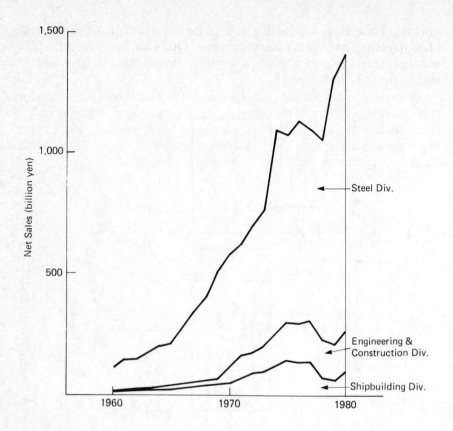

Figure 24.7
NKK's sales by divisions

ing research and development activities, and by becoming a more interesting enterprise on technology and science. At the same time, efforts are being made to improve the working environment. Intensification of pollution control measures has resulted in improved working conditions. The automation of facilities also tends to improve the work environment. We have created working conditions very different from the dusty and hot working environment of conventional blast furnace shops.

Another important factor is improvement of the working morale of employees. As you are aware, Japanese customs and practices of employment are rather unusual. Lifelong employment, and management by the Japanese seniority system, help basically to enhance loyalty to employers. In addition, Japan's steel companies actively promote mutual understanding between labour and management by constant dialogue to promote conditions for further cooperation. The voluntary small group activities among blue collar workers, called 'J-K' (*Jishu Kanri* = Quality circles – QC – or zero defect – ZD – management) activities also contribute to enhancing the will and motivation of employees.

An extensive survey was conducted in 1981 by the Japanese Federation of Iron and Steel Workers' Unions concerning the workers' job consciousness. (See Figure 24.8.) We trust our measures in this respect are having satisfactory results as expected. The results for NKK are well above the average for the steel industry at large.

1. Job motivation:	
Answer	%
Highly motivated	21.5
Fairly motivated	55.7
Hardly motivated	18.7
Not motivated	2.2

2. Loyalty to company:	
Answer	%
Wants to serve until retirement	71.1
Wants to change jobs soon	2.1
Wants to change jobs if a good job available elsewhere	6.9
Undecided	18.9

3. Attitude to J–K (quality circles) activities:	
Answer	%
Very much interested	19.2
Interested to an extent	53.1
Participating unwillingly	23.0
Not interested	2.4

Source: Inquiry survey by Iron and Steel Workers Union: June 1981

Figure 24.8
Job consciousness of steelworkers

Relations between government and steel industry

Up to this day, Japan's steel industry has received almost no assistance from the government in the industry's various measures to cope with the recent socio-economic changes. Nevertheless, it cannot be denied that this is because we in Japan are in a relatively favoured position. Namely, many of the facilities of the Japanese steel industry were constructed during the period of rapid growth of the Japanese economy in the 1960s and early 1970s to meet increasing domestic steel demand. Thus, Japan's steel mills are relatively modern and efficient. When steel demand declined, we were able to concentrate production in the more modern facilities.

Enormous funds are required for the construction and replacement of facilities in the steel industry. Besides, construction of iron and steelmaking facilities involves considerable investment for pollution control. It is very difficult for the industry to invest huge sums for replacement of old and obsolete facilities in the face of the low-cost production of newly developed countries where labour cost is comparatively low and facilities are new. To compete with steel from those countries, we think government support based on long-term policy is important. The method of such support may differ among countries according to their historical background as well as their economic and political systems. But personally, we would like to look for measures which make the best use of the vitality of private businesses to enable the pursuit of efficient management though it may involve some risks.

The Ohgishima Project

The Ohgishima project is an example of the revitalisation of a steel mill. This project was aimed at replacing our old Keihin steel works and took ten years before it was completed in 1979.

The birth of the Keihin works dates to 1912, when an open-hearth furnace and a seamless pipe mill were constructed in Kawasaki city. The steel mill was expanded gradually and by the late 1960s it had become a large integrated steel mill with seven blast furnaces and three LD converter shops with eight vessels to produce five and a half million tons of crude steel a year.

However, being located in the centre of an industrial zone close to Tokyo, it was difficult to procure adequate space. As seen from Figure 24.9, the mills and facilities were scattered in many separate areas and, besides, the facilities were small. Transportation between mills and shops was very complicated and productivity was low. With intensification of environmental regulations, it became clear that the existence of the works itself was threatened because the facilities were old and there was no space available for installation of anti-pollution equipment.

Various plans were proposed and studied. Finally a decision was made to expand a manmade island named Ohgishima and build a steel mill to replace the old facilities. The main reasons were to ensure continued close relations with customers in the area and to make efficient use of our skilled workers and affiliated companies also in this area.

The basic concept concerning the facilities was to concentrate production of hot metal and crude steel on the manmade Ohgishima island by installing fewer but larger facilities than those at the old plants and to realise the same total capacity. Also, hot rolling facilities such as hot strip mill and plate mill, and billet mill for seamless tube mills were to be constructed as downstream lines, and the facilities at the old plants were to be discarded. On the other hand, production facilities for seamless tubes and welded pipes, and the cold strip mill and related facilities at the old plants were to be improved and modernised at the existing places. Billets and hot coils, for these mills, were to be transported through an exclusive tunnel which was constructed under the sea to connect Ohgishima and the old plants. This tunnel was to be used also for connecting utilities such as electric power, gas, water and steam at the two areas under centralised control. (See Figure 24.10.)

The implementation of this project was very carefully planned in all conceivable respects. For example, labour productivity was planned to be more than doubled so this involved reduction of about 9,000 workers. During the ten years, natural attrition through retirement numbered about 4,000 and another 4,000 moved voluntarily as cadre workers to our new Fukuyama works, our company's other integrated steel mill. Thus, no dismissals were required in achieving the goal.

In conclusion, the Ohgishima project was a tremendous burden on the company. However, if the project had not been carried out, the Keihin works would not have been able to enjoy its present scale of production in view of intensification of environmental regulations, the cost aspects, and hiring an adequate number of qualified employees. In that case, perhaps, we would have decided on a different solution, but at any rate, we firmly believe that our decision for the Ohgishima project was one of the best possible solutions to meet the challenges of the decade.

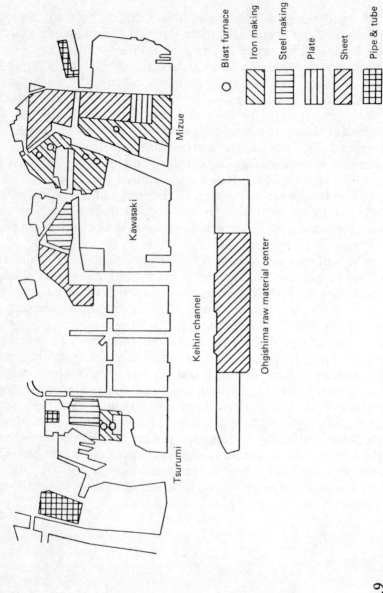

Figure 24.9
Ohgishima before

Blast furnace
Iron making
Steel making
Plate
Sheet
Pipe & tube

Mizue

Kawasaki

Tsurumi

Keihin channel

Ohgishima raw material center

458

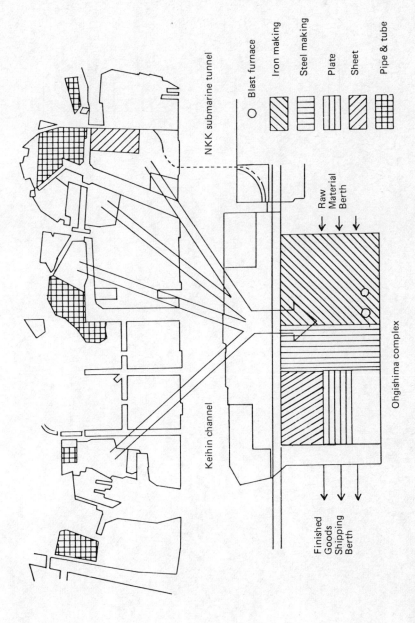

Figure 24.10
Ohgishima after

459

25 Transformation tendencies in the world steel industry and adaptive strategies

Bela Gold[1]

The painful competitive pressures experienced by most steel producers during the past decade are more likely to intensify than to ease during the 1980s. Resulting efforts to survive may be expected to reshape the structure of steel production among countries, within countries and within individual steel companies as well, while also posing major challenges to company managements, trade unions and governments.

In reviewing the major sources of pressures for change and resulting adjustment tendencies, attention will be given first to developments in input factor markets and next to developments in steel product markets. Technological innovations will then be examined both as additional sources of pressure for changes and also as means of responding to input and product market pressures. The first section of the chapter will conclude with a similar discussion of governmental policies as both sources of additional pressure and sources of responsive help. After briefly summarising the patterns of restructuring likely to result from the preceding developments, the second half of the discussion will focus on the readjustments to be considered by individual managements in order to strengthen future competitiveness.

Although this analysis is based on more than 30 years of research on the steel industries of the United States, Western Europe and Japan,[2] its purpose is merely to explore rather than forecast the readjustments likely to be encouraged by technological and economic conditions. Such prudence is counselled by the lessons of my past experience, which emphasise that the speed and extent of even seemingly necessary adaptations will probably continue to be restrained by the inertia attributable to the huge scale of established steel facilities and, even more, by their intricate interweaving with the social and political interests of their communities, regions and nations.

Major pressures reshaping the steel industry – input pressures

Availability and prices of energy inputs

There have been significant changes in the availability and price of all inputs required by the steel industry, including ore, energy, labour, capital, transportation and research. But the single most important one among these recently has undoubtedly been sharp increases in the price of energy. Such increases have been generating gradual shifts in the location of smelting operations, changes in the basic iron-producing processes employed and alterations in resulting products. And these adjustments are tending, in turn, to affect the size of steel mills, their own locations and the processes used by them to convert iron into steel.

Because of the huge energy requirements of smelting, one may anticipate a continuing gradual increase in the proportion of iron smelting operations located in areas possessing cheap energy, especially natural gas, and within reasonable proximity to large iron ore reserves, such as Australia and Venezuela. Although resulting savings have not been large enough to compel the transfer of established smelting operations, the availability of cheap oil and gas has encouraged the inclusion of smelting operations in some new steel mills – in the Middle East, for example – despite dependence on iron ore imported from distant sources.

Access to cheap natural gas has also triggered a second train of significant readjustments in the steel industry through its encouragement of increasing reliance on 'direct reduction' processes in place of blast furnaces. By eliminating the need for blast furnaces, sinter plants and coke ovens, such processes yield several major advantages, including very much lower investment requirements, less pollution and smaller plant sites. In addition, such processes achieve reasonably high levels of operating efficiency with smaller scale units and are also far more adjustable in output rates than blast furnaces, which usually run continuously for 5–8 years and are substantially rebuilt after shutdowns. The magnitude of such benefits has intensified efforts to improve direct reduction processes using other fuels. Developing effective processes using coal instead of gas, for example, would obviously improve the competitive positions of such countries as the United States and Sweden,[3] while Japanese efforts to utilise byproduct gases from nuclear power plants for direct reduction would clearly help to ease their severe energy disadvantages.

The resulting increase in the availability and price competitiveness of direct reduced iron also encourages increased reliance on electric arc furnaces in place of basic oxygen or the older open-hearth furnaces.

Such cold iron-electric furnace mills offer high productive efficiencies in relatively small capacity units, reduce investment requirements per ton of capacity by 75–85 per cent below those for large integrated mills, enable managements to vary scrap usage more widely to take advantage of fluctuations in scrap prices, and also permit more exact control of steel qualities.[4] This development facilitates increasing the proportion of steel output supplied by smaller plants, which may also be more widely dispersed to better service localised markets and which can also concentrate on narrower product mixes to gain the advantages of specialisation.

Of course, increases in the price of energy have increased pressures to conserve energy in the remaining steel production processes as well. One major contribution has involved adoption of continuous casting machines, which not only permit fuller utilisation of the heat contained in the molten steel to facilitate the following shaping process, but which also increase the yield of resulting products by reducing the losses common to the traditional processes which they replace.[5] The development of continuous billet casters has further enhanced the competitiveness of relatively small steel mills. And new developments, including horizontal continuous casting machines, promise to enlarge the array of products which can be made economically in small mills. In large steel mills, billet casting machines have been supplemented by slab casters, which offer similarly significant improvements in energy conservation and product yields in comparison with the long prevailing reliance on the teeming-reheating-slab mill sequence of producing larger shapes. It should be noted, however, that the shift to slab casters has been significantly slower than the growth in reliance on billet casters.

Availability and prices of labour inputs

The primary pressures on the steel industry from the labour market have been of three kinds. The most influential has been the continuing pressure for increasing wage rates and associated benefits. This has more than offset concomitant reductions in manhour requirements per unit of output, thus increasing unit wage costs in most countries, and especially in the United States and Western Europe.[6] In these areas, there has also been increasing labour resistance to reductions in employment. This has resulted in overmanning even what would otherwise be reasonably competitive facilities – as in England's modern Anchor Project – and in continuing the subsidised operations of submarginal mills in various parts of Europe. A further effect of such resistance has been to discourage the adoption of technological innovations which would increase competitiveness, but only if the reduced labour requirements which they make possible could be realised.[7] A third form of pressure from the labour market has involved increasing unwillingness to man night shifts, to accept hot or dirty jobs, or even to accept responsibility for the increasingly precise and timely responses required to ensure efficient, continuously high quality production by modern technologically advanced facilities.

Such pressures have obviously stimulated efforts to reduce labour requirements through increasing mechanisation, instrumentation and computerisation. Resulting advances have tended to reduce labour requirements per unit of output. They have also tended to reduce average skill requirements by reducing those required for most of the remaining direct labour, while necessitating relatively small increases in the number of more highly skilled operators of sophisticated equipment and in the size of maintenance staffs. The enormous magnitude of the net changes which may be required to ensure market competitiveness is suggested by the fact that the output per man in major Japanese plants is more than double the levels in the top 50 per cent of US and Western European plants, and the record of the leading Japanese plant of 1,800 tons of finished product per worker in 1980 is more than three times that of the best integrated steel plants in the US and Western Europe – and these cost advantages are further enhanced by lower wage rates and social benefits.[8] Small steel mills have also benefited from such labour-saving innovations, as indicated by the fact that, despite the long-vaunted scale economies of giant steel mills, Nucor Corporation, one of the leading 'mini-mills' in the United States, is currently approaching an annual output of 1,000 tons of finished product per worker, far above that of the largest and best steel mills in the country. Another feature of such plants which helps to attract and keep a desirable quality of labour is that electric furnaces readily permit two-shift operations and weekend shutdowns in contrast to the necessity for the uninterrupted operation of blast furnaces.[9]

It would be very difficult, of course, to effect the far-reaching changes in technology and facilities needed to achieve such sharp decreases in manhour requirements per unit of output in long-established mills, or to gain labour acceptance of resulting massive reductions in employment levels. Hence, the drive to regain competitiveness in countries subject to such labour pressures has involved considerable emphasis on building new mills in new regions, thus avoiding the burdens of established labour restrictions as well as of old facilities. Such efforts have certainly contributed to the construction of the few large new mills in Western Europe and the growth in the number of small mills in the US as well as elsewhere during recent years. Whether these undertakings remain within their previous national boundaries, as in Northern Italy, or involve the development of facilities abroad, may well depend, in the case of privately financed projects, on managerial evaluations of whether national and regional governmental requirements and trade union pressures undermine realistic opportunities to match the achievements of foreign competitors.

Availability and prices of capital inputs

Continuing sharp increases in the prices of capital goods and in interest rates, combined with increasing competition among industries and countries for available investment funds, have tended to inhibit the major additional investments in the steel industry needed to take advantage of recent technological advances as well as to restrain cost increases. In the absence of viable alternatives for avoiding an accelerating erosion of competitiveness, however, continuing investments have been necessary everywhere. And the associated need to maximise resulting benefits have encouraged reliance on widely similar strategies. In most long-established steel-producing countries, these have centred around three components: closing the oldest plants; partially modernising the better large plants; and adding some new plants.

Eight to ten years ago, new plant construction reflected substantial commitments to imitating, or even duplicating, the large-scale facilities so successfully pioneered and utilised by the Japanese. But such efforts have been sharply curtailed more recently in favour of smaller plants requiring smaller investments, energy and labour inputs per ton of finished goods capacity, through utilising the cold iron-electric furnace-continuous casting model combined with extensive computerised controls and automated materials handling.

This does not mean that large-scale plants are no longer economic. On the contrary, modern large Japanese steel mills are still the world's most efficient in virtually all respects. What needs to be more clearly recognised is that the margin of superiority of such plants is greatest in respect to producing huge quantities of large products involving long production runs of most orders, such as heavy plates and large coils of sheet. In respect to smaller, simpler products such as wire rod, the superiority of large mills is more open to question. In short, scale economies vary widely among different processes and different products. Current capital market pressure may be expected to lead to much more intensive analyses of the magnitude of such advantages for different steel operations and to result in more highly specialised plants representing combinations of processes and product mixes which promise maximum competitiveness in the particular markets for which they are designed.

Despite the increasing cost and reduced availability of capital, it should be recognised that prospective adjustments in steel plant operations are likely to involve increases rather than decreases in the ratio of capital to labour requirements. This need not, however, result in higher average unit costs and may not even involve a substantial increase in cost inflexibility. Such deviations from common expectations may be due, in part, to the fact that continuing technological progress tends to enhance the contributions to production of facilities and equipment far more than those of labour, thus making capital goods inputs progressively more economical relative to labour inputs.[10] Moreover, although capital goods prices and wage rates both rise during inflationary periods, the prices paid for the former stop rising as soon as they are purchased, while wage rates obviously continue to rise after workmen have been hired. On the other hand, the depreciation charges of new equipment may decline steadily under some capital recovery formulae and the net effects of any increased capital costs on total unit costs may be offset by the extent to which former labour inputs have been replaced.[11] Still another factor tending to increase the relative economy of capital inputs is the seemingly irreversible trend towards increasing payments to labour for non-working time, including: lay-offs; holidays; vacations; sickness; and pensions.

It is also important to recognise the need to modify traditional concepts of capital charges as 'fixed costs' and wages as 'variable costs', meaning that the former are unaffected by fluctuations in output, while the latter vary directly with them. Labour costs have obviously become much less 'variable' as a result of trade union resistance to reductions in employment and wage rates during recessions combined with the increasing penalties for lay-offs resulting from 'social benefit' requirements. At the same time, growing attention has been given to increasing the variability of capital charges, partly through allocating equipment costs at a fixed rate per machine-hour used and partly by adjusting depreciation rates according to the percentage of capacity utilisation. In order to correct some long-distorted perspectives concerning the prospective effects of these tendencies toward increasing the relative role of capital inputs, it should be emphasised that the widespread characterisation of the steel industry as 'capital-intensive', meaning that capital charges account for a major proportion of total costs, is quite mistaken. It is true, of course, that huge investments are required, but it is also true that these provide enormous productive capacity. As a result, capital charges tend to account for a much more modest proportion of total costs than is generally realised, averaging well under 9 per cent in the United States, for example.[12] A more meaningful, as well as a more correct, characterisation of the steel industry would be to call it a 'capital-dominated' industry, a term which I have used to emphasise that, even though its capital charges are relatively small, the scale and technological modernity of its capital facilities and equipment are the overwhelming determinants of a mill's production capacity and also of the effectiveness with which it can utilise labour and materials.[13]

Major pressures reshaping the steel industry – other major pressures

Pressures from customers

The most commonly recognised pressures from customers have traditionally emphasised minimising the costs of their purchases, meeting their quality requirements and assuring reliability and convenience of supply. But each of these has also assumed some modified forms in recent years. In respect to the first of these, the demand for minimum prices has been broadened to include a search for improvements in material characteristics which permit using smaller quantities to meet the same needs through improvements in strength or rigidity or other characteristics. In respect to the second, the increased automation and speed of manufacturing processes has intensified customer demands for highly consistent quality, lest variations interrupt the continuity of production operations through jamming machines, damaging tools or increasing product reject rates. In respect to the third, customer requirements have become more stringent as a result of efforts to reduce material inventories without jeopardising the continuity of production – the ultimate form of which has emerged in the Japanese 'Kanban system' of having suppliers deliver inputs on an almost continuous basis.

A second, and still growing, form of pressure involves seeking increasing contributions from material suppliers towards reducing the customers' own manufacturing costs. This frequently involves efforts to reduce processing needs by seeking material inputs which are more closely in accord with the product's final dimensions and configurations. Because of the recent development of many new processes for shaping and fabricating metals, customers are also pressing for improvements in such qualities as ductility and weldability to permit the utilisation of such faster, cheaper, and sometimes more precise, technologies.

Still another form of customer pressure involves seeking to improve the qualities of their own products by expanding the specifications to be met by materials suppliers. Long-standing demands for continuing improvements in durability and corrosion resistance have been supplemented in many cases by requirements for better surface quality and other service capabilities.

Increasing competition

Competition in the world steel industry has been intensifying during the past 15 years as capacity has expanded more rapidly than consumption except for relatively brief periods. The initial major increases in capacity came from the extraordinarily rapid growth of the Japanese industry. This was followed by the construction of some giant new mills in Western Europe seeking to copy the unexpectedly great competitive advantages of the large-scale Japanese mills. Although such modern additions rendered many older facilities submarginal, political pressures prevented their closure. Excess capacity was further increased as newly industrialising countries began to build steel mills of their own, partly as a basis for developing steel-using manufacturing industries and partly in the hope that their lower wage rates and, in some cases, advantageous access to needed inputs would enable them to gain some export markets as well. The disparity between capacity and consumption has also been increased as a result of successful efforts by substitute products, such as aluminium and plastics, to invade some steel markets, including automobile parts, piping and beverage containers. And the most recent addition to these competitive pressures, as noted earlier, has been the growing number of 'mini-mills' in the major steel-producing countries as well as in the more recently developing areas.

As a result of such developments, the world steel industry is likely to face three major changes in its international markets. First, the ratio of finished steel exports to world steel consumption will probably decline significantly because of increasing production for home use by newly industrialising countries. This trend is already under way and may well be intensified. Second, the composition of exports is likely to change in favour of its most and least sophisticated products. Exports of crude iron and steel as well as of relatively simple intermediate products from newly industrialising countries are already increasing and likely to increase much further for reasons discussed above. In addition, one may also anticipate substantial increases in the ratio of speciality steel exports to the total for several reasons. Perhaps the most important influences are the continuously rising technical requirements in virtually all of the long-established steel-using industries, including transportation, construction, machine building, extraction and manufacturing of consumer durable goods. Additional rapidly growing sources of pressure in this direction are the challenging requirements of a variety of relatively new industries, including coal gasification, deep oil well drilling, pollution control and others. Moreover, even the newly industrialising countries are likely to need progressively more sophisticated steels to support their development programmes.

Third, competition in international steel markets will grow more intense than ever. In part, this will result from the desperation policies of firms on the edge of bankruptcy and in part from government subsidised operations in older producing areas seeking to minimise domestic unemployment. But it will result in even greater measure from the combined impacts of new plants embodying more economical technologies for serving various market segments, as well as of new mills in developing countries which seek to export the excess of productive capacity over their, only gradually expanding, domestic consumption. The consequences for at least the next several years are likely to involve both product prices yielding only limited profit margins to most producers and also continuing resort to indirect, if not overt, restrictions on imports deemed to be gaining excessive proportions of domestic markets.

Pressures for pollution control

The primary targets of governmental pressures to reduce air and water pollution from steel mill operations have been blast furnaces, sinter plants, coke oven batteries and steelmaking furnaces. Because of the large scale of such units in integrated steel mills, minimising resulting pollution requires very large capital expenditures. In view of the poor financial performance of many of their plants, Western European governments have apparently been less demanding in the standards which have been enforced than in the United States. But even the latter have been surpassed by the demands which have had to be met by modern Japanese mills. It has been estimated that the unexcelled pollution control facilities in Ohgishima, the most recently completed integrated steel works in Japan, accounted for up to 20 per cent of total plant investment.

Efforts to meet such pollution control requirements have not only absorbed significant proportions of capital expenditures which might have been used for plant modernisation and expansion, but have also led to the re-direction of much research and development activity. Such commitments have been unavoidable for older plants in regions which have experienced the raising of these standards in recent years. But the prospect of having to allocate large amounts of scarce capital to facilitate which do not increase the volume of marketable products is clearly have a major influence on management planning in respect to the location and component operations of new facilities. The most extreme re-adjustment, encouraged by considerations of energy costs and investment requirements as well as pollution control considerations, as was noted earlier – involves buying direct-reduced iron and thus shifting the entire complex of iron-making operations elsewhere. More limited expedients of this kind to ease the problems of existing integrated mills may be illustrated by increases in the importation of coke and the successful experience of Kawasaki Steel in building a sinter plant in the Philippines to supply its Chiba Works near Tokyo and thus respond to the strong local agitation to reduce pollution emissions.

Major pressures reshaping the steel industry – technological pressures and responses

Pressures for increasing scale

Until the early 1970s, most important advances in steel industry technologies tended to yield increasing economies of scale. One of the most dramatic examples after 1950 was the Japanese pioneering of blast furnaces several times larger than had earlier been regarded as optimal.[14] Although this effort was motivated largely by the fact that the Japanese were under earlier and much greater pressure to economise on fuel, in view of the virtual lack of such domestic resources, continuing efforts to maximise the efficiency of blast furnaces also yielded economies of investment and labour and space. And the resulting increasing flow of molten iron was accommodated by increasing sizes of basic oxygen steelmaking furnaces and faster rolling equipment.

The consequent emergence of a balanced sequence of individually large capacity operating units, including parallel improvements in materials handling and finishing operations, yielded substantial economies in comparison with smaller scale integrated steel mills. Hence, western steel producers seeking to overcome the sharply enhanced competitiveness of Japan's new giant steel mills undertook to move in the same direction through huge new plants in Britain, France, Germany and Italy as well as through the addition of large units to American plants.

Contributions to the economies of smaller plants

Perhaps the single most important contribution to improving the competitiveness of relatively small steel mills was the adaptation and utilisation of electric arc furnaces to make carbon steels. Originally utilised to make alloy steels, and in wide use for that purpose for decades, such furnaces were increased in size and power and provided with improved refractories to become cost-effective producers of carbon steels from inputs of steel scrap, except during extraordinary peaks in scrap prices. Because such furnaces are readily adapted to using inputs of direct reduced iron as well, their role in carbon steel produc-

470

tion has been increasing steadily since the early 1960s – quadrupling such output in the United States, for example.

A second major contribution to the economies of small steel mills was the development of highly efficient multi-strand billet casters which, like their larger counterparts among continuous casters, yield substantial benefits in energy savings and yields. These intermediate products are then put through progressively improved rolling mills which produce small cross-section products such as bars, rods and simple structural shapes that can be cost-competitive with similar products made by large mills, as well as comparable in quality. Indeed, only the US Steel Corporation, the largest in the United States, produced more wire rod in the 1980s than Nucor Corporation, which operates three scrap-based mini-mills averaging less than 600,000 tons of annual capacity per plant. Of course, such basic facilities have been increasingly integrated, like large mills, by continuing improvements in materials handling and in production management systems.

As a result, the competitiveness of such mini-mills, which average about one-twelfth the capacity of leading integrated steel plants in the US, is indicated by the facts, which bear repetition, that they can be built for less than one-fifth the investment cost per ton of capacity for integrated mills and may require only half as many manhours per ton of finished product output.

It should be emphasised, however, that technological advances are continuing to increase the capabilities and cost-effectiveness of large scale integrated plants as well as of small-scale plants using cold iron or scrap inputs. Thus, the Japanese are continuing to increase the economies of giant steel mills by progressively reducing energy and labour inputs and also safeguarding their major advantages in investment requirements per ton of capacity. But the economies of such plants seem to be most marked in respect to a relatively limited array of large products, such as heavy plate, large structurals and big coils of sheet, to be provided in substantial tonnages. Hence, the construction of such additional giant plants is likely to be quite limited in numbers over the next decade outside the USSR and China.

On the other hand, the advantages of smaller plants which can be more highly specialised in product mix and dispersed more broadly to ensure closer access to each market, without incurring serious cost disadvantages, have stimulated intensified technological efforts to enhance the range of products which can be made on a small scale. Thus, horizontal continuous casters are being developed to make small plates and some steel sheet and strip products. Other processes are being developed to produce powder from direct reduced iron and to form it into ribbons, wire and other small products.[15] Finally, the development of relatively efficient small-scale steel plants may encourage renewed exploration of the possibilities of building 'captive' mills designed to serve the specific product requirements of one or a few major customers with closely similar needs.

Contribution to improving production management and products

The most important technological contribution to improving the effectiveness of steel mill management has been the development of increasingly comprehensive systems of computerised controls. Although the initial applications were limited to process controls applied to individual pieces of equipment, continuing advances in computers, instrumentation and software have encouraged the integration of successive operations, the development of increasingly effective production planning, control and inventory systems, and the linkage of such operating systems with accounting, order-entry, delivery and other management controls.

The major results have been to facilitate progress towards the increasingly continuous flow of work through successive operations, to improve product quality through increasingly knowledgeable and precise control of all processes, and to reduce energy and labour requirements. Such benefits have been available to small as well as large plants, although the greater complexity of the latter has meant slower progress towards achieving centralised control, especially in western countries.[16] Another important dimension of technological development has involved the design of increasingly efficient facilities for producing specialised products, especially in smaller capacity units during recent years. These include pipe and tube mills, rail mills, rod and bar mills, and various specialised facilities for coating and heat treatment. Combining any of these with the cold iron-electric furnace-continuous casting sequence already discussed has given impetus to an accelerating trend towards building such small specialised mills in the various regional markets with the greatest demand for each type of product. By also computerising the operation of such mills, major strides will be made towards realising the increasingly automatic factory – an advance that would be further facilitated by increasing the length of production runs through greater emphasis on the use of standardised products within each product line.

In the past, product development efforts concentrated primarily on improving the commonly required attributes of steel products, such as strength, hardness, flexibility and corrosion-resistance. Additional advances in such service capabilities continue to be needed by steel users seeking to further improve the performance of their products while also reducing the costs of making them. Thus, the pressure for weight reduction was never greater in the transportation industry. Similarly,

472

the introduction of new production and fabrication methods in manufacturing has increased the need for qualities such as greater ductility and weldability. And new application in nuclear power generation, medical devices, high temperature processes, ocean resource development, and space vehicles have also presented new challenges in respect to durability and corrosion-resistance.

These considerations call attention to two opposing directions of needed developments. One involves the need to re-emphasise commitments to basic research so as to more fully explore the scientific foundations of the potentials and limitations of steel and other ferrous materials, as well as the potentials of combining them with other metals and non-metallic materials. Such efforts would obviously tend to lead to an increasing proliferation of special materials to meet the highly differentiated needs of older, as well as newly emerging, industries. But the second issue involves renewed efforts to harness one of the fundamental principles of economy in all manufacturing industries: the periodic consolidation of wildly multiplying numbers of product types, qualities, configurations and sizes into a more limited array of standardised categories.

Standardisation tends to reduce manufacturing costs through facilitating the specialisation of equipment and increasing their utilisation through longer production runs. And the buyers of such standardised products get the benefits of lower prices for such inputs as well as assurance concerning their minimum service capabilities. This obviously need not curtail user efforts to determine the optimal specifications for their steel requirements, but it would encourage more careful evaluation of the prospective benefits of requiring non-standardised products relative to the higher prices to be paid for them. Thus, these dual pressures for greater standardisation as well as for greater differentiation of products may encourage some decreases in the product mix range of even large plants, while also increasing the degree of product specialisation of smaller plants – with the greater number concentrating on standardised products, and the remainder providing the wider array of specialised products needed for more distinctive applications.

Major pressures reshaping the steel industry – governmental pressures and responses

Government policy objectives

In most countries, government policies affecting their steel industries have been dominated by four objectives. Maintaining as much employment and foreign exchange earnings as possible have been paramount in most long-established western steel-producing countries, while newly industrialising countries have sought to increase such employment and earnings. Most countries have also sought to maintain or build a domestic steel industry to support their defence requirements and to assure the continuing availability of at least an acceptable minimum level of supplies for their steel-using industries. In addition, some governments have also sought to support labour demands for higher wage rates and larger social benefits, while others have actively resisted increases in steel prices on grounds of controlling inflationary pressures.

Implementation of policy objectives

Efforts to promote such objectives have led to a wide range of measures, some of which have tended to be in conflict with one another.

Obviously, the basic approach to increasing or to maintaining employment is to increase sales, which requires offering competitive or better products at competitive or lower prices. But it is precisely the firms which were unable to match the terms of new offerings in foreign and domestic markets that confronted decreases in sales and earnings. In order to help such firms, governments relied on a two-step strategy. In the short run, this called for reducing prices to whatever levels were necessary to retain some targeted market share, even if this involved selling below cost, in the expectation that governmental aid would ease resulting financial repercussions. Over the longer run, it called for reducing production costs and improving product quality through advances in technological capabilities. This involved urging existing firms to combine in the hope of gaining economies of scale and also providing the established sectors of the industry with investment funds at little or no cost in order to modernise some facilities in older plants and to build some new plants.

But such expedients have generally failed to achieve their intended basic employment and earnings objectives. In part, this resulted from the fact that merging failing companies seldom produces a more effective firm, unless massive cuts in employment and older facilities are combined with changes in personnel and policies as well as substantial new resources. Moreover, increases in the capacity of the modernised and new plants together with their improved performance potentials further intensified the pressure to close down the least competitive plants. It was also apparent that the cost advantages offered by the technologically upgraded facilities could not be realised unless resulting decreases in labour requirements were effected. Faced by powerful political pressures to minimise such employment reductions, both in old plants and in recently improved plants, and also to support demands for wage increases to compensate for inflation, governments continued to support sales at unprofitable prices and to support excessive manning levels. Indeed, the net cost of losses of steel sales is regarded by some governments as less than the cost of the unemployment benefits which would have to be paid to those dropped from their jobs and therefore producing no saleable products at all. At any rate, the political risks of increased unemployment to incumbent governments in democratic countries seemed even more obvious. Hence, when even the foregoing forms of government help proved inadequate to prevent imminent bankruptcies, the governments have assumed varying degrees of control and ownership to maintain operations.

It is also apparent, however, that the pursuit of such policies by several governments would tend to intensify the problems faced by each through maintaining excessive production and thus encouraging even more costly price reductions. This has stimulated intergovernmental efforts to negotiate agreements to limit imports and, in the case of the European Community, to fix minimum prices, to reduce subsidies and to close obsolete facilities. Such efforts may help, albeit slowly, to reduce inefficient capacity among the large plants producing traditionally broad arrays of products on which governmental support efforts have been concentrated. Whether they will also help to restructure their steel industries effectively to the pressures which have been reviewed is less clear.

Prospective responsive strategies for steel companies

By what means could steel producers most effectively strengthen their competitive positions within the complex of pressures which has been reviewed? It would certainly be presumptuous for an academician to recommend particular policies and actions in respect to the critical issues which must be confronted. But little harm can be done by relying on three decades of research on the steel industry to review relevant alternatives within broader perspectives than commonly dominate the thinking of officials under the constant hammering of immediate urgencies.

One of the most important requirements of a programme to regain and safeguard competitiveness is to clearly identify the magnitude of needed improvement goals over the next 5 to 10 years. Inadequate goals can only encourage inadequate policies of implementation. Yet it is my impression that such is currently the case in major sectors of the Western European, as well as the American, sectors of the industry. The reasons given for such limited goals are understandable enough. They include: the high cost and insufficient availability of capital; the threat of powerful trade union and broader political resistance to advances likely to entail substantial reductions in employment and even plant closures; and the likelihood that more efficient plants would be prevented from gaining significantly larger market shares through governmental aids to less efficient producers both in domestic and in foreign markets.

Planning within such long-prevailing constraints is bound to counsel concentration on essentially modest improvement targets. And this may well be 'practical', in the sense of helping to minimise current resistance to such targets from the variety of groups who might feel threatened by bolder objectives. But policies designed to achieve such minor gains are less likely to restore international competitiveness than to decrease it further – and thus intensify the burdens which have already been imposed. A programme to achieve major gains in competitiveness would have to begin by no longer asking: how much more can we accomplish within the restrictions and perceptions which have long placed a ceiling on the practical limits of planned changes? Instead, planning should begin with the question: what performance levels must we reach in order to match our major competitors in domestic markets, and in selected foreign markets, in 1990 and 1995?

I hope that my efforts to provide general guidelines for responding to this question will be recognised as intended to help clarify the nature and magnitude of the tasks to be confronted, rather than as ill-informed criticism of current responses under all-too-apparent urgencies, or as ignorance of the laudable efforts of some sectors of the industry.

What should be the magnitude of improvement targets? Performance levels obviously differ among steel mills depending on such factors as the modernity of technology, the scale of facilities, the product mix and capacity utilisation rates. Moreover, measures of physical performance must be adjusted for factor prices in order to estimate differences in cost competitiveness. It is also necessary to identify which competitors are to be the foci of performance comparisons. After all, the critical competition is not between entire national industries, but among the most efficient current and prospective producers seeking to increase shares in the specific regional and product markets of greatest concern to each management. Moreover, such appraisals must reach beyond the current capabilities of prospective competitors in order to make allowances for the continuing improvements which many of them are actively striving to achieve.

In order to avoid excessive generality in the following discussion, I shall concentrate on the improvement targets which would have to be reached by leading American integrated steel producers merely to avoid falling further behind leading Japanese competitors. Planning to surpass them does not seem to be a feasible target within the perspective of the 1980s. It is recognised, of course, that improvement targets resulting from such a narrowly oriented analysis may be of varying relevance to different steel producers.

Setting a 'labour productivity' target

A study in 1977 of the five largest integrated steel mills in Japan found that they averaged almost 600 tons of finished product per worker (including 'contract labour') at 90 per cent of capacity. This was well over twice the US average. Moreover, the best Japanese plant approximated 1,000 tons per worker, again well over double the record of the best US plant.[17] But the continuing intensive drive of the Japanese for further improvements is illustrated by the recent report that their newest integrated mill (Ohgishima), which had had a target of 1,000 tons per worker in 1977, actually produced 1,800 tons per worker in 1980.[18] This was about three times the level at Burns Harbor, one of the best plants in the US. Thus, despite recent domestic improvement efforts, the gap is not declining.

What improvement targets would be required for the US in order to prevent further declines in this sector of competitiveness? Even 900 tons per worker in 1986 would probably approximate less than one-half the Japanese level, and a target of 1,200 tons in 1991 might similarly yield no reduction in the gap. But effecting such increases in the US would require enormous changes in the industry, to say nothing of the serious social problems which would be generated if employment in this major industry were to be reduced by more than one-third within five years and by nearly two-third in ten years.

Even such improbable reductions in manhour requirements, however, would probably keep unit wage costs in the domestic steel industry higher than in Japan, because their total annual earnings and benefits per worker are likely to remain significantly below comparable levels in the US despite their very much higher output per worker.

Setting targets for improving energy consumption and capital requirements

The 1977 study cited above also reported estimates of total energy consumption per ton of output for the integrated Japanese steel industry averaging 30 per cent below US levels. But intensive efforts have also been under way in Japan to keep reducing such inputs further, and they have also made greater headway in replacing expensive oil and gas consumption. Hence, although US energy consumption rates have also improved in the steel industry since 1977, competitive targets for domestic producers may well require additional restrictions of at least

25 per cent within the next 5 years.[19]

Japan's integrated steel mills were built for about 40 per cent of comparable investment requirements per ton of capacity in the US. Even the latest mill built (Ohgishima) cost only about $666 per ton of capacity,[20] including about 20 per cent for what may well be the best pollution control system of any steel mill in the world. This compares with current US estimates for greenfield plants of $1,200–1,500 per ton of capacity for integrated mills. The Japanese investment advantage has been further enhanced by their lower interest rates and faster depreciation allowances. Hence, reductions of at least one-third in investment requirements per ton of modernised and new capacity would seem necessary to prevent increasing disadvantages in this sector.

Implications of these speculative targets

The preceding improvement targets are shockingly large, especially when compared to the modest gains achieved by the industry in recent years. In view of the current cost disadvantages of the domestic integrated steel industry, however, and in view of the continuing advances being made by foreign producers,[21] even the targets which have been suggested may prove insufficient to strengthen the industry's competitive position. At any rate, the critical need is to recognise the urgency of such farreaching improvement targets as the starting point for planning a revitalisation programme of adequate proportions. Continued acceptance of recent improvement rates can only endanger the future competitiveness of individual steel companies and of the industry.

The powerful potentials of commitment to major advances is well illustrated by the Japanese steel industry. In the late 1950s, despite their lack of iron ore, fuel, modern plants and engineers experienced in the latest technologies, and despite their severe shortage of capital, the Japanese undertook to match or surpass the world leaders in steel within 15 years – and drove to a spectacular success. By way of contrast, it may be recalled that the Council on Wage and Price Stability advised the President of the US in 1977 that it would be uneconomical for the domestic integrated steel industry either to modernise or to expand[22] – thus consigning to progressive decline an industry which was probably still the second most efficient in the world, which possessed abundant supplies of raw materials, and excellent scientific and technological specialists, and whose domestic market was still the best in the world.

If the nation depends on such foresight and leadership, the outcome is not difficult to predict. Hence, if the energetic, sustained and farreaching efforts necessary to strengthen the industry's international competitive position are to be mounted, the requisite leadership must come from company managements.

How can the suggested targets be reached?

The enormous magnitude of the improvement targets which have been suggested makes it impractical to seek their achievement through separate programmes for each. Instead, a comprehensive programme offering interrelated contributions to each is needed. In reviewing possible elements of such approaches, the following analysis will not cover needed supportive actions by government. Because such needs have already been the subject of prolonged and voluminous discussion,[23] the primary concentration here will be on what could be done by company managements.

Increasing the cost-competitiveness of current production operations

The most promising approach to increasing the cost competitiveness of production operations by 25 per cent within 5 years would seem to require closing the least cost-competitive 20 per cent of current capacity within 3 years, while increasing the capacity of remaining facilities by 25 per cent within 5 years. Lest such an improvement rate be considered unfeasible, it should be noted that gains of that order are not at all uncommon in Japanese mills.[24] Accordingly, attention will be turned to some of the sources which would contribute to achieving such a target.

The single most promising means of generating such an improvement may be to increase the size and upgrade the quality of the steel companies' engineering staffs and to set them annual improvement targets of 5 per cent in plant capacity, energy use, product quality and output per manhour *without* major changes in technology or facilities. This would require such staffs to undertake newly intensified analyses of current constraints on each of these aspects of performance along with careful evaluations of alternative means of easing them. It would also require more intensive efforts to bring performance in all plants up to the best instead of continuing to tolerate long-prevailing differentials. Some headquarters officials may feel that both of these proposals have long been carried out. My own plant level studies, however, raise doubts about the adequacy of such efforts.

A second, promising, source of needed improvements would involve the introduction of systematic and continuing programmes to upgrade the understanding of, and support for, productivity improvement programmes by intermediate levels of management and senior technical personnel as well as labour. It is all too often overlooked that production managers and their associated technical staff may feel that proposed innovations in processes or practices threaten their hitherto respected expertise and experience, that any associated organisational adjustments could undermine their status and authority, and that disruptions to production during the introduction of innovations would be likely to impair their performance, with attendant penalties. Resulting insecurities could engender resistance in the sophisticated and influential forms of offering senior managements, presumably expert, judgements which tend to discourage the adoption of proposed innovations and, at a later stage, of highlighting both the shortcomings of resulting acquisitions and attendant problems of gaining effective cooperation from labour in utilising them. The importance of gaining labour support, and some of the means for doing so, will be discussed later.

In order to help reinforce the preceding two means of improving the effectiveness of operations, most firms would benefit from the introduction of a more comprehensive productivity measurement and analysis system than is currently widespread. It should cover the interacting contributions of improvements in the utilisation of materials, energy and capital as well as of indirect and direct labour, in addition to identifying the effects of substitutions among such inputs. Such a system would also facilitate determining the location within the plant of all significant changes in productivity relationships, identifying their causes and tracing their effects on costs and product quality – along with their eventual impacts on profitability. Resulting evaluations at the level of cost centres and product lines, as well as at the level of plant aggregates, would enable management to uncover the operations which are progressing and those which are lagging, along with the factors responsible for such outcomes, as the basis for setting future improvement targets and appraising subsequent progress towards their realisation.[25]

A fourth potential means of increasing the cost-effectiveness of production might involve reducing the range of product lines made in each plant and also increasing the concentration within each product line on standardised categories. Maximum utilisation of the large investments in steel mills has been hampered both by the long-prevailing efforts of producers to supply as many as possible of the variety of steel products required by customers and also by the increasing tendency of buyers to submit orders with highly individualised specifications. But both of these run counter to two of the major sources of rising efficiency in most manufacturing industries: increasing the specialisation of equipment and processes; and contracting endlessly proliferating product specifications into a relatively limited number of standardised categories.

Steel mills have, of course, harnessed some of the benefits of specialisation through utilising a wide array of specialised facilities within most integrated mills, including rod, bar, rail and tube mills and various types of coating lines. But, as was noted earlier, the possibility should be explored that further economies might be achieved through the greater dispersion of such specialised operations in the form of separate plants located closer to the primary markets for their respective products. As for increasing standardisation, it is conceivable, of course, that every customer could justify each of his distinctive specifications. However, it seems even more likely that offering products which fall into standardised categories at significantly reduced prices might induce many buyers to modify their former requirements – thereby tending to increase the length of production runs, to reduce equipment downtime and to lower finished goods inventory requirements.[26] Finally, in addition to seeking such improvements in existing operations, urgent consideration needs to be given to hastening the utilisation of already available technological innovations capable of major contributions to cost-effectiveness. Foremost among these are continuous casting and the increasing computerisation of steel mill operations and related management controls. Planning for such additions to present mills might then be supplemented by evaluating prospective additions to capacity, including the possibilities of adding some small and intermediate-scale plants with narrower product specialisations and located closer to relevant markets.

Increasing the effectiveness of non-production operations

Effective efforts to advance firms' technological capabilities requires that measures such as those suggested above be reinforced by several concomitant changes in other parts of their organisations.

One of the most important of these would involve increasing the scope and effectiveness of corporate planning. In broad terms, this would require supplementing its long-prevailing emphasis on financial measures and targets, as well as its more recently added concern with market prospects, with a detailed analysis of prospective changes in the competitiveness of the firm's products and production capabilities. More specifically, the setting of sound future improvement targets for each of these should be preceded by:

a)   determinations of the firm's current advantages and disadvantages in respect to the competitiveness of its technologies, production facilities, operating costs and products;

b)   appraisals of the effects on these determinants of competitiveness over the next 5 years of prospective advances by competitors, of prospective pressures in product and factor markets, and of prospective changes in relevant government policies; and

c) estimates of the prospective benefits and costs of alternative improvement efforts in respect to each of these determinants of future competitiveness.

In short, it is difficult to understand how the most promising bases for improving future competitiveness can be identified without prior determination of the nature and magnitude of the firm's current and prospective advantages and disadvantages. Yet, it my impression that few major domestic steel companies have made thorough and realistic evaluations of their current competitiveness, and of the adequacy of the contributions to be expected of these planned improvements.

A second critically important area for reinforcing efforts to improve technological competitiveness involves improving the quality of the managerial and techonological manpower being recruited. There seems to be a serious imbalance between the heavy involvement of top management in attracting and guiding the allocation of capital resources and the relatively low-level and essentially routine efforts relied on to recruit most additions to the technical and managerial staffs. Yet the benefits of financial commitments depend very heavily on the aspiration levels, abilities and energy of the personnel entrusted with them. Surely, lower productivity levels, increasing reliance on foreign technological developments and more limited success in utilising such innovations[27] must be attributable, in significant measure, to limitations of technical and managerial performance.

But domestic steel companies have been severely weakened, relative to more rapidly growing and seemingly more exciting industries, in seeking to attract a reasonable share of the most promising younger (and even not-so-young) scientists, engineers and management specialists. As a result, it may require top management involvement to develop programmes offering intellectual challenges, financial incentives and promotional opportunities sufficiently attractive to bring in enough aggressive, imaginative and high-powered human assets to maximise realisation of the potentials embodied in the capital assets.

It should be noted, however, that any resulting success in recruiting may be quickly liquidated if the newcomers are then expected to endure the same long drawn-out sequence of slowly progressing and narrowly restricted job assignments as has been customary – and if they are expected to accept the long-established explanations of why they must keep waiting patiently until the time becomes ripe to make any of the changes that they may suggest.

A third essential means of reinforcing efforts to improve technological competitiveness involves increasing allocations to research and development programmes, especially those seeking major advances in production technologies and in product capabilities. Surely the decline in technological competitiveness has its roots in the inadequacy of past commitments to longer-term and recognisedly more risky basic research. Unfortunately, this seems to be continuing, with attendant increasing reliance on buying foreign technology and related consulting help.

Such policies have two serious weaknesses, however. They ensure continuing significant lags behind competitors because of the compounding of: delays in making new advances available to others; additional delays before newly available advances are evaluated and arrangements for their acquisition are made; and still further delays before such innovations can be installed and brought to effective levels of utilisation.[28] However, the ability of domestic steel company staffs to develop major advances tends to erode unless they remain actively engaged in challenging existing frontiers. Research and development capabilities cannot be maintained at peak levels by merely reading about foreign achievements, or despite being forced to concentrate on relatively short-term projects seeking only modest incremental improvements in current applications. Hence, the policy of minimising R and D costs and risks by relying on adoptions of foreign advances ensures the continuation, if not the expansion, of substantial lags in technological competitiveness.

Sight should not be lost, however, of the fact that the basic objectives of improving technological capabilities is to strengthen market competitiveness. Hence, efforts to achieve the former must be reinforced by increasing marketing aggressiveness and by ensuring greater responsiveness to customer needs. Our current survey of marketing efforts by foreign and domestic steel suppliers in major regional markets within the US suggests that the long-vaunted superiority of American marketing has also been seriously undermined. Specifically, we were astonished by the frequency with which customers reported,[29] and some sales representatives of domestic producers admitted (or complained), that efforts to sell foreign steel have been more continuous and energetic; been more quickly responsive to customer needs, including flexibility of delivery; ensured more consistent quality; and often offered more attractive price and financing arrangements. As a result, one cannot help wondering how effectively domestic salesmen are probing and reporting back to their companies both the reasons for recent customer shifts to other suppliers and the resulting trends in buyer attitudes which may foreshadow longerterm changes in market shares.

Finally, additional powerful contributions could be made to increasing the technological competitiveness and marketing strength of domestic steel companies through improving the quality, and broadening the scope, of the economic analyses guiding managerial decisions. Few steel companies have professionally outstanding economic staffs. Moreover, such staffs are seldom used to evaluate major options involving steel operations, being concentrated primarily on forecasting Gross National Product and steel demand – neither of which they do very well – and on replying to government inquiries, attacks and proposals. Specifically, despite my extended concern with the economics of the steel industry, I am not aware of any analytically persuasive evaluations by steel companies of such critical focuses of policy decisions as:

a) the economies of scale for different processes;
b) the economies of product specialisation or standardisation;
c) the magnitude, causes and effects of changes in productivity relationships;[30]
d) the prospective effects of major technological innovations, or even their actual effects after adoption (e.g., basic oxygen furnaces; increasing shift to iron ore pellets as blast furnace feed; continuous casting, computerisation);[31]
e) the economics of research and development, or of various segments of such programmes;
f) or even the effects of price adjustments on the level and composition of demand.

If economic issues are as important to business success as technological issues, one cannot help wondering whether steel companies can afford to depend on a lower level of expertise in the former than in the latter, or to limit the potential contributions of economists in considering alternative decisions affecting the magnitude and direction of capital allocations to improve technological capabilities.

Increasing labour contributions to improved competitiveness

Needed advances in productive efficiency are likely to involve not only the sharp reductions in total labour requirements per unit of output discussed earlier, but also major shifts in the composition of labour skills and further decreases in the flexibility of employment levels. Thus, increased utilisation of more advanced technologies and more comprehensive computerisation are likely to result in substantial displacements of semi-skilled and unskilled tasks, while increasing the need for more highly skilled control, maintenance and set-up personnel. Employment levels are also likely to be more stable for the latter because they are determined primarily by the continuing services required to ensure effective functioning of these more highly automated production systems rather than by variations in their output.

In order to help maximise the potential contributions of such influential operating personnel, as well as their motivations, considerations should be given to altering their terms of employment as well as their work assignments. Possible changes might include shifting the basis of their payments from hourly rates to salaries, and guaranteeing at least 40–45 weeks of employment annually. In addition, staff capabilities might be improved through providing periodic new training programmes to update knowledge of emerging developments and also through organising regularised rotations in assignments, which would not only engender a fuller grasp of interrelationships among various operations, but would also facilitate cooperation in coping with a wide array of possible disruptions.

Efforts to keep increasing the effectiveness of highly integrated operations are also likely to require some additional measures affecting the operating labour as well as the technical servicing personnel, engineers and managers. One of these might involve introducing group incentives to encourage the mutually reinforcing contributions which are essential in such systems. And another might involve periodic meetings of the entire production staff – as well as more frequent meetings of the technical service personnel, engineers and supervision – in order to review current problems and achievements as well as suggestions for improvement.

Some implications of suggested changes

Prospective changes in the structure of national steel industries

There is likely to be a progressive increase in the proportion of total domestic capacity supplied by small and medium size mills relative to large integrated plants. This will result partly from the closing, or paring down, of some of the large old mills in locations which have been declining in competitiveness, and partly from the addition of 'mini-mills'.

The latter trend derives from the substantial advances which have been made in recent years in the economic competitiveness of such smaller units, originally with annual capacities of 250,000–350,000 tons, but gradually rising until some are now in the 600,000–750,000 ton range. By combining high-powered electric furnaces to melt scrap or direct reduced iron with continuous casters, which offer the double rewards of energy conservation and higher product yields, such mills can reduce investment requirements by 75–80 per cent, as was noted earlier. And less complex facilities, narrower product mixes and more effectively integrated operations also make possible substantially greater output per manhour.

Increases in the proportion of capacity supplied by mini-mills will also be associated with a greater geographical dispersion of steelmaking. Because such mills are not dependent on low-cost bulk transport facilities to supply huge quantities of coal and iron ore, and because scrap and electricity are widely available, new mills can locate closer to the markets they intend to serve. Moreover, because of their smaller capacity, they can focus on supplying smaller and more specialised markets than would be economical for large integrated mills.

The foregoing obviously does not mean that larger mills no longer offer significant scale economies, but rather that their advantages are limited to certain products and processes instead of applying to all steel operations. Mini-mills, too, are competitive only within a limited product and volume range, although these have been widening gradually as they have adopted new, and somewhat larger, equipment and simultaneously explored improved operating practices. At any rate, the key point here is that continuing developments in products, processes, equipment and facilities urge more effective analyses of changing patterns of scale economies.[32] There would seem to be a reasonable basis, however, for hazarding the judgement that there is likely to be little further copying of giant Japanese facilities in the US or Western Europe during the next decade.

Another change in the structure of steel operations may involve increases both in product specialisation by plants and in the proportion of standard products within major product lines. Both are likely to shape the modernisation programmes in existing plants as well as the planning of new plants. So far, progress in these directions seems to have resulted from limited experimental explorations. But the potentials involved may well be significant enough in terms of their prospective bearing on competitiveness to warrant more intensive and more comprehensive analyses of the benefits and burdens of decreasing a plant's range of product offerings, and also of offering price inducements to encourage a greater concentration of orders within standard categories. It should be noted in this connection, however, that correct analyses are quite likely to result in different conclusions for different plants, depending on their present production capabilities, market competition in various product lines and prospective customer requirements.

Still another change in the structure of steel production may involve a renewed interest by some large steel customers in establishing full or partial ownership of steel mills. Because of the high levels of efficiency now attainable in small mills, and the possibility of gaining additional special benefits through having such mills concentrate on the particular products needed by the customer – including unique metallurgical specifications as well as the shapes and dimensions which would minimise additional manufacturing costs – there may well be significant benefits to a customer whose average (or minimum consistent) requirements are sufficient to absorb all, or a very large proportion, of the the output of a mini-mill operating at continuously high levels of capacity utilisation.

Prospective changes within steel companies

The preceding analysis also suggests that a variety of changes may have to be considered within steel companies. In already established multiplant firms, these would include closing the least competitive units, modernising the best plants and eventually adding advanced new plants or units. Modernisation programmes would tend to centre around the addition of electric furnaces, continuous casters, units to permit increasingly effective production of specialised products, and the organisational and equipment requirements for increasing the computerisation and automation of all operations. The new facilities to be added are likely to be predominantly smaller in scale, relatively narrow in product mix, more widely dispersed among regional markets and based on the cold iron-electric furnace-continuous casting model.

All operations are likely to be subjected to more intensive and continuous pressures to improve competitiveness through reducing energy requirements, increasing product yield, maintaining higher average utilisation rates, maximising capacity per dollar of investment and decreasing unit wage costs. Increasing commitments are also likely to be necessary to improve the qualities and service capabilities of products in basic, as well as in incremental, ways.

Effective evaluation of such varied and far-reaching options may well require a more balanced consideration of technological, as well as of marketing and financial viewpoints, along with sounder analytical foundations for each.

Modernising existing plants and adding new plants would obviously require substantial increases in investment. Raising such capital may necessitate selling off some of the company's long-hoarded natural resource holdings in addition to demonstrating increased profit capabilities. And in order to allocate sufficient capital to such long-term improvement programmes, it may be necessary to evaluate them within the perspectives of 'strategic investments designed to achieve long-term competitiveness and profit' rather than within the essentially short-term maximising 'net present value' criteria of common capital budgeting methods.[33]

In order to become truly competitive with foreign competitors, and in order to help preserve steel markets under pressure from various substitutes, steel companies will also have to undertake technological development programmes on a scale far surpassing past commitments. Major advances in many fields of science and technology offer new insights and potentials affecting every aspect of steelmaking. Exploration of such developments requires new kinds of specialists, new kinds of research and new criteria for determining the adequacy of allocations to such efforts.

In short, safeguarding its future would seem to require each company's management to supplement continuing catch-as-catch-can responses to an endless succession of immediate urgencies with a major effort to develop a long-term improvement programme. To be effective such an undertaking would probably require significant commitments of funds, personnel and time in order to provide:

a) an objective evaluation of its relative competitiveness in each of its current and prospective product and geographical markets – including specific identification of its strengths and weaknesses;

b) estimates of what it would take to catch up with competitors sufficiently to hold, or regain, a desired share of such markets, after allowing for continued improvement efforts by others as well; and

c) evaluations of the proportion of such gains it could realistically strive to achieve on the basis of expected resource availabilities.

Such appraisals would then provide a sound basis for planning an integrated programme of technological changes, facilities improvements, product mix changes, organisational readjustments and alterations in managerial policies.

Although many managements claim that they already have such information, I have found few persuasive evidences of such assessments which are both comprehensive and objective. In some cases, this means that the prospects are so frightening that managements seek to avoid coming closer to grips with them – perhaps hoping that catastrophe can be fended off until they retire. In other cases, it reflects a fundamental refusal to accept the need for far-reaching changes from the familiar arrangements of the past and hence emphasises instead that all will be well as soon as competitors, governments and trade unions begin behaving 'sensibly' once again.

In any case, the fact is that a significant proportion of current steel operations will not survive, and those which change least are the least likely to survive. The result will be changes in the structure of steel operations among countries, as well as within countries. And far-reaching changes are also inevitable with steel companies and even within steel mills.

Notes

1 William E. Umstattd Professor of Industrial Economics and Director of the Research Program in Industrial Economics, Case Western Reserve University, Cleveland, Ohio, USA.

2 The most recent publications summarising such findings are Gold, B., *Productivity, Technology and Capital: Economic Analysis, Management Strategies and Government Policies*, D.C. Heath-Lexington Books, Lexington, Mass., 1979, 1982; and Gold, B. *et al, Evaluating Technological Innovations: Methods, Expectations and Findings*, D.C. Heath-Lexington Books, Lexington, Mass., 1980.

3 For a brief introduction to Swedish efforts in this area, see Eketorp, S., 'Thoughts about Metallurgy facing the Year 2000', *Stahl u. Eisen*, July 1981.

4 For further discussion, see *Fortune* (6 April, 1981, pp.43–46) and Gold, B., 'Interactions between Technological Innovations and Factor Prices', *Revue d'Economie Industrielle*, Paris, Spring 1980.

5 For more detailed discussion, see Rosegger, G., 'Continuous Slab Casting', in Gold *et al*. op.cit. 1980.

6 For example, see US results in Gold, B., 'Tracing Gaps Between Expectations and Results of Technological Innovations: The Case of Iron and Steel', *Journal of Industrial Economics*, September 1976.

7 See Gold, B., 'Economic Perspectives in Evaluating Technological Development Alternatives', *Iron and Steelmaker*, December 1978.

8 For detailed analysis of 1976, see Gold, B., 'Steel Technologies and Costs in the US and Japan', *Iron and Steel Engineer*, April 1978. The 1980 estimate was published in the *Wall Street Journal*, 7 April 1981.

9 For a fuller discussion of Nucor achievements and policies, see *Fortune*, 6 April 1981, pp.43–6.

10 For a more detailed analysis, see Gold, B., op.cit. 1976.

11 Resulting labour cost savings may be less than expected, how-

ever, because wage rates are often increased further as a result of claims on the basis of the very gains in 'labour productivity' which are entirely attributable to the investments in additional modern equipment.

12 See Gold, B., op.cit., 1976.

13 See Gold, B., op.cit., 1979, 1982, Chapters 5 and 10.

14 For further discussion, see Gold, B., 'Evaluating Scale Economies: The Case of the Japanese Blast Furnaces', *Journal of Industrial Economics*, September 1974; and Gold, B., op.cit., April 1978.

15 For further discussion, see Eketorp, S., op.cit., July 1981.

16 The more effective Japanese approach already applied in large mills is discussed in Gold, B., 'Factors Stimulating Technological Progress in Japanese Industries: The Case of Computerization in Steel', *Quarterly Review of Economics and Business*, Winter (December) 1978.

17 For a more detailed analysis, see Gold, B., op.cit., September 1974.

18 *Wall Street Journal*, 7 April 1981.

19 Ibid.

20 Ibid.

21 For a review of Japan's rapid progress in the computerisation of steel operations, for example, see [16].

22 For a summary of the Council's report and a critique of its findings, see Gold, B., op.cit., April 1978.

23 My own suggestions for governmental support were summarised in Gold, B., op.cit., 1979, 1982, Chapter 17.

24 For a reference to comparable pressures on Japanese engineers, see Ibid, p.278.

25 For a fuller discussion of such a system along with some illustrative empirical results of applying it, see Gold, B., op.cit., 1979, 1982, Chapters 3, 5, 9 and 10.

26 It should be noted that effective evaluation of the potential benefits of specialisation and standardisation may well require modifications in the prevailing accounting allocations of cost which were developed during periods characterised by limited sensitivity to these additional sources of potential economies.

27 This may be illustrated by US experience with continuous casting, see Rosegger, G., op.cit., 1980.

28 For further discussion, see Gold, B., 'Managerial Considerations in Evaluating the Role of Licensing in Technological Development Strategies', *Managerial and Decision Economics*, 1982.

29 Part of a study for the National Science Foundation of the

factors affecting the international competitiveness of five domestic industries, 1982.

30   For example, it might be of interest to note that our findings of long-continued declines in the productivity of fixed investment (i.e., in productive capacity per dollar of net fixed investment) in the steel industry after 1947 occasioned astonishment within the industry repeatedly with each successive report between the earliest (Gold, B., *Explanations in Management Economics: Productivity, Costs, Technology and Growth*, Macmillan, London 1971; Basic Books, New York 1971, pp. 63–6)and the latest (Gold, B. *et al.*, op. cit., 1982, pp. 284–5).

31   For an examination of some of these effects, see Gold, B. *et al.*, op. cit., 1982, Chapters 6–13.

32   There is, in any case, considerable mythology associated with widespread beliefs about the sources and magnitudes of scale economies. For further discussion, see Gold, B., op.cit., September 1974; and Gold, B., 'Changing Perspectives on Size, Scale and Returns: An Interpretive Survey', *Journal of Economic Literature*, March 1981.

33   For further discussion, see Gold, B., 'On the Shaky Foundations of Capital Budgeting', *California Management Review*, Winter (March), 1977; and Gold, B. and Boylan, M.G., 'Capital Budgeting, Industrial Capacity and Imports', *Quarterly Review of Economics and Business*, Autumn 1975.

PART VII
CONCLUSIONS

PART VIII

26 Conclusions

Introduction

The conclusions to be drawn, in brief, from the analyses of the steel industries of major steel-producing (mature) countries will be cast into the shape of an overview of strategic options available at the levels of the firm, the region, and of (national or supranational) government.

Strategy at firm level

Given the international competitive situation in the steel industry, only firms being prepared to, and being capable of, continuously striving for improvement with the aim of reaching and staying at the top will have a chance to survive in the industry. Such firms must be prepared to meet a number of requirements.

They must be prepared to maintain a high level of physical investment, and to continuously stay in the forefront of the industry, being prepared to meet the competition on the terms of the industrial leaders, which, for the foreseeable future, will be the leading Japanese firms. Such high levels of physical investment must be maintained, despite the high levels of inflation and high levels of cost of capital. Firms wishing to play a role in steel must be prepared to invest in order to continuously reduce energy costs, as well as increasing labour costs (which increasingly are to be regarded as a fixed cost), but also to stay prepared for environmental protection requirements and environmental recovery raised against the steel industry. Most of all, a high level of investment is needed in order to meet the continuously increasing qualitative requirements stemming from the market where the firm has to face, and to stay ahead, of competition. The qualitative requirements (and there is little difference between carbon steel and alloy steel, since carbon steel will continue to take market shares from, and intrude into, the traditional domains of alloy or special steel) already discernible concern: corrosion resistance; capacity to withstand even extreme hot and cold temperatures (in the latter case, for cryogenic processes); improved chemical properties (up to non-crystalline alloys); improved maintenance, durability, formability properties; the deep drawing qualities of thinner, lighter and corrosion protected steel; and increased elasticity and reduced weight combined with other maintained or improved qualities. In summary, the increased qualitative requirements imply improvements in metallurgic as well as in finishing processes, as a basis for higher value added. It would be very hard for a firm to find one or several niches in which it may stay safely over longer periods of time. In this respect, the character of the industry has changed in almost revolutionary ways since the mid-1960s.

In order to meet the above and other requirements, a steel firm must prepare to increase the size, and to continuously upgrade the quality of, its engineering staff, not only to meet these qualitative requirements but also productivity targets which must be pushed steadily forward, to ensure the proper performance of the high quality equipment the firms must employ.

A steel firm aiming at superiority will have to reduce its range of products, achieve scale economies, cut inventories, standardise but also maintain flexibility, at competitive costs. If the firm is operating in the integrated sector, it must also be prepared to take advantage of mini-mill cost efficiencies and flexibility, and to adopt mini-mill market orientated behaviour. It should be remembered that mini-mills are now approaching sizes lying in the medium range of integrated mills. The firm must also be prepared to speed up its adoption of technological innovation in many aspects, for example, in the field of flexible automation. It must be prepared to reduce the internal inertia against technological change not only from the financial criteria employed by the firms, but also, to an important extent, from behavioural obstacles in the minds of middle and upper management and of technicians who in requirements for continuous innovation see threats to their competence and thus employment security.

The firm will have to improve its corporate planning features in order to identify its competitive advantages, their likely duration and how they may be improved, and it will have, at frequent intervals, to decide whether to change radically or if random or piecemeal adjustment would suffice. Similarly the firm must overcome its weaknesses and vulnerability by deliberate improvement.

As customers' quality requirements rise steadily because of the superior performance of Japanese plants, in particular, the firm is requested to maintain high consistency in quality, which not only requires the utilisation of top modern equipment in metallurgy and finishing, but also places high emphasis on control processes as well as on the behavioural requirements of all levels of personnel.

The firm will have to take an integrated view on all the sources of productivity by keeping continuous control of its staff productivity, capital, technology, material, energy and so on. Specific and integrated productivity analysis is a must, and the firm must continuously ask itself 'what would happen if...?', where economies of scale are to be harvested, on what are they based, where small-scale economies are to be achieved, and how the cost and income properties of innovation can be assessed.

Perhaps one of the most thorough changes needed is in marketing. Salesmen have to be trained not only to take and transmit orders, but also to listen and to actively solve their customers' problems together with them. To be competent partners in problem-solving for their customers', researchers, developers and designers, these salesmen will have to have a high level of competence.

The firms have also to improve their R and D objectives and performance, not only for their own processes and products, but also for those of their customers, as well as for processes to be invested in, and processes which may have to be developed together with sellers of technology. R and D will have to extend into customers' problems but also into control and monitoring systems for metallurgy and finishing processes.

Steel firms will have to continuously improve their manpower relations, *inter alia*, implying continuous motivation and engagement of all their personnel in problem-solving, productivity and quality improvement, cost-saving progammes and so on.

Steelmaking firms will have to extend their environmental awareness, not only to new challenges and problems in the physical environments. They will also have to be alert when it comes to assessing the specific requirements imposed on the infrastructure, on the administrative systems in the environment, of the economic and social structure of the environment they act in, and increasingly become prepared to help, assist but also guide local and regional government when it comes to protecting the economic and social structure against the problems of monostructuring attendant on the presence of the steel industry.

Most of all, steel firms will have to improve their management to meet the above, and other, requirements. The first task of management would be to identify the improvement goals, the targets at which survival at profit is possible. It will have to set integrated as well as local or suboptimisation goals and do this in cooperation with many different members of the firm and its environment, in particular, its customers. Management will have to improve its capacity for finding new and original solutions to many problems which may not yet have been identified.

A few comments may be added to the above account of major firm strategies to be pursued. When it comes to the mini-mill sector, perhaps a most amazing feature is that practically none of the large integrated producers have entered into this sector. At present, approximately 300 mills are operating in the sector, which was practically non-existent some 15 years ago. The sector is today as large as the steel sector in the newly developing countries, that is, accounting for approximately one seventh of the world's capacity, and having a considerable growth potential. It is a most interesting low-cost, high-flexibility alternative, also from the point of view of siting flexibility, which is one of the new requirements for continued competitiveness in the steel industry.

The readiness to move to new sites – site optimisation – matters nowadays; site tradition does not. Steel production is either moving close to the centres of the consumption and demand, or to deep-sea locations, if dependent on bulk transportation of raw materials. The situation concerning scrap supplies is different, as scrap is usually available in metropolitan agglomerations and is also traded internationally. In recent years, trade obstacles have occasionally been raised against exports of scrap in particular (e.g. from the United States of America. There is, however, an alternative to scrap: sponge iron or direct reduced iron even if the booming cost of natural gas has reduced the attractiveness of DRI).

The problem of which product portfolio to offer the market is different for large and small firms, although the differences are shrinking, both because of the diminishing size gap or size difference between big and small, but also due to the development of so-called flexible automation, whatever this broad term may cover.

One prediction, a rather plausible one, is that world trade will be reduced, because new countries have been steadily entering the market as suppliers, at least of their home market or of neighbouring markets. Newly industrialised countries and less developed countries will, most likely, grow in importance as a market segment to which the old industrialised countries have only limited access. Second, trade of special steel will increase *inter alia* because of product differentiation and consequent specialisation. Third, low-quality steel will be offered at low prices on the world market by the newly industrialised, or even the less developed countries.

The problem of qualities and specialisation is one to be carefully watched. Many of the world's foremost experts claim that only firms employing the best available technology, at competitive cost, will survive. But it is not in the ties to (firms') technology that survival capacity is anchored. It is rather in ties to the market, in the firms' readiness to assist and to develop competence in problem-solving together with the customers, that interesting development is to be found.

As an example of the qualities syndrome, it may be mentioned that approximately four-fifths of all nuclear energy plants' costly standstills and temporary closedowns are caused by problems in the steel vessels, in piping (leaks, cracks) and in material failure or in fatigue, for example in turbines. Obviously then, some of the world's leading nuclear energy equipment manufacturers have not had access to capable steel suppliers or have been unable to foresee or specify the necessary qualitative requirements when it comes to the steel needed under the circumstances prevailing in nuclear energy plants, and to communicate the pertinent requirements to the steel producers.

Firms not prepared or able to invest cannot survive in the competition of tomorrow. The projected steel shortage, of which there was much talk during the first years of the crisis, is a myth, mainly used to postpone unpleasant decisions to close plants or to liquidate firms.

Logistic services, as a part of marketing, customer services and participation in problem-solving of large customers' flows of material, is an area which should be paid attention. It is not a typical problem for which average steel firms normally have developed high competence. Nevertheless, today or tomorrow, it is one of the means of competition, to be able to serve the customer at a low internal cost of logistics.

Managers at different levels must be willing, again, to commit themselves to their firms as much as their Japanese counterparts do. The steel industry must be able to attract top level managerial capacity again, in order to stay alert in competition. Management style, in particular when it comes to labour relations, needs careful reconsideration, mainly on one of the basic questions: if workers and management have to be afraid of losing their jobs, if they are successfully pushing productivity (and if productivity is not improving by approximately 5 per cent a year, a plant is doomed to be closed within 5–8 years' time), then they will do whatever they can to withhold their potential constructive contributions. One of the claims is that investment in steel takes a long time to mature, in particular in integrated plants. One should, however, consider that change in managerial thinking also takes time, mainly because good solutions to urgent problems of the type mentioned are not easily at hand.

Firms in general, and large firms most of all, do not diversify well outside their basic business or technology (in terms of industrial statistics, they are not overly successful in diversifying outside their two-digit ISIC groups). One of the questions which has been raised in connection with the steel industry is whether it should integrate further, as some firms (for example, the large German producers) have been doing. This may or may not be a good strategy, depending very much on whether one gets into competition with one's own major customers.

The situation in Japan is quite different from the other highly developed industrialised countries: the centres of economic activity are the *Zaibatsus*, the trade houses, which are financial centres, and which are market-orientated. The *Zaibatsus* are often engaged in widely different businesses, of which steel may be one. In the Japanese case, it is not a problem of the steel industry diversifying into other industries, but rather a question of *Zaibatsus* getting involved in different types of business. That is an entirely different problem, implying different solutions.

The Japanese have discovered that they will have to increase their indirect steel exports much more, as the direct steel export is stagnant. One must foresee consequent action to be implemented within a few years' time.

The above short summary of firms' strategies is by no means complete. How could it be? Good management is always ahead of generalisations offered in books or in articles.

Regional policy

As briefly discussed in Chapter 1, the term 'region' is not easily defined. A region may be equal to a state but, more often, it is not. The region which interests in the present context is a spatial area more or less heavily influenced by the existence of one or several large, integrated steelworks or, as often is the case in traditional steel locations, an area in which coal and steel have been dominating the economic life. If looked upon from this point of view, a 'region' will hardly ever be identical to an administrative decision making unit in public administration. Regions most often cut across borders of such units. Sometimes municipalities, cities, or districts in such regions join together in problem-orientated assemblies, regional boards, or whatever they may be called, in order to discuss or coordinate their joint problems and opportunities stemming from the existence of dominant industry (industries) in their 'region'. Occasionally, this is made difficult because such relationships may cut across political borders of state or country type (as has been, and still is, the case in some European regions for instance).

A basic problem facing most regions is that they cannot and do not select the industries within their boundaries. At least they have comparatively little influence on what industry will settle there. In far too many cases, the awareness of dependence on one or a few industries emerges only when the industries are running into crises. This is rather unfortunate, because it is usually too late to do anything about the situation, for several reasons, one of the most important ones being that the regional administration, if there exists any such body, has not only little competence as a rule, but very few instruments at its disposal to influence the regional economic structure for the better. If regions had an influence, and they are increasingly becoming aware that they ought to have such an influence, on the economic structure of their domain, they ought to be interested in a differentiated structure of economic activities. The advantage of a differentiated structure, over a monostructure, is that it represents a kind of 'insurance': it is less likely that all economic activities and all firms will be affected by economic crisis at the same time, and to the same degree.

As we have seen in Part V of this volume, the presence of big steel in the region not only gives the region certain structural characteristics, which are usually called 'monostructural'; steel also requires infrastructural characteristics to be present, or to be provided, perhaps more than other industries do. Because of their size, steel plants are quite often in a position to dominate the local or regional political administration. Dominance tends to increase with 'tradition', that is, by the length of time the industry has been present and dominant.

What one would recommend regional bodies of the type discussed above to do during 'normal' times, that is before a crisis turns up, is to analyse the region's dependence on different kinds of economic activities and, if necessary and possible, to induce a development towards differentiation. This is certainly not easy to achieve and is something which does not come about by command or prohibition, but rather by deliberate long-term orientated programmes aimed at fostering a good economic climate in the region, and also at providing a wide range of 'infrastructure', in traffic, schooling, research, cultural and social development and so on. The aim is to make the region an interesting and attractive one to settle in for both people and enterprises.

Regions would also do well to analyse how small and medium-scale business is developing, most of all when large industry is heavily represented. As has been mentioned before, the presence of a differentiated structure of small and medium size enterprises in different lines of business may constitute a most desirable and valuable safety net in case of a crisis in big industry. As also has been claimed, the problem of large firms, large workplaces or plants is that, if in trouble, they are not easily replaced by equally sized plants or by employers requesting the same type of skills and knowledge from the people, who were employed by the faltering industry or firm.

This leads to the conclusion that one of the most important and long term tasks of regional bodies is to analyse long term labour market trends, with respect to both requirements and supply, quantity and quality-wise, within the region, but also to induce strategies both to close gaps, if they become apparent, and to cater for differentiation wherever possible.

Whereas diversification or differentiation may not be the optimal strategy for a firm – success more often comes from specialisation and concentration – the situation for the region is different, even if it may appear most attractive to build up copies of 'Silicon Valley' wherever possible. But who knows how durable the life of a singular rapid-growth industry may be over time? The region cannot harvest so-called windfall profits and accumulate reserves to be used in times of crisis. Its decision structure is quite simply different from that of private business.

Another type of regional strategy to be applied in crisis management is to stimulate enterprise mobility whenever this is possible and feasible. If enterprises do not, or cannot, move into a certain region, then it may become necessary to move people, with all the problems that this may imply in suffering, loss of income, loss of economic base, and for the family.

Strategies at national level

When discussing policies for the ailing steel industry at national level, one must keep in mind that the national level is a political level. If, thus, the political opinion in the country is that an industry, of whatever type, should be owned by the public, the major question of public policy is answered in principle, and it is therefore left to us to discuss options available of governmental policy *vis-á-vis* ailing steel (or faltering industries at large).

In countries which some parties call free, and others may call capitalist, government is usually involved in some kind of economic policy at large, the aim of which is to regulate the business climate in one way or the other. When the business cycle is overheated, the government or some other body, usually the national bank, may take steps to prevent the economy from overheating, that is, running into inflation. The level of inflation to be permitted has shifted over time and certainly shifts between different countries. Many countries during the 1950s and 1960s felt that inflation should not exceed 2 or 3 per cent a year, but the 1970s have let us become accustomed to much higher

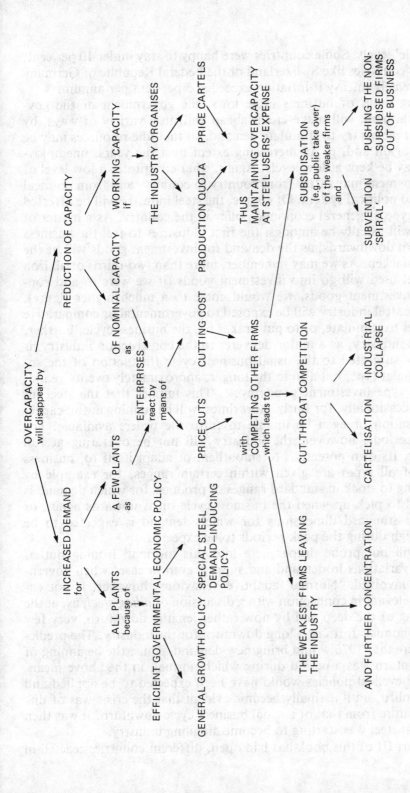

Figure 26.1
Steel market problems — a schematic overview (H. Wienert)

503

'tolerable' levels. Some countries were happy to stay under 10 per cent, whereas countries like Switzerland or the Federal Republic of Germany felt extremely uneasy if inflation exceeded 5 per cent per annum.

During times of business cycle lows, the government or the governmental bank will make credit cheap and, in a variety of ways, by different means try to stimulate demand, so that idle resources may be utilised again and, to an increasing extent over the years, unemployment may be kept at low levels. Again, what constitutes a low level of unemployment may differ from country to country, and from political system to political system. Of course, the steel industry will be affected by this type of general economic policy in the country. As a matter of fact, it will usually be amongst the first industries to feel the business cycle turn downwards, as the demand for investment goods will be the first to slacken. As we may remember, more than two-thirds of all iron and steel used will go into investment goods (if we were to add consumer investment goods, we would come to a much higher figure). Thus the steel industry will be exposed to governmental or comparative measures to stimulate, or to put brakes on, the business cycle. Further, the steel industry, as a major deliverer to the construction industry, is not only subjected to the usual business cycle fluctuation of the so-called Juglar type, but also to the longer, approximately twenty years', Kusnetz type investment goods cycle. This implies that the steel industry occasionally, for fairly short times, will be running at full capacity utilisation or even be unable to fulfil the orders available. For longer periods, however, the industry will not be operating at full capacity. Its own potentials to smoothen, or adapt itself to, business cycles of all types are given within certain ranges, for example by producing to stock in standard ranges of products for which demand is expected to pick up when the business cycle turns upwards again, or even for standard dimensions for which demand is expected to be rather high during the peak periods to be expected.

We will not probe deeper here into this 'normal' ironing out of cyclical variations in demand and supply, even in cases when government is involved. 'Normal' adaptive behaviour, however, is not entirely irrelevant in connection with a discussion of *ailing steel*, as, at the beginning of the deep and by now rather lengthy downturn, very few steel economists foresaw a long downturn for the industry. The predictions were that 1976 would bring new demand. Thus, the beginning of the downturn was a period during which normal, in the above meaning, anti-cyclical policies would have been expected to be applied and were applied, until it finally became evident that the crisis was of different nature from that of a usual business cycle downturn. It was then clear that steel was starting to become an ailing industry.

As Part III of this book has laid open, different countries reacted in

504

quite different ways to the emerging crisis symptoms.

What options would then be available to cope with a crisis? In the first place, so-called temporary aid for *adaptation* was used as an instrument. As steel firms used up their reserves, labour market authorities stepped in with, to beging with traditional labour market instruments: unemployment relief, retraining, for jobs outside the industry or sometimes even within the industry, temporary employment programmes and so on. In some countries the slack period was used to give the industry an opportunity to improve its environmental characteristics, that is to install equipment for cleaning exhausts into air and water. Later steps were also taken in energy-saving measures.

The above measures towards necessary or desirable improvement of the industry's environmental and energy utilisation criteria were, of course, implemented – often earlier than required by legislation or regulation – in order to keep up employment in the industry. There were, however, interesting differences in the financial arrangements. In some countries the money for the investment was given to the industry, to cover the full cost, or the labour-cost part, as part of unemployment relief, that is the company had only to carry a certain share of the burden of investment for environmental improvement. In other countries (for example, Sweden) business cycle smoothing measures imply a skimming off of 'over-profits' gained during a boom period, by temporarily freezing them to deposit funds at zero interest in a government bank, and unfreezing them during a slack period, to be used for certain types of investment, such as improving productive capacity, or improving the environment. In this case it is the firm itself which pays, by being permitted or induced to use money that it was forced to save (rather than to give it away to its employees or its shareholders). Mixes of using reserves plus unemployment relief have also been reported.

Another type of financing is by loans from Merchant banks, with the public sector, government or a similar institution, acting as guarantor (earmarked special-purpose loans against state guarantee), in some cases also 'conditional' loans.

Similar programmes to those just reported on environmental or energy problems have later been implemented for restructuring purposes, which, in many cases, have meant the scrapping of old capacity and the installation of new modern capacity, sometimes also implying rather substantive capacity increases.

Other means to stimulate demand, like government orders or 'buy domestic laws', should not be forgotten, even if they fall into the ranges of protective measures. In some cases industries were gradually or (depending on the severity of the crisis) totally transferred into public ownership, that is by the public taking over shares or contributing new capital against shares, or by nationalising the industry, for example, in order to safeguard it from bankruptcy and closure, or to make possible investment, or to compensate the owners for scrapping over-capacity (whatever overcapacity may mean in this context). Some states have been travelling in the opposite direction, for example, Italy and, to some extent, even Great Britain have decided to sell off public-owned enterprises, although in both cases it has been most difficult to find financiers prepared to invest in ailing steel.

A new practice has emerged in a range of countries implying a government preference to keep labour employed in industries suffering from slack in demand and from consequent structural overcapacity, or from their inability to meet market requirements, rather than put it on unemployment relief. The financial arrangements are often wage subsidies during periods of guaranteed employment. Thus the public is taking over, fully or partly, the wages to be paid. The objective of such programmes is to keep unemployment figures at 'reasonable' levels, mainly in monostructured regions, as heavy unemployment would be a testimony to the ruling government's inability to manage the economy or to control industry. Such measures are sometimes taken when hopes are still being nurtured that the tide may turn quite soon and that the problem is therefore just to overcome a few weeks or a few months of unemployment.

These programmes are rather problematic, as they spoil productivity, by having people idle often for a considerable time. Of course, this may be regarded as a better option than the unemployed totally losing their workplaces and therefore not even having access to some rights and privileges which may be maintained by 'artificial' employment.

From the point of view of international trade and subsidy rules, this becomes a questionable policy because one never knows where the frontier line is drawn between unemployment relief and subsidisation of wages and salaries for unprofitable firms which are unable, in the short, as well as in the long run, to meet the market's requirements. A crucial problem is, again, that nobody is actually in a position to know what the truth is about real cost, for what the money is spent, or what causes the bad market situation for the firm. In this respect, it makes absolutely no difference if it is a government-owned firm or not.

A cleaner way, from the point of view of trade policy agreements, is to give troubled firms loans, under strict conditions, or, as has been done in a few cases in the Federal Republic of Germany, to have the government to act as a guarantor, under strict conditions.

One type of programme which has been applied with some success in the Federal Republic of Germany is to give certain public support to research into innovative ventures, that is projects furthering risky technology, for example, to set up pilot plants with new technology or with technology as yet untested in larger-scale applications. Innovation or research programmes of this type usually are run on a conditional lending basis: if the innovation turns out to be successful, the money is to be paid back; if it was unsuccessful the loss is shared between the firm and the government according to an agreed formula. It should be remembered from Rosegger's statement in Chapter 17, that innovation is very much needed for the industry but that innovation alone will hardly be able to save the industry from drastic capacity reduction.

A policy not infrequently adopted is that of stimulating mergers between ailing steel firms. This is a programme for better or for worse, more often for worse: merging two or more bad firms will never produce a good firm. Merging good firms[1] may produce monopoly and this does not, in the long run, help the industry or the employed. The problem would not be very serious, if one could trust in markets given freedom of trade. Monopolies would not be able to operate if imports were permitted. They could only monopolise certain niches in which they were superior (and in some regions in which they were protected by cost of transportation thresholds). It would also be a poor strategy to sink healthy firms by forcing them to take over sick firms (as was the rule in France for a time). This, at least in the case of the steel industry, would not only mean an unnecessary draining of capital from firms with a survival potential to those who have no chance; it may bring the healthy part, which is fit for competition, into danger, and, last but not least, it would, in times of crisis, divert management's attention from necessary market orientation and activity on the markets to internal problems, settling the many disputes and controversies brought about by the merger imperative.

Government may, if the industry itself, because of competition, is unable to assemble into forums for discussing its problems and their potential solutions, create such opportunities, acting as a neutral party. It also may appoint independent chairmen of investigations, or studies of the industry to be made.

The European Community, under the Davignon plan, has created a temporary cartel, not only to avoid price competition, but also to raise the money for the necessary scrapping of overcapacity and the consequent premiums to be paid for certain firms which will have to give up substantial parts of their productive capacity. The German innovator, Dr Willy Korf, has proposed a similar programme along slightly different lines, which, however, would not only help big steel but also pay due regard to the most vigorous mini-mill sector on the European market. Korf claims that the governments should ensure that the most competitive firms are not being weeded out because of the crisis, and that they should ascertain that only the technically most advanced and competent firms will get a chance to survive. Korf has sued the state for subsidising ailing steel corporations, as this hurts the firms which have to compete without support.

Policy summary

In attempting to summarise, a number of policy alternatives available may be put against each other.

Public or private industry

The history of steel, as we know it by now, has proved that government owned industries have, as a rule, been less capable of coping with the crisis than private industry, although the record of failures, if examined uncritically, would speak for government industry at least as providing secure workplaces – in the short run.

For justice's sake, it should be mentioned that the state-owned steel industry of Austria, as well as the state-owned German subset of the steel industry, have withstood the crisis in a remarkable way, although even they have lately been shaken and driven close to the brinks of the crisis. On the whole, government takeovers so far have not proved to be a superior, in most cases not even an equivalent, alternative for coping with the crisis, compared to private steel industry.

Cartelisation or non-cartelisation[2]

The history of cartels is full of cases which demonstrate that cartels transfer money from consumers to shareholders or other interested parties in the industry. For that reason most countries maintain anti-trust or cartel legislation. As has been demonstrated in the section on the US steel industry, the United States steel industry has been unusually successful in operating an oligopoly, of almost duopoly type, over many years, a fact which now may be regarded as being one of the major causes of the recurrent crisis of the industry.

In order to reduce the European steel industry's capacity to levels needed to only, in principle, serve the European steel market, it has been temporarily cartelised. Its termination date, end of 1985, even if seriously meant, is highly questionable. Cartelisation has drastically increased steel prices and hit the steel users, mainly the machinery industry, very hard, and may certainly have contributed to a slower than necessary recovery from the deep business cycle crisis of the late 1970s and early 1980s. It is 'legitimate' because it has been concluded with the consensus of nine industrial nations. It is compatible with the rules of the General Agreement on Tariffs and Trade, GATT.

It may well be suspected that the existence of cartels in the medium, but most of all in the long, run will hit the industry in two ways: it will lose, faster than it otherwise would have done, market shares to both mini-mills and to low-priced imports from developing countries. It

508

may, as the American example has demonstrated, be losing in the high quality ranges to the more competent and cost-wise most competitive producers, at present in principle, and almost exclusively, the Japanese steelmakers.

Government policy guidance for steel – or not

A major question to be looked into is if government guidance is helpful and, if it is, what type of government control or guidance may be most helpful.

In recent history, from the late 1960s onwards, the record of government interference is not very good. In the first place, many governments decided to invest in the steel industry because they felt they should be strong in steel, either to be able to compete in the world market, or (as in the case of Italy) to produce steel for their own industrial development. In a few cases government involvement has been motivated by the non-availability of sufficient private capital for heavy greenfield investment. The history of this investment boom time demonstrates that governments have peshaps been poorer forecasters than industry, although taken on the average, industry has not been too clever either.

Seen in the longer view, the case of Japan may be of most relevant interest. The history of Japanese involvement in steel started with the initiative of the Ministry of Agriculture and Commerce to launch Yawata Steel Company as a government enterprise, to provide steel for the Imperial Army and Navy. It was a quite typical decision to be taken in a nation developing industrially and, as the case was in Japan of the late nineteenth and early twentieth century, building an empire. The continued history until the 1930s shows a slightly mixed picture. Private steel was able to develop side-by-side with Yawata Steel, and it was only during the preparations for the next imperial step that the Yawata–Fuji merger produced the leading enterprise, Japan Steel, in an attempt to establish industrial leadership.

After the dissolution of the government-owned steel industry and the Yawata–Fuji merger as well, MITI firstly did what was natural in a devastated country: to get its basic industries going, and to have them develop without undue interference from the occupational power. MITI was most successful in channelling money into the steel industry and helping it grow, being able to fall back on a long tradition of grooming the industry, even if it now was a private industry. Remarkable is the decline of MITI's influence after the Sumitomo incident. Although it induced the Yawata–Fuji (re-)merger to Nippon Steel, and thus actually won a victory, after the incident it has to turn to a milder tone of 'administrative guidance'.

MITI has been, and still is, most influential now that the Japanese steel industry is labelled as an ailing industry. It has done remarkably well in guiding the industry, at the same time keeping up vigorous competition within it. This is the most remarkable and positive case of government guidance.

The question whether it is to be big steel or little steel has already been answered: it will be both, although little steel will be gaining, at the expense of big steel, as it has been doing remarkably well during its fairly short existence. The US forecast for the industry, achieving a 25 per cent share of the market by 1990, may very well not be the final growth limit of the industry. Most likely it will grow considerably beyond this share, at the same time as big steel will be losing part of its market to imports from newly industrialised or less developed countries as well.

Certainly, little steel will make government dealing with steel much less necessary, as generally is the case for smaller industry. Governments' concern has regularly been with big industry, mainly because of the many jobs concerned and because of the heavy impact failing big industry used to have on monostructured regions. Having a sound and competitive mini-mill sector in steel to grow to considerable strength will lessen the need for government interference with the industry and will lessen the regional problems when coping with ailing large-scale industries, that is, not only steel, but also shipyards the aircraft industry, and the automobile industry.

Notes

1   CF Goldberg, W., *Mergers, Motives, Modes and Methods*, Gower, 1983.
2   In most cases, the steel industry is an oligopoly or at least it behaves like an oligopoly, even given the growth of the mini-mill sector, which was following the pricing behaviour of large steel quite well, and thus running at profit, when big stell was already in trouble.

Glossary

ADL: Arthur D. Little, Inc., Cambridge, Massachusetts, Consultants.

AISI: American Iron and Steel Institute.

Alloy Steel: speciality steel, not treated in this volume.

Alloy steelmakers: produce special products of different types. They may or – more often – may not engage in reduction of iron from raw materials. This segment of the industry is not dealt with in this volume. It will however be mentioned repeatedly.

BAT: best available technology (state of the art).

BATEA: best available technology economically achievable (mainly used in connection with environmental protection against industrial polution).

BOP: The basic oxygen process will often be mentioned as the major and widely used post-World War II steelmaking process, in which steelmaking furnaces (BOF) to which oxygen is injected are used.

BPT: best possible technology (used in setting criteria for the measurement of compliance with industrial environmental protection measures).

CC: The continuous casting process is a major innovation in finishing operations of steel.

COMECON: (or, in Russian, SEV) Commission for Economic Co-operation of the Socialist Counctries.

Continuous casting: A process for solidifying steel or other material in the form of a continuous strand rather than individual ingots.

Davignon plan: Plan for the restructuring of the steel industry in the member countries of the European Community as developed by Etienne Vicomte Davignon, Industry Commissioner of the European Community. The plan was developed and revised during the late 1970s and is valid until the end of 1985.

DHHU: Dortmund – Hörder Hüttenunion, predecessor (part) of Hoesch AG, German steel-producing firm.

Downstream *see* Vertical integration.

DR: direct reduction. A family of processes for making iron from ore without exceeding the melting temperature. No blast furnace is needed.

DRI: Iron produced by direct reduction, see DR.

Dumping: The sale of exported goods at less than the price charged by the manufacturer in his home market, or in some cases at less than cost. Dumping is restricted under the GATT as an 'unfair' pay practice.

EAF: electric arc furnaces. These are often used in mini-mills.

ECU: European Currency Unit.

EPA: United States Environmental Protection Agency.

Escape clause: Section 201 of the US Trade Act of 1974, which permits temporary restrictions on imports in the absence of prohibited practices, such as dumping, if sudden surges of these imports substantially injure a domestic industry.

Fe: ferrum, iron (as chemical element).

fob: free on board – factory prices.

FRG: Federal Republic of Germany.

FTC: US Federal Trade Commission.

GATT: General Agreement on Tariffs and Trade: An international organisation, based in Geneva, that provides a forum for trade negotiations. Member countries are committed to reducing the barriers to world trade, and extending its volume.

GDP: Gross Domestic Product. The total value of goods and services produced by an economy over a given period, usually one year.

GNP: Gross National Product. GDP plus the income accruing from foreign investment, less payments made to investors in foreign countries.

IISI: International Iron and Steel Institute.

Integration: Integrated steelmaking. An indication of the extent to which a given firm produces the materials, components or substances that are inputs to end products. An integrated steel firm begins by making iron from ore, then converts the iron to steel, and later to finished products.

ITC: International Trade Commission. An independent agency of the US Government, which investigates and issues rules on trade related matters, primarily concerned with imports.

KVR: Ruhr District, part of NRW. KVR is the acronym for an assembly of municipalities and towns in the district.

Marginal costs: The incremental costs associated with an increase in volume of production.

Mini-mill: A small non-integrated steel mill, typically scrap-based and using electric furnaces to produce a limited range of products.

MITI: Ministry of International Trade and Industry, Japan.

Non-integrated: Steelmaking firms that do not reduce iron from ore, but make finished products starting with steel scrap or refined ore.

NRI: Nomura Research Institute, Tokyo, Consultants.

NRW: North Rhine-Westphalia, one of the states of the F.R.G.; the Ruhr district is located in this state.

OECD: Organisation for Economic Cooperation and Development. An international organisation composed of industrial countries. Its aims are to encourage economic growth and employment and to promote the development of industrialising countries.

OHF: open-hearth furnaces. These are an older technology for the reduction of steel from iron. It has largely been replaced by BOF or EAF in most modern steel works.

Pellets: refined ore, used in non-integrated steel making.

Productivity: Output per unit of input-used, usually used to mean labour productivity, the physical quantity or value of goods produced per unit of labour input. Labour input is usually measured in workerhours.

Quality: A statistical measure of the extent to which devices, products etc. meet design specifications. For steel, quality has a complex meaning. It might be expressed in terms of surface characteristics, physical properties or chemical composition.

RWI: Rheinisch-Westfälisches Institut für Wirtschaftsforschung, one of the large German institutes for economic research and forecasting.

Sinter: refined ore, used in non-integrated steel making.

Solomon Plan: A programme for revitalising the American steel industry, prepared by a commission headed by Anthony Solomon. The commission's report was issued in 1977.

Special steelmakers: produce special products of different types. They may or – more often – may not engage in reduction of iron from raw materials. This segment of the industry is not dealt with in this volume. It will however be mentioned repeatedly.

SRI: Stanford Research Institute, Palo Alto, Ca., consultants.

Steel capacity: The figures for nominal, effective and production capability differ significantly from each other.

Nominal capacity: Production capacity of main equipment: converter, electric furnace etc., disregarding factors of material supplies, imbalances between processes. Equipment not in use is disregarded.

Effective capacity: Maximum possible production during a year under normal working conditions, attention paid to repairs, maintenance, holidays; attention also paid to reductions and additions taking place during the year. Assumption of raw material availability.

Production capability: Tonnage capability to produce raw steel at full demand. No bottlenecks concerning raw materials, fuels, supplies; no bottlenecks in supplies of steel industry processes like coking, iron and steelmaking, rolling and finishing. Due attention paid to current environmental and safety requirements.

Ton: Three different weight measures used in steel statistics often contribute to confusion.

US (short) ton = 2,000 lb. = 907 kg

Metric ton (tonne) = 2,205 lb. = 1,000 kg (1,103 short ton)

(0,9843 long ton)

GB (long) ton = 2,240 lb. = 1,016 kg

International statistics, e.g. as compiled by the International Iron and Steel Institute, (IISI) most frequently use the metric ton (tonne). US domestic statistics very often use short tons, British statistics long tons.

514

TPM: trigger price mechanism. Sets a floor price for steel imports in the United States, the price being based on the production costs of the lowest producer (Japan) plus transportation charges, adjusted for currency fluctuations. Steel imports entering the United States below this price automatically 'trigger' an accelerated antidumping investigation.

TSP: Total Suspended Particulars, an air pollution criterion.

UNCTAD: United Nations Conference on Trade and Development.

UNIDO: United Nations Industrial Development Organisation.

Upstream *see* Vertical integration.

Vertical integration: An indication of the extent to which a given firm produces the materials, components etc. that are inputs to its end products. An integrated steel firm begins by making iron from ore, then converts the iron to steel etc. The terms 'downstream' and 'upstream' are used as shorthand for 'forward' (towards the end user) and 'backward' (towards raw material supplies) vertical integration respectively.

VRA: voluntary restraint agreement. A negotiated limit on imports similar to a so-called orderly marketing agreement, a negotiated limit on imports of a certain type of product from a particular country. VRAs on steel negotiated by the United States with the EEC and Japan were in effect from 1969 to 1974. They limited steel imports to specific tonnages plus 5 per cent annual growth. The agreements were not strictly 'voluntary'.

Bibliography

Abel, K. (1981): *Corporate Organizational Design and Its Effect on Innovation*, England, Oxford University, Dept. of Engineering Science.

Ács, J. (1967): *Auswirkungen der wissenschaftlich-technischen Revolution auf die Produktion und den internationalen Handel von Schwarzmetallen*, Dissertation an der Hochschule für Ökonomie, Berlin-Karlshorst, GDR.

Ács, J. (1981): *Price Behaviour and the Theory of the Firm in Competitive and Corporate Markets: A Study of the U.S. Steel Industry*, London, University Microfilms International.

Adams, W. and Dirlam, J.B. (1966): 'Big Steel Inventions and Innovation', in *The Quarterly Journal of Economics*, vol. LXXX, no. 2, pp.167–189.

Adams, W. and Mueller, H. (1982): 'The Steel Industry', in Adams, W. (ed.), *The Structure of American Industry*, New York, MacMillan.

American Iron and Steel Institute, Yearbooks.

American Iron and Steel Institute (1961): *The Competitive Challenge to Steel*, New York.

American Iron and Steel Institute (1970): *Steel Imports – A National Concern*, Washington DC.

American Iron and Steel Institute (1980): *Steel at the Crossroads The American Steel Industry in the 1980's*, Washington DC.

American Iron and Steel Institute (1980): *Annual Statistical Report*, Washington DC.

American Iron and Steel Institute (1982): *Steel and America: An Annual Report*, Washington DC.

Anonymous (1982): 'Neues Verfahren zur Eisengewinnung – VÖEST macht Kohlereduktion serienreif' in *Die Presse*, 1 September.

517

Assider (1980): *Prospettive e problemi della siderurgia italiana* .

Aylen, J. (1981): *Plant Size and Efficiency in the Steel Industry: An International Comparison*, University of Salford.

Batelle-Institute: (1980) *Der Arbeitsmarkt in Nordrhein-Westfalen und im Ruhrgebiet. Analyse der Entwicklung bis zum Jahre 1995*, Frankfurt-Main.

Bömer, H. (1977): 'Internationale Kapitalkonzentration und regionale Krisenentwicklung am Beispiel der Montanindustrie und der Montanregionen der Europäischen Gemeinschaft', *Dortmunder Beiträge zur Raumplanung*, Bd. 5, Dortmund.

Bömer, H. (1979): 'Regionale Strukturkrisen im staatsmonopolistischen Kapitalismus und marxistische Raumokönomie am Beispiel der Ruhrgebietskrise', in *Marxistische Studien* (Jahrbuch des IMSF), February, pp.138 ff.

Bömer, H. and Schröter, L. (1974): 'Ursachenanalyse regionaler Krisen anfälligkeit. Zur Anwendung der Theorie der Überakkumulation/ Entwertung auf regionale Probleme', *Seminarberichte*, no. 10, Gesellschaft für Regionalforschung, p.35 ff.

Bowden, P. (1979: *Regional Development in Action*, Anglo-German Foundation for the Study of the Industrial Society, London.

Boyce, R.L. (1981): 'The Structure of Steel Markets in the United States', paper presented at the Twelfth Atlantic Economic Conference; New York, 8–11 October.

Brune, R., Hennies-Rautenberg, H. and Löbbe, K. (1978): *Wirtschaftsstrukturelle Bestandsaufnahme für das Ruhrgebiet, 1. Fortschreibung, Essen* .

Brune, R., Heilemann, U., Karrenberg, H. and Löbbe, K. (1979): 'Überlegungen zu regionalpolitischen Maßnahmen für das Ruhrgebiet' in *Mitteilungen des Rheinisch-Westfälischen Instituts für Wirtschaftsforschung*, Heft 1, Essen.

Bundesforschungsanstalt für Landeskunde und Raumordnung (BFLR) (1980): 'Arbeitslosenquote als Indikator der Regionalpolitik?' Schwerpunktheft der *Informationen zur Raumentwicklung*, Heft 3-4.

Burckhardt, H. (1981): *25 Jahre Kohlepolitik*, Baden-Baden.

Bureau of Census: *Annual Survey of Manufacturers, Industry Profiles*, various years.

Carney, J., Hudson, R. and Lewis, J. (eds) (1980): *Regions in Crisis*, London.

Census of Manufacturers: *Summary Statistics*, various editions.

Chase Econometrics (1980): *Steel Long Term International Forecast*, September (reserved to subscribers).

Chenery, H.B. (1960): 'Patterns of Industrial Growth', in *American Economic Review*, vol. 5, no. 4, September.

Cockerill, A. and Silberston, A. (1974): *The Steel Industry: International Comparison of Industrial Structure and Performance*, Cambridge.

Controller General Report to the Congress (1981): *New Strategy Required for Aiding Distressed Steel Industry*, Washington DC, 8 January (GAO Steel Report).

Congress of the United States, Office of Technology Assessment (1980): *Technology and Steel Industry Competitiveness*, Washington DC, June (OTA Steel Report).

Congress of the United States, Office of Technology Assessment (1981): *U.S. Industrial Competitiveness: A Comparison of Steel, Electronics and Automobiles*, Washington DC.

Crandall, R.W. (1980): 'Analyse de la crise actuelle de la sidérurgie dans les pays membres de l'OECD' in, *OECD Steel in the 80s*, Paris.

Crandall, R.W. (1980): *The U.S. Steel Industry in Recurrent Crisis: Policy Options in a Competitive World*, Washington DC, Brookings Institute.

Davignon, E. Count (1978): 'Restructuring in the Eighties', Metals Society, International Conference *The Steel Industry the Eighties* .

De Menton, J. (1980): *Restructurations Industrielles en Europe dans le Secteur de l'acier*, Ministère de l'Industrie, Centre d'Etudes et de Prévision Paris.

Dicke, H. (1983): 'Krise in der Stahlindustrie: Markt oder Politikversagen' in *Die Weltwirtschaft*, pp.110–13.

Dielmann, H.J. (1981), 'U.S. Response to Foreign Steel: Returning to Trigger Prices', *Columbia Journal of World Business*, vol. 16, no. 3, pp.32–42.

Dilley, D.R. and McBride, D.L. (1967): 'Oxygen Steelmaking – Fact vs Folklore' in *Iron and Steel Engineer*, October.

Dirlam, J.B. and Mueller, H.G. (1981): *Import Restraints and Reindustrialisation: The Case of the U.S. Steel Industry*, Murfreesboro, Middle Tennessee State University Conference Paper, series no. 67.

Dobrov, G.M. (1976): *The Dynamics and Management of Technological Development as an Object for Applied Systems Analysis*, Luxemburg, Report to IIASA.

Domröse, W. and Koch, K. (1976): 'Metallurgie und Verfahrenstechnik der kontinuierlichen Stahlerzeugung' in *Stahl und Eisen*, October.

EEC – GD III (1981, 1982): Confidential Papers on Restructuring of the Steel Industry.

Eketorp, S. (1981): 'Thoughts about Metallurgy Facing the Year 2000' in *Stahl und Eisen*, July.

Eurostat (1974–1981): *Iron and Steel Yearbook*, varrious issues.

Eurostat: *Wages and Incomes*, various issues.

Fama, E.F. and Miller, M.H. (1972): *Theory of Finance*, New York, Holt, Rinehart and Winston, Inc.

Federal Trade Commission (1977): *The United States Steel Industry and its International Rivals: Trends and Factors Determining International Competitiveness*, Washington DC.

Foy, L.W. (1975), 'We Can Do It', presentation at the Annual Meeting, American Iron and Steel Institute, New York, 21 May (mimeo).

GATT: *Yearbooks*, various issues.

General Accounting Office (1981): New Strategy Required for Aiding Distressed Steel Industry, Washington DC, 8 January.

Gold, B. (1964): 'Industry Growth Patterns: Theory and Empirical Findings' in *Journal of Industrial Economics*, November.

Gold, B. (1969): 'The Framework of Decisions for Major Technological Innovations', in Baier, K. and Rescher N. (eds), *Values and the Future*, New York, Free Press.

Gold, B. (1971): *Explorations in Managerial Economics: Productivity, Costs, Technology and Growth*, New York, Basic Books.

Gold, B. (1974): 'Evaluating Scale Economies: The Case of Japanese Blast Furnaces' in *Journal of Industrial Economics*, September.

Gold, B. (1975): *Technological Change: Economics, Management and Environment*, Oxford, Pergamon Press.

Gold, B. (1976): 'Tracing Gaps Between Expectations and Results of Technological Innovations: The Case of Iron and Steel' in *Journal of Industrial Economics*, September.

Gold, B. (1977): 'On the Shaky Foundations of Capital Budgeting' in *California Management Review*, Spring.

Gold, B. (1978): 'Interactions Between Technological Innovations and Factor Prices' in *Revue d'Economie Industrielle*, Paris, Spring.

Gold, B. (1978): 'Economic Perspectives in Evaluating Technological Development Alternatives' in *Iron and Steelmaker*, December.

Gold, B. (1978): 'Factors Stimulating the Technological Progress of Japanese Industries: The Case of Computerization in Steel' in *Quarterly Review of Economics and Business*, Winter.

Gold, B. (1978): 'Steel Technologies and Costs in the U.S. and Japan' in *Iron and Steel Engineer*, April.

Gold, B. (1978): 'Prospective Changes in the World Steel Industry during the 1980's', Metals Society, International Conference *The Steel Industry in the Eighties* .

Gold, B. (1978): 'Long Run Prospects for the World Steel Industry: Some Internal Potentials for Improvement', Metals Society International Conference *The Steel Industry in the Eighties*.

Gold, B. (1979): Analyzing the Structure of Technological Impacts: From Physical to Economic Effects' in Gold, B. (ed.), *Productivity, Technology and Capital*, Lexington, Mass., D.C. Heath-Lexington Books.

Gold, B. (1981): 'Changing Perspectives on Size, Scale and Returns: An Interpretive Survey' in *Journal of Economic Literature*, March.

Gold, B. (1982): 'Managerial Considerations in Evaluating the Role of Licensing in Technological Development Strategies', *Managerial and Decision Economics* .

Gold, B. and Boylan, M.G. (1975): 'Capital Budgeting, Industrial Capacity and Imports' in *Quarterly Review of Economics and Business*, Autumn.

Gold, B., Peirce, W.S. and Rosegger, G. (1970): 'Diffusion of Major Technological Innovations in U.S. Iron and Steel Manufacturing', *Journal of Industrial Economics*, July.

Gold, B., Rosegger, G. and Boylan, M.G. (1980): *Evaluating Technological Innovations: Methods, Expectations, and Findings*, Lexington, Mass., D.C. Heath-Lexington Books.

Goldberg, W.H. (1983): *Governments and Multinationals: The Strategy of Conflict vs Cooperation*, Cambridge, Mass., Oelgeschlager, Gunn and Hain.

Goldberg, W.H. (1983): *Mergers, Motives, Modes and Methods*, Aldershot, Gower.

Gott, E.H. (1970): 'The Economic Importance of Continuous Casting of Steel Slabs' in *International Iron and Steel Institute 1969 Report of Proceedings*, Tokyo.

Hall, P. (1966): *The World Cities*, London, Weidenfeld and Nicholson.

Hall, W.K. (1980): 'Survival Strategies in a Hostile Environment' in *Harvard Business Review*, September-October, pp.75–85.

Hamermesh, R.G. and Silk, S.B. (1979): 'How to Compete in Stagnant Industries', *Harvard Business Review*, September.

Hellen, J.A. (1974): 'North-Rhine-Westphalia', in *Problem Regions of Europe*, Oxford University Press.

Hiebler, H. (1982): 'Neue Technologie in der Stahlindustrie', paper presented at the International Symposium *Automation in the Steel Industry*, Vienna, European Coordination Center for Research and Documentation in Social Sciences.

Hinterhuber, H. (1975): *Innovationsdynamik und Unternehmensführung*, Vienna – New York, Springer Verlag.

Hinterhuber, H. (1977): *Strategische Unternehmensführung*, Berlin New York, de Gruyter.

HMSO (1981): *National Income and Expenditure 1980*, HMSO® London, Central Statistical Office.

HMSO (1981-2): *Effects of BSC's Corporate Plan*, HMSO, London, Fourth Report from the Industry and Trade Committee (2 vols), vol 2, p.6.

Hosoki, S. and Kono, T. (1978): 'Japanese Steel Industry and its Rate of Development', Metals Society, International Conference *The Steel Industry in the Eighties* .

Hosoki, S., Kono, T. and Imai, K. (1975): *Iron and Steel*, Japanese Economic Studies.

Imai, K. (1975): *Iron and Steel*, Japanese Economic Studies.

International Iron and Steel Institute (IISI): 'Working Group on Steel Demand Forecasting', Provisional papers.

International Iron and Steel Institute (1972): *Projection 85*, Brussels.

International Iron and Steel Institute (1974-9): *World Indirect Trade in steel*.

International Iron and Steel Institute (1977): Committee on Economic Studies, *Projection 90*, Draft, Brussels.

International Iron and Steel Institute (1978): *A Handbook of World Steel Statistics* .

International Iron and Steel Institute (1980): *Causes of the Mid-1970's Recession in Steel Demand*, Brussels, p.155.

International Iron and Steel Institute (1981): *World Steel in Figures*.

Irvine, K.J. (1978): 'Developing Steels for the Market', Metals Society, International Conference *The Steel Industry in the Eighties* .

I.S.C.O.: *Conguintura italiana, 1970-1980*, Rome.

Japan Iron and Steel Federation: *The Steel Industry of Japan, 1974-1980* .

Johnson, C. (1982): *MITI and the Japanese Miracle: The Growth of Industrial Policy, 1925-75*, Stanford CA, Stanford University Press.

Kaiser Engineers, *L-D Process Newsletter*, occasional.

Kawahito, K. (1972): *The Japanese Steel Industry with an Analysis of the U.S. Steel Import Problem*, New York, Washington, London, Praeger Publishers.

Kawahito, K. (1978): *Sources of the Differences in Steel-making Yield between Japan and the United States*, Middle Tennessee State University, Murfreesboro, monograph series no. 20, Business and Economic Research Center, July.

Kawahito, K. (1982): 'Japanese Steel in the American Market: Conflict and Causes' in *The World Economy*, vol. 5, pp.259-78.

Kawahito, K. (1982): 'Steel and the U.S. Antidumping Statutes' in *Journal of World Trade Law*, vol. 16, pp.157-64.

Kawata, S. (ed.) (1980): *Japan's Iron and Steel Industry,* Tokyo.

Kern, W. (1976): 'Innovation und Investition' in *Investitionstheorie und Investitionspolitik privater und öffentlicher Unternehmen,* Wiesbaden, Gabler.

Kommunalverband Ruhrgebiet (1980): 'Statistik der tätigen Personen NW und Gebiet des KVR', Essen, Manuskript.

Kono, T. (1980): 'Évolution de l'industrie sidérurgique mondiale jusqu'en 1975: Demande, échanges et capacités de production' in *OECD, Steel in the 80's,* Paris.

Krupp Stahl AG (1982): *Die Entwicklung der Krupp-Stahl AG 1966–1982,* Essen.

Lamberts, W. (1972): 'Der Strukturwandel im Ruhrgebiet – eine Zwischenbilanz, 1. Folge: Die Bedeutung des Montan-Komplexes für die Ruhrwirtschaft', *Mitteilungen des Rheinisch-Westfälischen Instituts für Wirtschaftsforschung,* Essen, Heft 3.

Lauffs, H.W. and Zühlke, W. (1976): *Politische Planung im Ruhrgebiet,* Göttingen.

Leminsky, G. (1978): 'Labour, Capital and Management in the Eighties in the Steel Industry', Metals Society, International Conference *The Steel Industry in the Eighties* .

Liestmann, W.D. (1975): 'Stranggiessen von Stahl als Verfahren zwischen Schmelzbetrieb und Fertigwalzwerk' in *Stahl und Eisen,* January.

Maizels, A. (1963): *Industrial Growth and World Trade,* Cambridge, Cambridge University Press.

Marcus, P. (1978): *World Steel Dynamics,* New York, Paine Webber Mitchel Hutchins, quarterly issues.

Margulies, F. (1982): *Optional Study – Steel Industry: A Cross-National Comparative Research,* Vienna, European Coordination Center for Re search and Documentation in Social Sciences.

Mattessich, R. (1978): *Instrumental Reasoning and Systems Methodology,* Dordrecht, Holland, Reidel.

McCormack, G. (1980–81): 'The Re-instated Steel Trigger Price Mechan ism' in *Ford International Law Journal,* vol. 4, no. 2, pp.289-339.

McManus, G.J. (1967): 'Slab Casting: Caution Gives Way to Action' in *Iron Age,* 16 February.

Mensch, G. (1971): 'Zur Dynamik des technischen Fortschritts' in *Zeitschrift für Betriebswirtschaft,* vol. 41.

Mensch, G. (1972): 'Basisinnovationen und Verbesserungsinnovationen' in *Zeitschrift für Betriebswirtschaft,* vol. 42.

Mensch, G. (1975): *Das technologische Patt,* Frankfurt am Main, Umschau-Verlag Breidenstein KG.

Memorandum (1981): *Demokratische Wirtschaftspolitik gegen Marktmacht und Sparmaßnahmen,* Cologne.

Miles, T. (1980): 'La Croissance de la production d'acier au cours des années 80', OECD, *Steel in the 80s,* Paris.

Minister für Arbeit, Gesundheit und Soziales (1982): 'Runderlaß, Abstände zwischen Industrie bzw. Gewerbegebieten und Wohngebieten im Rahmen der Bauleitplanung (Abstandserlaß)' in *Ministerialblatt für das Land Nordrhein-Westfalen* 35, no. 67, pp.1376–1984.

Ministry of Labour, *Monthly Survey of Labor Statistics,* Japan.

Mönnich, H.: *Aufbruch ins Revier. Aufbruch nach Europa. Hoesch 1871–1971,* Munich, 1971.

Monopolkommission (1977): *Hauptgutachten 1973/75, Mehr Wettbewerb ist möglich,* Baden-Baden, 2nd ed.

Monopolkommission (1978): *Hauptgutachten 1976/77, Fortschreitende Konzentration bei Großunternehmen* Baden-Baden.

Morton, S.K. *et al.* (1973): 'Continuous Casting of Steel' in *Iron and Steel Institute Journal,* January.

Moser, G. (1978): *Die Konzernorganisation der VÖEST-ALPINE AG: als Antwort auf eine im Umbruch stehende Stahlwelt',* Linz, manuscript.

Mueller, H.G. (1977): The United States' Steel Industry and its Principal Foreign Rivals: Past Performance and Outlook for the 1980's, Middle Tennessee State University, Murfreesboro, Conference Paper Series No. 16, Business and Economic Research Center, February.

Mueller, H.G. (1979): *Factors Determining Competitiveness in the World Steel Market,* Middle Tennessee State University, Murfreesborough, Conference Paper Series No. 56, Business and Economic Research Center, October.

Mueller, H.G. (1980): *Competitiveness of the U.S. Steel Industry after the new Trigger Price Maintenance System,* Middle Tennessee State University, Murfreesboro, monograph no. 25, Business and Economic Research Center.

Mueller, H.G. (1982): 'The Steel Industry' in *The Annals of the American Academy PPS 460,* March, pp.73–82.

Mueller, H.G. (1982): *A Comparative Analysis of Steel Industries in Industrialized and Newly Industrializing Countries,* Middle Tennessee State University, Murfreesboro, Conference Paper Series No. 72, Business and Economic Research Center.

Mueller, H.G. and Kawahito, K. (1978): *Steel Industry Economics: A Comparative Analysis of Structure, Conduct and Performance,* New York, Japan's Steel Information Center.

Mueller, H.G. and Kawahito, K. (1979): *The International Steel Market: Present Crisis and Outlook for the 1980's*, Middle Tennessee State University, Murfreesboro, Conference Paper Series No. 46, Business and Economic Research Center, May.

Mueller, H.G. and Van der Ven, H. (1982): 'Perils of the Brussels–Washington Steel Pact of 1982' in *The World Economy*, vol. 5, pp.259–78.

Myers, J. and Nakamura, L. (1978): *Saving Energy in Manufacturing: The Post Embargo Record*, Ballinger Publishing.

Neumann, M. (1979): *Vertikale Integration and Konzentration zwischen Unternehmen der Eisen und Stahlindustrie und des Maschinenbaus,* Gutachten erstellt für die Monopolkommission.

Neumann, M. (1982): '"Predatory Pricing" by a Quantity Setting MultiProduct Firm' in *American Economic Review*, vol. 72.

Niethammer, L. (1979): *Umständliche Erläuterung der seelischen Störung eines Communalbaumeisters in Preußens größtem Industriedorf – oder die Unfähigkeit zur Stadtentwicklung*, Frankfurt, Syndikat.

Nippon Steel Corporation (1981): *How Japan's Steel Industry has Improv ed its Productivity*.

OECD (1980): *Steel in the 80s*, Paris.

OECD, *The Steel Industry in 1975, 1976, 1977, 1978, 1979*.

Office of Technology Assessment, U.S. Congress (1980): *Technology and Steel Industry Competitiveness*, Washington, DC, US Government Printing Office.

Oster, S. (1982): 'The Diffusion of Innovation Among Steel Firms: The Basic Oxygen Furnace' in *Bell Journal of Economics*, vol. 13, no. 1, pp.45–56.

Overbeck, E. (1977): 'Strukturwandel – Neue Chance für die Unternehmen' in *Zeitschrift für betriebswirtschaftliche Forschung*, vol. 29.

Paine Webber Mitchell Hutchins Inc. (1979): *Steel Prices, Costs and Profits*, New York.

Paine Webber Mitchell Hutchins Inc. (1981): *Steel Strategist*, New York.

Peco, F. (1980): 'La recherche d'une compétitivité internationale pour les entreprises sidérurgiques', OECD, *Steel in the 80s*

Peirce, W.S. (1980): 'The Effects of Technological Change: Exploring Successive Ripples' in Gold, B. (ed.) *Technological Change: Economics Management and Environment*, Oxford, Pergamon.

Prognos, A.G. (1980): *Sammlung, Aufbereitung und Schätzung der Daten zur Überarbeitung der Prognosen für die 75 Gebietseinheiten für das Bundesraumordnungsprogramm*, Basle.

Ray, G. (1982): 'The Management of Technological Change in the British Steel Industry', Lausanne, paper presented at the XXV Meeting of the Institute of Management Sciences.

Robbins, N.A. (1979): in *The American Steel Industry in the 1980's – The Crucial Decade,* Washington DC, American Iron and Steel Institute.

Rosegger, G. (1976): *'Diffusion Research in the Industrial Setting: Some Conceptual Clarifications'* in *Technological Forecasting and Social Change,* March.

Rosegger, G. (1979): 'Diffusion and Technological Specificity: The Case of Continuous Casting' in *Journal of Industrial Economics,* September.

Rosegger, G. (1980): *The Economics of Production and Innovation: An Industrial Perspective,* Oxford, Pergamon.

Rosegger, G. (1980): 'Exploratory Evaluations of Major Technological Innovations: Basic Oxygen Furnaces and Continuous Casting' in Gold, B. *et al., Evaluating Technological Innovations,* Lexington, Mass., Lexington Books.

Rosegger, G. (1982): 'Diffusion and Scale Dynamics: A Case Study' in *Technovation,* vol. 1, pp.201–304.

Schenck, H. (1972): *Probleme der Stahlindustrie,* Vienna, ÖIAG.

Siderurgia Latinoamericana, June 1981, p.7.

Steel Employment News, 11 March 1982.

Steel Tripartite Advisory Committee on the United States Steel Industry (1980): 'Report to the President', Washington DC, 25 September (mimeo).

Stegemann, K. (1977): *Price Competition and Output Adjustment in the European Steel Market,* Mohr, Tübingen, Kieler Studien.

Stoffaes, C. and Gaddoneix, P. (1980): 'Steel and the State in France' in *Annalen der Gemeinwirtschaft,* no. 4, pp.404–21.

Takacs, W.E. (1976): *Quantitative Restrictions on International Trade,* Baltimore, Ph.D. dissertation, Johns Hopkins University Press.

UNIDO (1980): *Picture for 1985 of the World Iron and Steel Industry,* Vienna.

United Nations (1968): *Economic Commission for Europe, Economic Aspects of Continuous Casting of Steel,* New York, Document no. ST/ECE/-Steel/23, United Nations.

UNO, ECE (1980): *The Steel Market in 1980,* New York.

US Congress, Subcommittee on Trade of the Committee on Ways and Means (1978): *Administration's Comprehensive Program for the Steel Industry,* Washington, DC, US Government Printing Office.

US Council on Wages and Price Stability (1977): *Report to the President on Prices and Costs in the United States Steel Industry*, Washington, DC, US Government Printing Office.

U.S. Federal Trade Commission (1978): *The United States Steel Industry and its international Rivals: Trends and Factors Determining International Competitiveness*, US Government Printing Office.

Väth, W. (1980): 'Industrielle Restrukturierung und Strukturpolitik – das Beispiel Saarland', paper on *Probleme der Stadtpolitik in den 80er Jahren*, Essen 1.-2, 10.

Vogel, E.F. (1979): *Japan as No. 1*, Cambridge MA, Harvard University Press.

Wegener, M. and Vannahme, M. (1981): 'Regional Unemployment and the Housing Market: Spatial Effects of Industrial Decline', Paper, 13th Annual Conference: Regional Science Association (British Section), University of Durham, September.

Willim, F. (1975): 'Einplanung von Stranggießanlagen bei Blasstahlwerken' in *Stahl und Eisen*, February.

Witte, E. (1975): 'Erfolg von Investitionen in neuen Technologien durch richtige Auswahl der Mitarbeiter' in *Fortschrittliche Betriebsfuhrung und Industrial Engineering*, vol. 24, p.47.

Witte, E. (1973): *Organisation für Innovationsentscheidungen – Das Promotorenmodell*, Göttingen, O. Schwartz.

Wyman, J.C. and Gold, B. (1979): *Technology and Steel*, New York, Shearson Hayden Stone Inc., 21 February.

Zarnowitz, V. and Moore, G.H. (1977): 'The Recession and Recovery of 1973-1976', p.508 in NBER *Explorations in Economic Research*, vol. 4, p.471 ff.

Author index

Subject index

531